Chemical Kinetics and Transport

Chemical Kinetics and Transport

Peter C. Jordan
Brandeis University
Waltham, Massachusetts

Plenum Press · New York and London

Library of Congress Cataloging in Publication Data

Jordan, Peter C
Chemical kinetics and transport.

Includes bibliographical references and index.
1. Chemical reaction, Rate of. 2. Transport theory. I. Title.
QD501.J7573 1979 541'.39 78-20999
ISBN 0-306-40122-3

First Printing – March 1979
Second Printing – January 1980

A Division of Plenum Publishing Corporation
227 West 17th Street, New York, N.Y. 10011

Printed in the United States of America

Preface

This book began as a program of self-education. While teaching undergraduate physical chemistry, I became progressively more dissatisfied with my approach to chemical kinetics. The solution to my problem was to write a detailed set of lecture notes which covered more material, in greater depth, than could be presented in undergraduate physical chemistry. These notes are the foundation upon which this book is built.

My background led me to view chemical kinetics as closely related to transport phenomena. While the relationship of these topics is well known, it is often ignored, except for brief discussions of irreversible thermodynamics. In fact, the physics underlying such apparently dissimilar processes as reaction and energy transfer is not so very different. The intermolecular potential is to transport what the potential-energy surface is to reactivity.

Instead of beginning the sections devoted to chemical kinetics with a discussion of various theories, I have chosen to treat phenomenology and mechanism first. In this way the essential unity of kinetic arguments, whether applied to gas-phase or solution-phase reaction, can be emphasized. Theories of rate constants and of chemical dynamics are treated last, so that their strengths and weaknesses may be more clearly highlighted.

The book is designed for students in their senior year or first year of graduate school. A year of undergraduate physical chemistry is essential preparation. While further exposure to chemical thermodynamics, statistical thermodynamics, or molecular spectroscopy is an asset, it is not necessary.

As a result of my perspective, the choice of topics and organization of the material differs from that of other kinetics texts. The introductory chapter, on equilibrium kinetic theory, is mainly review. The next two chapters, treating transport in both gas and solution, introduce many chemically useful concepts. Chapter 4 focuses on phenomenology—the experimental methods used to determine a rate law, the meaning of rate laws and rate constants. The next three chapters are devoted to mechanisms of thermal and photochemical reaction, first in systems that are stable to perturbation and then in those for which explosion or oscillation is possible. Much of this material is in the standard repertory. However, the sections devoted to heterogeneous catalysis, infrared laser photochemistry, and oscillating chemical reactions incorporate the results of recent, occasionally speculative, research. Chapter 8 treats single-collision chemistry—first, to relate the reaction cross section and the rate constant, and then, to demonstrate how mechanistic information can be deduced from scattering data.

The first of the theoretical chapters (Chapter 9) treats approaches to the calculation of thermal rate constants. The material is familiar—activated complex theory, RRKM theory of unimolecular reaction, Debye theory of diffusion-limited reaction—and emphasizes how much information can be correlated on the basis of quite limited models. In the final chapter, the dynamics of single-collision chemistry is analyzed within a highly simplified framework; the model, based on classical mechanics, collinear collision geometries, and naive potential-energy surfaces, illuminates many of the features that account for chemical reactivity.

I have tried to present both the arguments and the derivations so that they can be readily followed. However, some steps have been intentionally omitted. I believe that unless a student works through the material individually, the subject matter cannot be mastered. Examples have been chosen to illustrate general principles, not to be exhaustive. The problems, which are an indispensable adjunct to each chapter, provide further illustrations.

While the arrangement of the material is not traditional, each chapter, and many of the sections, are, to a great degree, self-contained. Topics may, if the instructor so desires, be presented in any order.

I am indebted to many individuals for their assistance. Foremost among these are my colleagues, Michael Henchman and Kenneth Kustin, who clarified my ideas and suggested many of their own; without their help this book would never have been written. The reviewers, Lawrence J. Parkhurst and Robert E. Wyatt, made many helpful and perceptive suggestions. Students at Brandeis University have rendered considerable assistance, especially in pointing out ambiguous or unclear passages. Finally I am grateful to Evangeline P. Goodwin for her rapid and accurate typing and to Virginia G. Steel for her excellent draftsmanship.

Peter C. Jordan

Contents

6. *Photochemistry*

7. *Nonstationary State Mechanisms*

8. *Single-Collision Chemistry*

Notation

Those symbols which are only used once in the text do not appear in this listing. Familiar mathematical functions are also not included, nor are quantities that are simply intermediates in a derivation. Capital letters are listed before lowercase letters. Greek letters are listed at the end.

A	Arrhenius A-factor, reaction affinity, energy acceptor in excitation transfer reaction
ABC	Product of principal moments of inertia of a nonlinear molecule
C_i	Normality of the ith component
C_v	Constant volume heat capacity
D	Energy donor in excitation transfer reaction
D	Dissociation energy for dynamical model of Chapter 10
D_i	Diffusion coefficient of the ith component
E	Energy
E_a	Arrhenius activation energy
$\mathbf{E}$	Electric field
F	Faraday's number
$F_{\mathrm{A}}(\mathbf{c}) = F_{\mathrm{A}}(u, v, w)$	Three-dimensional velocity distribution function for molecules of species A
$F_{\mathrm{A}}(c)$	Speed distribution function for molecules of species A
$\Delta G_0^{\ddagger}$	Gibbs free energy of activation
$G_x(u)$	x component of molecular flux as a function of x component of velocity

$\Delta H_0^\ddagger$	Enthalpy of activation
I	Ionic strength, moment of inertia of a linear molecule
I_A	Intensity of a beam of molecules of species A
$I(\nu)$	Normalized fluorescence intensity at frequency ν
$I(\lambda, x)$	Position-dependent light intensity at wavelength λ
$I_A(x)$	Position-dependent intensity of a beam of molecules of species A
J	Rotational quantum number
$\mathbf{J}_i$	Diffusive flux of the ith component
$\mathbf{J}_q$	Heat flux
$K(T)$, $K(T, p)$, $K(T, p, I)$	Equilibrium constants
$K^\ddagger$	Equilibrium constant for activation
K_m	Michaelis constant
M	Total mass
M_A	Molecular weight of species A
N_0	Avogadro's number
N_A	Number of particles of species A
$N^\ddagger$	Hypothetical number of activated complexes
$N(\epsilon_\alpha)$	Density of states for the αth degree(s) of freedom of an activated molecule with energy ϵ_α in the specified degree(s) of freedom
P_x	Momentum flux in x direction
$P(\epsilon)$	Probability density for states of energy ϵ
Q	Enthalpy release per mole of reaction
Q_A	Molar partition function for species A
$Q^\ddagger$	Reduced molar partition function for the activated complex
$Q(\epsilon, i, j)$, $Q(\epsilon)$, $Q(E)$	Reaction cross sections as functions of molecular states i, j, collision energy ϵ, or molar collision energy E
$\partial^2 Q(\epsilon, \theta, \phi)/\partial\theta\, \partial\phi$	Differential cross section for molecules with collision energy ϵ scattering into solid angle θ, ϕ
R	Gas constant, total reaction rate
R_f (R_r)	Rate of forward (reverse) reaction
S_i	The ith singlet state of an electronic manifold
$\Delta S_0^\ddagger$	Entropy of activation
T	Absolute temperature
T_i	The ith triplet state of an electronic manifold
U_A	Potential energy of a molecule of species A

V Volume, limiting velocity of an enzyme-catalyzed reaction
V_{coll} Collision volume of crossed molecular beams
$\Delta V_0^\ddagger$ Volume of activation
X, Y Relative coordinates for collinear motion in the dynamical model of Chapter 10
Y_i Symbol denoting the ith reactant
$[Y_i]$ Concentration of the ith reactant
Z_A Total collision frequency for molecules of species A
a_i Activity of the ith component
$a^\ddagger$ Hypothetical activity of the activated complex
b Impact parameter
c Molecular speed
c_0 Velocity of light
c_i Concentration of the ith component
$\mathbf{c}_i$ Velocity of the ith particle
$\mathbf{c}_M$ Velocity of the center of mass
$\tilde{c}_V$ Constant volume heat capacity per molecule
d (d_{AB}) Hard-sphere diameter (for an A–B collision)
e Electronic charge, energy per mole
$\mathrm{erf}(x)$ Error function
$f_A(c)$ Three-dimensional velocity distribution for molecules of species A as a function of speed
$\mathbf{f}$ Frictional force on a particle
g_α Degeneracy factor for the αth degree(s) of freedom
h Planck's constant
$\mathbf{j}_i$ Current density attributable to the ith ionic species
k Boltzmann's constant, rate constant for reaction
$k^{(n)}, k_{\pm n}$ Rate constants for reaction
k_d Rate constant for irreversible decomposition
k_{diff} (k_{diss}) Rate constant for association (dissociation) in a diffusion-limited reaction
k_∞ (k_0) Limiting rate constant at high (low) pressure in unimolecular decomposition
$k(\epsilon)$ Energy-dependent rate constant for reaction
m_e Electron mass
m_i Mass of the ith particle

n_i	Number density of the *i*th component
$n^\ddagger$	Hypothetical number density of the activated complex
p	Pressure
p_i	Partial pressure of the *i*th component
$p(x)$	Position-dependent scattering probability for molecules in a beam
$p(b, \epsilon, i, j), p(b, \epsilon)$	Reaction probability as functions of molecular states i, j, collision energy ϵ, and impact parameter b
q_A	Molecular partition function for species A
$q^\ddagger$	Reduced molecular partition function for the activated complex
q_α	Partition function for the αth independent degree(s) of freedom
r	Distance in spherical polar coordinates
r_α	Rate of the αth step in a reaction sequence
$\mathbf{r}_i$	Position of the *i*th particle
$\mathbf{r}_M$	Position of the center of mass
t	Time
u	x component of velocity, mobility in an electric field
$u(r)$	Intermolecular potential function
v	y component of velocity, volume per mole, reaction velocity in an enzyme-catalyzed reaction, vibrational quantum number
$\mathbf{v}$	Drift velocity in an external field
w	z component of velocity
x, y	Position components for the equivalent particle in the dynamical model of Chapter 10
x, y, z	Position components in Cartesian coordinates
z_i	Charge of the *i*th ionic species
$z_A(c)$	Speed-dependent collision frequency per molecule of species A
$\Gamma(x)$	Gamma function
Δ	Reaction exo- or endoergicity for dynamical model of Chapter 10
Λ_i (Λ_i^0)	Equivalent conductance of the *i*th electrolyte (at infinite dilution)
Ψ	Time- and position-dependent wave function

$\alpha_i(\lambda)$	Molar absorption coefficient of the *i*th component at wavelength λ
β	$1/kT$, angle characterizing mass ratio in dynamical model of Chapter 10
γ	Thermal coupling constant for dissipation of heat to surroundings
γ_i	Activity coefficient of the *i*th component
$\gamma^\ddagger$	Hypothetical activity coefficient for the activated complex
$\gamma(\lambda)$	Light attenuation coefficient per unit length at wavelength λ
ϵ	Energy per molecule, dielectric constant, energy parameter in Lennard-Jones potential
$\epsilon^\ddagger$	Classical barrier height per molecule in activated complex theory
$\epsilon_i(\nu)$	Molar extinction coefficient for the *i*th component at frequency ν
η	Shear viscosity
θ	Polar angular velocity in spherical polar coordinates, polar angle in spherical polar coordinates, angle of deflection, angle characterizing energy partitioning in dynamical model of Chapter 10
θ_i	Fractional surface coverage attributable to adsorption of the *i*th component
$\varkappa$	Thermal conductivity, transmission coefficient
$\varkappa_i$	Decay constant for the *i*th relaxation process
λ	Wavelength, decay constant for an independent kinetic process
λ_A	Mean free path of species A
λ_i (λ_i^0)	Equivalent conductance of the *i*th ionic species (at infinite dilution)
μ_i	Chemical potential of the *i*th component
$\tilde{\mu}_i$	Electrochemical potential of the *i*th component
μ_{ij}	Reduced mass for relative motion of particles *i* and *j*
ν	Frequency
ν_A	Number of atoms of type A in a compound
ν_i	Stoichiometric coefficient of the *i*th reactant, vibrational frequency of the *i*th normal mode

- $\nu^\ddagger$ Frequency defined by curvature at the peak of the reaction profile
- $\xi(t)$ Time-dependent progress variable
- $\xi(\epsilon)$ Depletion factor in the theory of unimolecular reaction
- ϱ_i Mass density of the *i*th component
- $\varrho_i(E)$ Density of quantum levels in electronic state *i* at energy *E*
- $\varrho_x(u)$ One-dimensional velocity distribution
- σ Specific conductivity, hard-sphere diameter for a collision, length parameter in Lennard-Jones potential, symmetry number of a molecule
- $\tau(\lambda)$ Intrinsic lifetime of a photoexcited state
- ϕ Azimuthal angular velocity in spherical polar coordinates, azimuthal angle in spherical polar coordinates, angle of deflection
- ϕ_λ Quantum yield at wavelength λ
- ϕ_f (ϕ_ϱ) Fluorescence (phosphorescence) quantum efficiency
- χ Polar angle in spherical polar coordinates
- ψ Electric potential, position-dependent wave function
- ω Oscillation frequency of a kinetic process, azimuthal angle in spherical polar coordinates

1

Kinetic Theory of Gases—Equilibrium

1.1. Introduction

The kinetic theory of gases represents the first truly successful effort to construct a model which provides a mechanical basis for understanding the properties of bulk matter. By combining the molecular hypothesis with Newtonian mechanics and using a statistical approach, the theory provides an explanation for the thermodynamic similarities between dilute gases, makes numerous verifiable predictions, and suggests a number of unifying principles. In its simplest form the theory is based upon the following premises:

(1) a gas is composed of an enormous number of molecules;
(2) in the absence of external forces particles move in straight lines[(1)];
(3) particles collide infrequently;
(4) collisions, whether with the walls or with other molecules, are elastic (momentum and kinetic energy are conserved).

In a dilute gas, collisions, although relatively infrequent, are very important. They insure that molecules move in all directions with a continuum of speeds. Rather than attempting to determine the velocity of each molecule (a hopeless undertaking considering that collisions continually alter the velocities), one describes the gas by a *velocity distribution*. Such a

[(1)] The interparticle forces actually cause trajectories to bend as molecules approach one another. However, quite accurate predictions follow if billiard ball dynamics are assumed.

description, which focuses upon the behavior of classes of molecules, eliminates reference to the properties of individual molecules. While the statistical picture provides only a limited amount of information about a gas, it is in fact a description that corresponds precisely to the available experimental data.[2]

1.2. The Maxwell–Boltzmann Distribution of Velocities

A macroscopic sample of a gas contains an enormous number of molecules continually colliding with one another and with the walls of the container. The effect of these collisions is to change the velocity of the colliding particles. The fundamental problem in kinetic theory is to determine the distribution of velocities of the molecules in a gas. The answer was first given by Maxwell for a gas in thermal equilibrium. Indicative of the importance of this problem is the fact that, over a period of years, Maxwell and Boltzmann presented four different derivations of the basic equation, each somewhat more sophisticated than its predecessor. We shall consider the two simplest ones.

The velocity of a molecule, $\mathbf{c}$, is specified by its components u, v, and w in the x, y, and z directions, respectively. If we were to attempt to find the number of molecules with a specific velocity $\mathbf{c}$, we would have an immediate problem of measurement. The more precisely $\mathbf{c}$ can be measured, the fewer the molecules that will be found with the prescribed velocity. Rather than attempt unnecessary accuracy,[3] we focus on the quantity $F(\mathbf{c}) = F(u, v, w)$, where $F(u, v, w)\, du\, dv\, dw$ is the *fraction of molecules* with velocity components between u and $u + du$, v and $v + dv$, w and $w + dw$; the total number of molecules in this velocity domain is then $NF(u, v, w)\, du\, dv\, dw$. If du, dv, and dw are chosen too small, we again encounter the problem that few molecules will lie in the velocity range of interest and the quantity $F(\mathbf{c})$ might not be a smooth function of $\mathbf{c}$. However, as long as the system of interest is macroscopic, the velocity increments may be prescribed as small as measurement permits; there is always an enormous number of molecules in the specified velocity domain and $F(\mathbf{c})$ is a smoothly varying function.

[2] Experiments using molecular beams provide a direct measure of the effect of interparticle collisions. Interpretation of such data requires the analysis of molecular trajectories (Chapter 8).

[3] In fact there are limits, given by the uncertainty principle, as to how precisely one may specify molecular velocities (Problem 1.3).

The quantity $F(u, v, w)$ is the velocity distribution function. It is not a probability but rather a *probability density* and represents the fraction of molecules with the velocity components u, v, and w per unit volume in a three-dimensional velocity space.[4] Since the total probability is 1, the distribution is normalized:

$$1 = \int_{-\infty}^{+\infty} \int_{-\infty}^{+\infty} \int_{-\infty}^{+\infty} F(u, v, w)\, du\, dv\, dw \tag{1.1}$$

By integrating over y and z components of velocity we can compute $\varrho_x(u)\, du$, the fraction of molecules with x component of velocity between u and $u + du$:

$$\varrho_x(u) = \int_{-\infty}^{+\infty} \int_{-\infty}^{+\infty} F(u, v, w)\, dv\, dw \tag{1.2}$$

Naturally $\varrho_x(u)$ is also normalized,

$$1 = \int_{-\infty}^{+\infty} \varrho_x(u)\, du \tag{1.3}$$

as is seen by combining (1.2) and (1.1).

As yet $F(u, v, w)$ is unspecified; it may be determined using Maxwell's original arguments.[5] Experience suggests that, as long as gravitational effects are unimportant, a gas at equilibrium is isotropic. The orientations of the coordinate axes are immaterial; the axes may be rotated arbitrarily. Thus the velocity distribution cannot depend upon u, v, or w in any way that distinguishes one direction from another. The only molecular property with no orientational dependence is the magnitude of the velocity $c = (u^2 + v^2 + w^2)^{1/2}$ so that an alternative description of the velocity distribution is as a function of c only, $f(c)$. The equivalence of the descriptions requires that

$$F(u, v, w) = f(c) \tag{1.4}$$

In addition, Maxwell introduced the simple hypothesis that information about the distribution of x components of velocity provides no insight into the distribution of either y or z components, i.e., the distributions in

[4] The concept is analogous to the familiar idea of spatial density. The number of molecules in a volume element limited by x and $x + dx$, y and $y + dy$, and z and $z + dz$ is $dN = \varrho(x, y, z)\, dx\, dy\, dz$; $\varrho(x, y, z)$ is the *number density*, i.e., the *number per unit volume*.

[5] J. C. Maxwell, *Phil. Mag*. **19**, 19 (1860).

u, v, and w are uncorrelated and $F(u, v, w)$ can be factored:

$$F(u, v, w) = \varrho_x(u)\varrho_y(v)\varrho_z(w) \tag{1.5}$$

The form of these functions is specified completely by (1.4) and (1.5).[6] Take logarithms and differentiate both sides of these equations with respect to u: the result, making use of the chain rule when differentiating $\ln f(c)$, is

$$\left(\frac{\partial \ln F(u, v, w)}{\partial u}\right)_{v,w} = \frac{d \ln \varrho_x(u)}{du} = \frac{u}{c}\frac{d \ln f(c)}{dc} \tag{1.6}$$

Dividing both sides of (1.6) by u yields

$$\frac{1}{u}\frac{d \ln \varrho_x(u)}{du} = \frac{1}{c}\frac{d \ln f(c)}{dc} \tag{1.7}$$

If u is fixed the left-hand side of (1.7) is a constant, $-2A$. However, c may still be altered independently by varying v or w; no matter what value c takes, A does not change so that the quantity $c^{-1}\, d \ln f(c)/dc$ is independent of c. Thus $f(c)$ is the solution of the differential equation

$$\frac{d \ln f(c)}{dc} = -2Ac$$

which is

$$\ln f(c) = -Ac^2 + \ln B$$

or in exponential form

$$f(c) = B \exp(-Ac^2) \tag{1.8}$$

The rationale for the negative sign in the exponential is now clear; in this way infinite velocities are forbidden. Factoring (1.8), the separate distributions in u, v, and w can be obtained; $\varrho_x(u)$ is

$$\varrho_x(u) = B^{1/3} \exp(-Au^2) \tag{1.9}$$

with similar expressions for $\varrho_y(v)$ and $\varrho_z(w)$. Evaluation of A and B requires further information. The normalization (1.1) or (1.3) provides one condition; in the next section the other is found.

Before determining A and B let us derive (1.8) in a way that does not presume the factorization property (1.5). Consider collisions in a dilute

[6] A more sophisticated derivation, not based upon (1.5), was formulated a few years later [J. C. Maxwell, *Trans. Roy. Soc.* **157**, 49 (1867)]. It is presented later in this section.

gas of identical molecules. As interparticle distances are large only a relatively few molecules are interacting at any instant; of these interactions the overwhelming majority involve pairs of particles. The trajectories of a colliding pair are determined by the intermolecular forces between them. As long as the forces are short range the details of the force law are unimportant.[7] The physical significance of such collisions in dilute gases is that they provide a mechanism for momentum transfer between molecules. In a typical collision molecules approach with initial velocities $\mathbf{c}_1$ and $\mathbf{c}_2$, interact, and then separate with final velocities $\mathbf{c}_1'$ and $\mathbf{c}_2'$. The details of the interaction need not concern us; we only need to know the initial and final velocities. Limiting consideration to structureless identical particles, energy conservation requires that[8]

$$c_1^2 + c_2^2 = c_1'^2 + c_2'^2 \tag{1.10}$$

For the process considered the rate must be proportional to the fraction of molecules of each velocity, i.e., to $F(\mathbf{c}_1)F(\mathbf{c}_2)$. The rate of the reverse process, in which molecules initially with velocities $\mathbf{c}_1'$ and $\mathbf{c}_2'$ emerge with velocities $\mathbf{c}_1$ and $\mathbf{c}_2$, is analogously proportional to $F(\mathbf{c}_1')F(\mathbf{c}_2')$. In a system at equilibrium the rate of any process and its inverse are equal.[9] Since, as we shall now demonstrate, the proportionality constants determining the two rates are equal, detailed balance requires that

$$F(\mathbf{c}_1)F(\mathbf{c}_2) = F(\mathbf{c}_1')F(\mathbf{c}_2') \tag{1.11}$$

which, with (1.10), has the unique solution (1.8).

To establish that the proportionality constants are the same, consider the collision processes as viewed by an observer moving with the center-of-mass velocity $(\mathbf{c}_1 + \mathbf{c}_2)/2 = (\mathbf{c}_1' + \mathbf{c}_2')/2$. Combining this with (1.10) we find that the relative speed is the same before and after collision, $|\,\mathbf{c}_1 - \mathbf{c}_2\,| = |\,\mathbf{c}_1' - \mathbf{c}_2'\,| = c_{12}$. Collision accomplishes nothing more than

[7] Short-range forces are those for which the integral $\int_{r_0}^{\infty} u(r)4\pi r^2\,dr$ is finite [here $u(r)$ is the intermolecular potential]; r_0 is arbitrary. Ordinary van der Waals interactions, which fall off as r^{-6}, are short range. On the other hand, coulombic forces, for which $u(r) \propto r^{-1}$, are long range. The consequence is that even in a dilute plasma one must consider the motion of the ions to be correlated.

[8] Such processes, in which *kinetic* energy is conserved, are known as elastic processes. In general, when the effect of molecular structure is considered, collisions are *inelastic*, i.e., there is interconversion of translational and rotational or vibrational energy. Only inelastic collision processes can lead to chemical reaction.

[9] The correctness of this assertion, known as the principle of detailed balance, is not immediately apparent. It will be discussed more fully in Chapter 4.

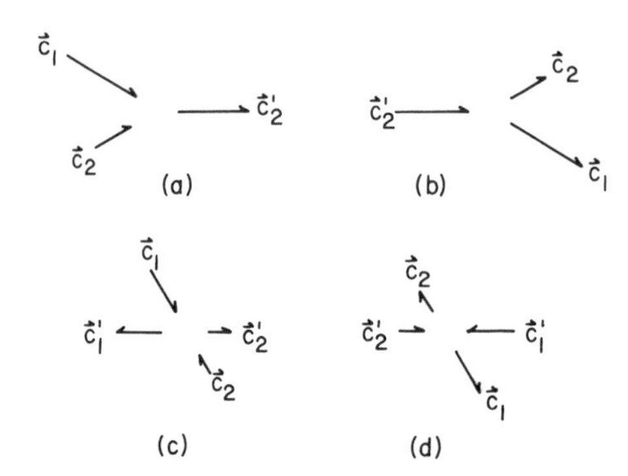

Fig. 1.1. Dynamics of a binary collision and its inverse as viewed in lab coordinates (a, b) and center-of-mass coordinates (c, d). In this example $m_2 = 2m_1$ and $\mathbf{c}_1' = 0$. Note that in center-of-mass coordinates the collision and its inverse are simply related by rotation about the center of mass.

to reorient the relative velocity vector. Thus, in the center-of-mass system two processes lead to the changes shown in Fig. 1.1. In each case the particles are deflected through the same angle. There is no physical difference between them. Thus, barring the effect of walls or interactions with a third particle, there is no reason for preferring either. The absolute rates must be equal and (1.11) is established.

To show that (1.10) and (1.11) are indeed equivalent to (1.8), specialize to the case that $\mathbf{c}_1' = 0$. Then $|\,\mathbf{c}_2'\,| = (c_1^2 + c_2^2)^{1/2}$ and with (1.4) we have

$$f(c_1)f(c_2) = f(0)f([c_1^2 + c_2^2]^{1/2}) \tag{1.12}$$

Using arguments similar to those which led from (1.5) to (1.8) again establishes the form of $f(c)$.

1.3. Determination of Pressure

The pressure of a gas is the force per unit area exerted due to molecular collisions with the walls of the container. Since force is momentum change per unit time, pressure is determined by computing the momentum transfer per unit time per unit area. Consider a wall at $x = L$; if at each collision there is perfect reflection the x component of velocity changes from u to $-u$ and the net momentum transfer is $2mu$. While perfect reflection is not a reasonable assumption we know that the gas has an isotropic velocity distribution and from (1.8), $F(u, v, w) = F(-u, v, w)$. Thus, on the average, the fraction of molecules leaving the walls with an x component of velocity $-u$ is the same as that striking the wall with x component of velocity u. The net effect is a mean momentum transfer of $2mu$.

To determine the contribution such collisions make to the pressure we must compute the number of molecules with x component of velocity

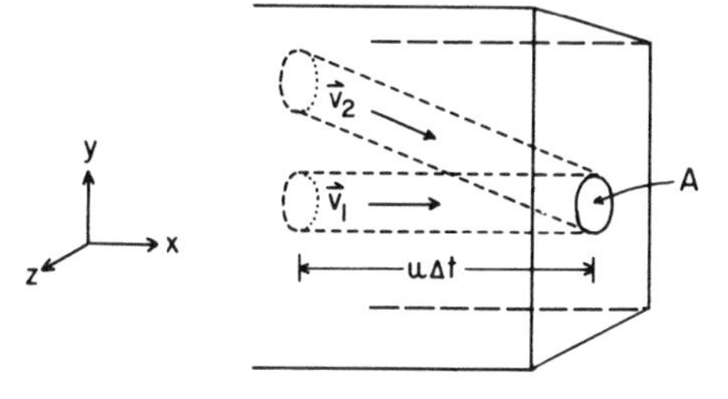

Fig. 1.2. Collision cylinders for molecules with different velocities $\mathbf{v}_1$ and $\mathbf{v}_2$, but the same x component of velocity, striking a target area A in a time Δt.

between u and $u + du$ that strike an area A in a time Δt. From (1.2) the total number of such molecules is $N\varrho_x(u)\,du$. Only a small fraction of these hit the wall during a limited time. From Fig. 1.2 we see that the molecules within the x distance $u\,\Delta t$ hit the wall during Δt. The values of v and w are irrelevant.[10] The fraction that hits the designated area is the ratio of volume, $Au\,\Delta t$, of the collision cylinder to the volume, V, of the container, so that the total number of molecules providing a momentum transfer $2mu$ to the area A in the time Δt is

$$\frac{Au\,\Delta t}{V}\,N\varrho_x(u)\,du \tag{1.13}$$

The momentum transferred by these molecules is

$$(2mu)\,\frac{Au\,\Delta t}{V}\,N\varrho_x(u)\,du$$

Integrating over all u greater than zero (only molecules moving toward the wall may collide) yields the total momentum transfer at $x = L$; dividing by A and Δt we obtain the pressure

$$p = \frac{N}{V}\,2m\int_0^\infty u^2\varrho_x(u)\,du$$

Since $\varrho_x(u) = \varrho_x(-u)$ the range of the integration can be extended to

[10] Molecules within this distance may collide with other molecules before striking the wall. Although kinetic energy and momentum are conserved (collisions are elastic) each molecule will generally change its x component of velocity. Simultaneously (i.e., during Δt) molecules with different x components of velocity will collide and one of the collision products may then have an x component u. The rate of these processes must be equal at equilibrium. Otherwise the distribution $\varrho_x(u)$ would be time dependent, which, as we have seen, it is not. Thus, on balance, for each molecule with component u initially within $u\Delta t$ of the wall, a molecule (not necessarily the same one) with component u will strike the wall during Δt.

minus infinity, which indicates that

$$pV = Nm \int_{-\infty}^{+\infty} u^2 \varrho_x(u)\, du = Nm\overline{u^2} \tag{1.14}$$

where $\overline{u^2}$ is the average value of u^2. As isotropy requires that $\overline{u^2} = \overline{v^2} = \overline{w^2} = \overline{c^2}/3$ an alternate form is

$$pV = \tfrac{1}{3}Nm\overline{c^2} = \tfrac{2}{3}N\bar{\epsilon} = \tfrac{2}{3}E \tag{1.15}$$

where $\bar{\epsilon}$ is the average kinetic energy per molecule and E the total kinetic energy. Dilute gases are ideal; thus $pV = NkT$, which, with (1.15), yields two fundamental results relating mean kinetic energy and mean molecular speed to temperature,

$$\tfrac{1}{2}m\overline{c^2} = \tfrac{3}{2}kT, \qquad E = \tfrac{3}{2}NkT \tag{1.16}$$

The first of these defines the root mean square (rms) speed

$$c_{\text{rms}} = (\overline{c^2})^{1/2} = (3kT/m)^{1/2} \tag{1.17}$$

To obtain the parameters A and B substitute (1.9) in (1.3) and (1.14). The resulting integrals are

$$\begin{aligned} 1 &= B^{1/3} \int_0^{\infty} \exp(-Au^2)\, du \\ pV/N = kT &= B^{1/3} \int_0^{\infty} mu^2 \exp(-Au^2)\, du \end{aligned} \tag{1.18}$$

which are examples of those commonly encountered in applications of kinetic theory. They are of the general form

$$2(A^{s+1})^{1/2} \int_0^{\infty} u^s \exp(-Au^2)\, du = \int_0^{\infty} x^{(s-1)/2} e^{-x}\, dx \equiv \Gamma\left(\frac{s+1}{2}\right) \tag{1.19}$$

Some basic properties of the Γ function are

$$\Gamma(x+1) = x\Gamma(x), \qquad \Gamma(1) = 1, \qquad \Gamma(\tfrac{1}{2}) = \pi^{1/2} \tag{1.20}$$

from which one can establish the important special cases

$$\begin{aligned} &\Gamma(n+1) = n!, \qquad n = 0, 1, 2, \ldots \\ &\Gamma(n+\tfrac{1}{2}) = (2n)!\ \pi^{1/2}/2^{2n}n!, \qquad n = 0, 1, 2, \ldots^{(11)} \end{aligned} \tag{1.21}$$

(11) For an extensive discussion of the Γ function including a tabulation of values of $\Gamma(x)$, for $1 \le x \le 2$, see M. Abramowitz and I. A. Stegun, eds., *Handbook of Mathematical Functions* (New York: Dover Publications Inc., 1965), pp. 253–293.

Using these results we find that

$$1 = 2B^{1/3}(\pi/2A)^{1/2}, \qquad kT = 2mB^{1/3}(\pi/2A^3)^{1/2}$$

which indicate that

$$A = m/kT, \qquad B^{1/3} = (m/2\pi kT)^{1/2}$$

Now, using (1.8) and (1.9) the velocity distribution function is completely specified:

$$\begin{aligned} \varrho_x(u) &= (m/2\pi kT)^{1/2} \exp(-mu^2/2kT) \\ F(\mathbf{c}) = F(u, v, w) &= (m/2\pi kT)^{3/2} \exp(-mc^2/2kT) \end{aligned} \tag{1.22}$$

1.4. Properties of the Maxwell Distribution

The one-dimensional velocity distribution is plotted for N_2 at 100 K and 300 K in Fig. 1.3. As a consequence of the isotropy of the gas it is an even function of u, i.e., $\varrho_x(u) = \varrho_x(-u)$. At low temperature the distribution is more sharply peaked, a corollary of the fact that higher molecular speeds are less probable at low temperatures.

The three-dimensional velocity distribution can be expressed in a number of ways. We have, so far, considered a Cartesian description of the velocity vector. If, however, we switch to spherical polar coordinates in which

$$u = c \sin\theta \cos\varphi, \qquad v = c \sin\theta \sin\varphi, \qquad w = \cos\theta$$

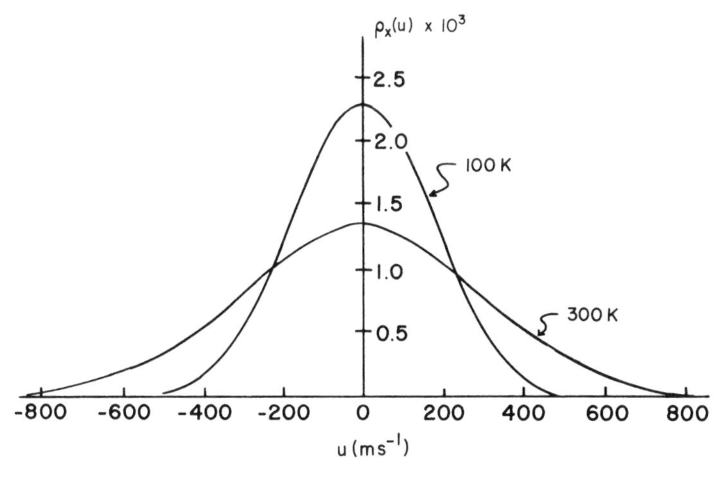

Fig. 1.3. One-dimensional velocity distribution function for N_2 at 100 K and 300 K.

the fraction of molecules with speed between c and $c + dc$ oriented in a solid angle between θ and $\theta + d\theta$ and between φ and $\varphi + d\varphi$ can be found using (1.22) and the relationship between differential volume elements in Cartesian and spherical polar coordinates

$$F(c, \theta, \varphi)\, dc\, d\theta\, d\varphi = (m/2\pi kT)^{3/2} \exp(-mc^2/2kT)c^2 \sin\theta\, dc\, d\theta\, d\varphi \qquad (1.23)$$

In many cases the molecular orientation is not of interest and we need consider only the distribution of *speeds*

$$F(c) = \int_0^{\pi}\int_0^{2\pi} F(c, \theta, \varphi)\, d\theta\, d\varphi = 4\pi(m/2\pi kT)^{3/2}c^2 \exp(-mc^2/2kT) \qquad (1.24)$$

This distribution is plotted in Fig. 1.4 in terms of the dimensionless parameter $c(M/2RT)^{1/2}$. Such a plot adjusts the speed scale to account for effects due to mass and temperature variation.

The maximum in the speed distribution, which occurs at the most probable speed $c_{\mathrm{mp}}(M/2RT)^{1/2} = 1$, corresponds to speeds of 422 m s^{-1} and 244 m s^{-1} for N_2 at 300 K and 100 K, respectively; plotted on the same scale the speed distribution at 100 K would be more sharply peaked than that at 300 K, just as was the case for the one-dimensional velocity distribution of Fig. 1.3. The maximum is due to two opposing effects. As the speed increases, the exponential factor $f(c)$ decreases, reflecting the consideration that a higher speed means a higher energy; on the other hand the factor $4\pi c^2$ increases, reflecting the fact that higher speeds can be attained in many

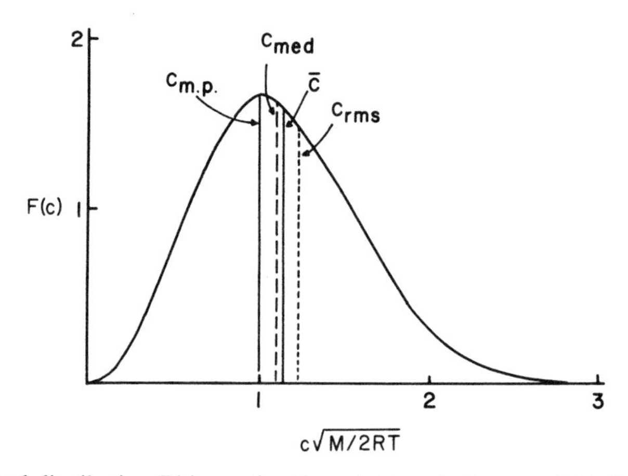

Fig. 1.4. Speed distribution $F(c)$ as a function of dimensionless speed. Various measures of molecular speed are indicated.

more ways than lower ones.[12] The result is the most probable speed, which is found by calculating $dF(c)/dc$ and setting it equal to zero. This measure of molecular speed is somewhat different from the rms speed already calculated, another consequence of having a distribution. A third common measure of speed is the average (or mean) speed, $\bar{c}$, given by the integral $\int_0^\infty cF(c)\,dc$; using the formulas (1.19)–(1.21) yields

$$\bar{c} = (8RT/\pi M)^{1/2} \tag{1.25}$$

intermediate between c_{mp} and c_{rms}.

In addition to the various averages that can be obtained, the speed distribution may be used directly to determine the fraction of molecules within a given speed range. As $F(c)\,dc$ is the probability that a molecule has a speed between c and $c + dc$, the fraction of molecules with speeds between c_l and c_u is

$$\int_{c_l}^{c_u} F(c)\,dc$$

In terms of the dimensionless parameter, $y = c/c_{\text{mp}} = c(M/2RT)^{1/2}$, the required fraction is

$$\frac{4}{\pi^{1/2}} \int_{y_l}^{y_u} y^2 \exp(-y^2)\,dy \tag{1.26}$$

demonstrating that the ratio of the molecular energies of interest to the thermal energy of the gas determines this property of the gas. The integral is related to a common transcendental function, the error function[13]

$$\operatorname{erf}(x) = \frac{2}{\pi^{1/2}} \int_0^x \exp(-y^2)\,dy$$

For many purposes the fraction with speeds above the cutoff c_l is of interest. Integrating by parts and setting $y_u = \infty$ the result is

$$\begin{aligned} &\frac{2}{\pi^{1/2}} \left[y_l \exp(-y_l^2) + \int_{y_l}^{\infty} \exp(-y^2)\,dy \right] \\ &\quad = \frac{2}{\pi^{1/2}}\, y_l \exp(-y_l^2) + [1 - \operatorname{erf}(y_l)] \end{aligned} \tag{1.27}$$

[12] Molecules with speed c have their velocity vector $\mathbf{c}$ on the surface of a sphere of surface area $4\pi c^2$. The larger the value of c the larger this area is.

[13] This function can be looked up in mathematical tables; it is defined so that $\operatorname{erf}(0) = 0$ and $\operatorname{erf}(\infty) = 1$. Using these tables is little more trouble than using ordinary tables of trigonometric functions. See M. Abramovitz and I. A. Stegun, eds., *Handbook of Mathematical Functions* (New York: Dover Publications Inc., 1965), p. 310.

which can be used to provide yet another measure of molecular speed, the median speed; by definition half the molecules move faster and half slower than the median. To compute the median refer to tables of the error function and find the value of y_l for which (1.27) equals 0.5; the result is $y_l = 1.0876$ or $c_{\text{med}} = 1.538(RT/M)^{1/2}$. The relative values of the four measures of molecular speed are then

$$c_{\text{mp}} : c_{\text{med}} : \bar{c} : c_{\text{rms}} = 1 : 1.088 : 1.128 : 1.225$$

which are also indicated in the plot of $F(c)$. When the cutoff speed is large, (1.27) is simplified. For 10% accuracy the integral term can be ignored whenever $y_l \gtrsim 3$; if 1% accuracy is required $y_l \gtrsim 7$.

The dimensionless parameter which characterizes the speed distribution is the ratio of the molecular kinetic energy to kT. Thus we may determine a Maxwellian energy distribution by defining $\epsilon = mc^2/2$ from which $d\epsilon = mc\,dc$ and (1.24) becomes

$$\begin{aligned} F(c)\,dc &= 4\pi(m/2\pi kT)^{3/2}(2\epsilon/m)^{1/2}e^{-\epsilon/kT}\,d\epsilon/m \\ &\equiv F(\epsilon)\,d\epsilon = 2\pi(1/\pi kT)^{3/2}\epsilon^{1/2}e^{-\epsilon/kT}\,d\epsilon \end{aligned} \tag{1.28}$$

which measures the fraction of molecules with energy between ϵ and $\epsilon + d\epsilon$.

A plot of $F(\epsilon)$ would be similar to that of $F(c)$ in Fig. 1.4. Of more interest is the fraction of molecules with energies greater than a specified value ϵ_0. This quantity, $F^*(\epsilon_0)$, is $\int_{\epsilon_0}^{\infty} F(\epsilon)\,d\epsilon$. Integrating by parts the result, expressed in terms of a dimensionless parameter $\alpha_0 = \epsilon_0/kT$, is

$$F^*(\alpha_0) = 2(\alpha_0)^{1/2}\exp(-\alpha_0)/\pi^{1/2} + [1 - \text{erf}(\alpha_0^{1/2})] \tag{1.29}$$

which could equally well be obtained from (1.27) by setting $y_l = (\alpha_0)^{1/2}$ and $y = \alpha^{1/2}$. It is plotted in Fig. 1.5; note how rapidly $F^*(\alpha_0)$ decreases for $\alpha_0 > 1$. The fraction of molecules with energy greater than 6.9×10^{-21} J (which is kT at 500 K) is indicated for a number of temperatures; it falls precipitously at the lower temperatures. While (1.29) and Fig. 1.5 precisely describe only a three-dimensional dilute gas, the qualitative behavior is general, extending even to the molecular energy distribution in liquids and solids, which has important ramifications, particularly in chemical kinetics. As only molecules with energies greater than a threshold energy, ϵ_0, can react this suggests that the reaction probability should be proportional to $\exp(-\epsilon_0/kT)$. Thus a basis for understanding the enormous temperature dependence of the rate of chemical reactions can be provided.

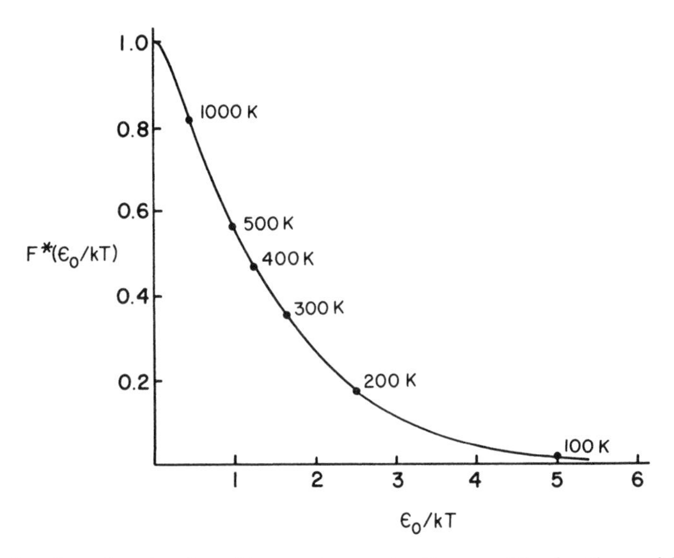

Fig. 1.5. Fraction of molecules with energies greater than ϵ_0. The fractions with energy greater than 6.9×10^{-21} J at various temperatures are indicated.

1.5. The Molecular Flux and Effusion of Gases

A direct measure of molecular speed can be obtained from the process of effusion. In such an experiment a pinhole is made in the wall of a container and gas leaks into an evacuated collecting vessel. The rate of gas flow is determined from the pressure in the collection chamber. If the pinhole is sufficiently small so that leakage does not perturb the gas in the container, the effusion rate is the same as the rate at which molecules collide with a wall. In our determination of the pressure we computed this quantity.

The molecular flux, $G(u)\,du$, is defined as the number of molecules with x component of velocity between u and $u + du$ which strike unit area in unit time, a quantity already calculated in Section 1.3. Using the same arguments which led to (1.13) yields

$$A\,\Delta t G(u)\,du = (Au\,\Delta t/V)N\varrho_x(u)\,du$$

or

$$G(u)\,du = (N/V)u\varrho_x(u)\,du \tag{1.30}$$

For effusion into vacuum there is no return flow and the total flux is found by integrating over all values of u greater than zero (only molecules with these velocities are moving toward the hole). Using (1.22) and (1.19)–(1.21) the total flux is

$$G = \frac{N}{V} \int_0^\infty u\varrho_x(u)\, du = n\bar{c}/4 \tag{1.31}$$

with $n = N/V$. If the area of a pinhole can be measured, an effusion experiment provides a direct measure of the mean speed; if not, effusion can be used to compare the mean speeds of different gases,[14]

$$\frac{G_1}{G_2} = \frac{n_1\bar{c}_1}{n_2\bar{c}_2} = \frac{n_1(T_1/m_1)^{1/2}}{n_2(T_2/m_2)^{1/2}} \tag{1.32}$$

Problems

1.1. Show that (1.12) implies (1.8).

1.2. A gas is in a *container* which undergoes uniform translational motion in the x direction with velocity V_x, e.g., the container is in an airplane. Assuming that the gas is at equilibrium, express the Maxwell distribution with respect to a stationary origin, e.g., the airport.

1.3. A 1-liter cube contains N_2 at 300 K and 10^{-3} atm. Using the uncertainty principle compute the minimum uncertainty in the specification of x component of velocity u. Accepting this quantity as the minimum possible value of Δu, compute the fraction of molecules with x component of velocity between u and $u + \Delta u$ for (a) $u = \bar{u} = 0$, (b) for $u = u_{\text{rms}} = (kT/m)^{1/2}$. How many molecules are in the two velocity domains?

1.4. Show that the speed distribution (1.24) and the energy distribution (1.28) are both normalized.

1.5. What fraction of molecules in He and CO have kinetic energies greater than 1.0 eV at 500 K and at 2000 K?

1.6. In the treatment of surface phenomena the speed and energy distributions in two dimensions are of interest. Transform to polar coordinates $u = c \cos \varphi$, $v = c \sin \varphi$ and show that

$$\begin{aligned} F(u, v)\, dv\, du &= \varrho_x(u)\varrho_y(v)\, du\, dv \\ &\equiv F(c, \varphi)\, dc\, d\varphi = (m/2\pi kT) \exp(-mc^2/2kT)\, c\, dc\, d\varphi \end{aligned}$$

---

[14] The effect was put to practical use in the Manhattan project. In order to construct an atomic bomb ^{235}U had to be separated from its far more abundant isotope ^{238}U. This was originally accomplished by building a huge gas-effusion system for the separation of gaseous $^{235}UF_6$ from $^{238}UF_6$.

1.7. Integrate $F(c, \varphi)\, dc\, d\varphi$ over all φ and show that the two-dimensional speed distribution is

$$F(c)\, dc = (m/kT) \exp(-mc^2/2kT)\, c\, dc$$

Now transform to energy as a variable and obtain the two-dimensional energy distribution

$$F(\epsilon)\, d\epsilon = (1/kT)e^{-\epsilon/kT}\, d\epsilon$$

1.8. For a two-dimensional gas what are c_{mp}, c_{med}, $\bar{c}$, and c_{rms}?

1.9. What is the average translational kinetic energy of particles effusing through a pinhole? (Assume that the idealized conditions for an effusion experiment are met.)

1.10. Given the following isotopic abundance

H: 0.99972	D: 0.00028
^{235}U: 0.00715	^{238}U: 0.99285

compute the isotopic composition when the following naturally occurring isotopic mixtures effuse through a pinhole:

(a) molecular hydrogen
(b) uranium hexafluoride

1.11. To construct an atomic bomb based on uranium, the fuel must be nearly pure ^{235}U. How many passes through an effusion apparatus are required to enrich naturally occurring uranium to 95% ^{235}U? to 99% ^{235}U? The effusing gas is UF_6.

General References

S. Golden, *Elements of the Theory of Gases* (Reading, Mass.: Addison–Wesley, 1964), Chapters 3 and 4.

W. Kauzmann, *Kinetic Theory of Gases* (New York: Benjamin, 1966), Chapter 4.

J. E. Mayer and M. G. Mayer, *Statistical Mechanics* (New York: John Wiley, 1940), pp. 1–18.

R. D. Present, *Kinetic Theory of Gases* (New York: McGraw–Hill, 1958), Chapters 1, 2, and 5.

2

Kinetic Theory of Gases—Transport

2.1. Introduction

According to the kinetic model, molecules move very rapidly. At 300 K the rms speed of gaseous bromine is 125 m sec^{-1}. Yet, if a small amount of bromine is added to nitrogen at 300 K and 1 atm the rate at which the bromine diffuses through the gas is much slower, about 10^{-3} m sec^{-1}. The presence of the nitrogen impedes the movement of the bromine, reducing the effective rate of travel by a factor of 10^5 from that which occurs in the absence of nitrogen. The most obvious possible cause for such retardation is the effect of molecular collision.

As long as the dimensions of the container are large enough so that molecules undergo collisions in traversing the vessel, a simple, although imperfect, analogy is found in the new amusement park ride of miniature hovercraft supported above a rink by air jets. Imagine that all drivers are blindfolded, the steering wheels of their craft locked so that they move in straight lines, and that they are required to cross the arena. If there are few craft the chances are excellent that no collisions will occur and that they will traverse the floor without incident. As more craft participate collisions become more likely and the distance traveled between them decreases. Each collision redirects the craft, which goes charging off on its new path almost surely to collide with another vehicle. The possibility of making headway is enormously reduced and the time required to cross the floor increases greatly. On the molecular level the difficulties are much the same. Thus, the time required for a molecule to traverse a container

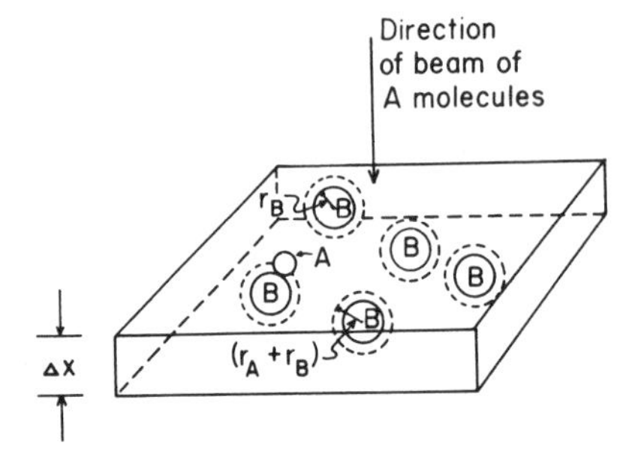

Fig. 2.1. Model for calculating the A–B collision probability when a beam of A molecules is directed at a gas of B molecules. One A–B pair is shown making a glancing collision. The dashed circles indicate the effective A–B collision diameters.

of gas is related to the distance traveled between collisions or, equally well, to the collision frequency.

2.2. Mean Free Path

An approximate determination of the *mean free path*, i.e., the average distance traveled between collisions, can most simply be done by considering the following experiment in which a beam of A molecules, originally of intensity $I_A(0)$, is attenuated as it passes through a gas containing B molecules. The beam intensity at any point in the gas is $I_A(x)$. To compute $I_A(x)$ requires determining the probability, $p(x)\,dx$, that an A molecule is scattered from the beam between x and $x + dx$. The decrease in intensity (the number *ejected* from the beam) in this interval is the product of the scattering probability and the intensity

$$dI_A(x) = -I_A(x)p(x)\,dx \tag{2.1}$$

To compute $p(x)$ consider a cross section of the gas of thickness Δx, as shown in Fig. 2.1. The B molecules provide the targets for the beam of A molecules. Assuming that the molecules are rigid spheres[(1)] of radii r_A and r_B, the effective area of each target, known as the *collision cross section*, is $\pi(r_A + r_B)^2$; stated differently, if the path of the center of an A molecule passes within a distance $r_A + r_B$ of the center of a B molecule, a collision occurs. The probability of a collision is the ratio of the area covered by target molecules to the area of the beam, a. If B is so dilute that individual target areas do not overlap, this probability is

$$p(x)\,\Delta x = (n_B a\,\Delta x)(\pi\,d_{AB}^2)/a$$

(1) The hard-sphere assumption is a considerable restriction, and most certainly an unrealistic one. See footnote (1) in Chapter 1.

where n_B is the number density of B and $d_{AB} = r_A + r_B$; the number of B molecules in the specified volume is $n_B a \,\Delta x$. Simplifying the expression yields $p(x)$, the probability per unit distance of a molecule in the beam colliding with the target gas,

$$p(x) \equiv \frac{1}{l} = n_B \pi \, d_{AB}^2 \tag{2.2}$$

Combining (2.1) and (2.2) we obtain

$$\frac{dI_A}{dx} = -\frac{1}{l} I_A(x)$$

which has the solution

$$I_A(x) = I_A(0) e^{-x/l} \tag{2.3}$$

In the beam experiment the mean free path, λ_A, is the average distance traveled by an A molecule before it undergoes a collision. Thus λ_A is the integrated product of the fraction of A molecules ejected between x and $x + dx$ multiplied by the distance the molecules have traveled:

$$\lambda_A = \int_{x=0}^{x=\infty} \left[\frac{-dI_A}{I_A(0)}\right] x$$

Using (2.1), (2.2), and (2.3) the result is

$$\lambda_A = \int_0^\infty p(x) \left[\frac{I_A(x)}{I_A(0)}\right] x \, dx = \int_0^\infty \frac{1}{l} e^{-x/l} x \, dx = l \tag{2.4}$$

If the target gas were a mixture of A and B the analysis would have to be generalized to account for A–A collisions. Then $p(x)$ and l would no longer be given by (2.2). Assuming the target gas mixture is dilute enough so that the collision probabilities are additive the appropriate generalization is

$$p(x) = \frac{1}{l_A} = n_A \pi \, d_{AA}^2 + n_B \pi \, d_{AB}^2 \tag{2.5}$$

with $d_{AA} = 2r_A$.

There is an obvious flaw in the arguments used to derive (2.4) even granting the assumption of rigid spherical molecules and the limitation to low-density gases. No account has been taken of the variation in the relative velocity of the molecules making up the beam and the target, an effect which alters the numerical factor in (2.4) but not the qualitative behavior; λ_A still decreases as the gas density or the collision cross section increases.

Since molecular diameters, determined from the volume of condensed matter, range between 0.2 and 0.5 nm, the mean free paths can be estimated. If we use (2.2) and (2.4) the mean free path of Br_2 molecules in N_2 gas at 300 K and 1 atm can be estimated. Assuming 0.4 nm for the N_2–Br_2 distance of closest approach we find $\lambda_{Br_2} \sim 100$ nm. Since at 300 K the mean speed of Br_2 is ~ 200 m sec^{-1}, the typical bromine molecule makes $\sim 2\times 10^9$ collisions per sec. Each collision redirects the molecule; thus it is no surprise that diffusion through a molecular maze is a much slower process than unimpeded effusion.

2.3. Collision Frequency

An accurate calculation of mean free path requires computing the average distance a molecule travels between collisions, a quantity which depends upon the speed of the molecule. The mean free path of stationary molecules is zero since they do not move until struck by other molecules. To very fast molecules the remainder of the gas appears as immobile target molecules and the calculation of the previous section is exact. We thus expect (2.4) to provide an upper bound for the speed-dependent mean free path.

In a gaseous mixture of A and B at equilibrium a representative A molecule with speed c travels a distance $c\,\Delta t$ in a time Δt. During this period such a typical molecule makes $z_A(c)\,\Delta t$ collisions[2] [$z_A(c)$ is the speed-dependent collision rate]. The average distance between collisions, $\lambda_A(c)$, is

$$\lambda_A(c) = c\,\Delta t / z_A(c)\,\Delta t = c/z_A(c) \tag{2.6}$$

and the mean free path is found by averaging over speeds

$$\lambda_A = \int_0^\infty F_A(c)\left[\frac{c}{z_A(c)}\right] dc \tag{2.7}$$

where $F_A(c)$ is the speed distribution function (1.24).

As a representative A molecule moves it sweeps out collision cylinders with respect to the other molecules in the gas. In Fig. 2.2 such a collision

[2] As previously noted, collisions alter the speed of the individual molecules. However, as the gas is at equilibrium, the total number of A molecules with a given speed is invariant. It is thus correct to consider the properties of molecules of a particular speed even though the members of this class are continually changing. See Section 1.2 and footnote (10) in Chapter 1.

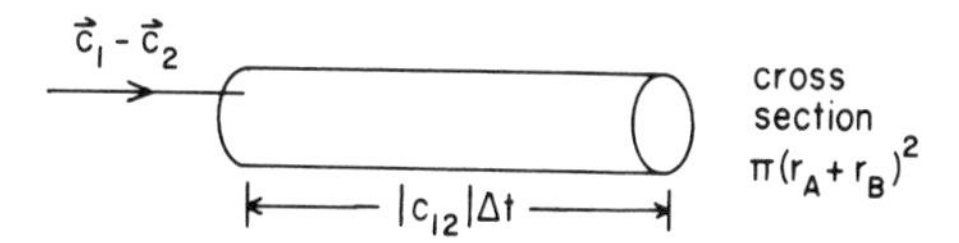

Fig. 2.2. Collision cylinder for A–B collisions.

cylinder is shown for interaction with B molecules with velocity between $\mathbf{c}_2$ and $\mathbf{c}_2 + d\mathbf{c}_2$.[3] The cross-sectional area is πd_{AB}^2, just as in the model discussed in Section 2.2. The height of the cylinder depends upon the *relative speed* of the colliding molecules; those approaching slowly sweep out a smaller volume and have a smaller chance of undergoing a collision. Denoting the velocity of the A molecule as $\mathbf{c}_1$, the collision cylinder described in a time Δt has the volume

$$\pi\, d_{AB}^2 c_{12}\, \Delta t \tag{2.8}$$

where $c_{12} = |\,\mathbf{c}_1 - \mathbf{c}_2\,|$ is the relative speed of the colliding pair. As the collision probability is the ratio of (2.8) to the total volume V and the number of B molecules in the specified velocity domain is $N_B F_B(\mathbf{c}_2)\, d\mathbf{c}_2$, the A–B collision rate is

$$n_B \pi\, d_{AB}^2 c_{12} F_B(\mathbf{c}_2)\, d\mathbf{c}_2 \tag{2.9}$$

By integrating over all values of the velocity $\mathbf{c}_2$, $z_{AB}(c)$, the collision frequency of a representative A molecule of speed c with B molecules of arbitrary velocity can be found. The calculation is left as a problem. The result is

$$z_{AB}(c) = n_B \pi\, d_{AB}^2 \left(\frac{2kT}{\pi m_B}\right)^{1/2} \left[\exp(-\alpha^2) + \frac{1+2\alpha^2}{\alpha} \int_0^\alpha \exp(-y^2)\, dy\right] \tag{2.10}$$

where $\alpha = c(m_B/2kT)^{1/2}$ and the integral is related to the error function introduced in Section 1.4. In a binary mixture the total collision frequency is

$$z_A(c) = z_{AA}(c) + z_{AB}(c) \tag{2.11}$$

which, when combined with (2.10) and (2.6), determines $\lambda_A(c)$. The speed-dependent mean free path is an involved function of the relative masses of the two components, their densities, and their collision cross sections.

[3] This is a shorthand notation; it indicates a velocity vector with components in a restricted domain. In Cartesian coordinates the x, y, and z components lie between u and $u + du$, v and $v + dv$, and w and $w + dw$, respectively. In spherical polar coordinates the speed is between c and $c + dc$, and the solid angle is bounded by θ and $\theta + d\theta$ and φ and $\varphi + d\varphi$.

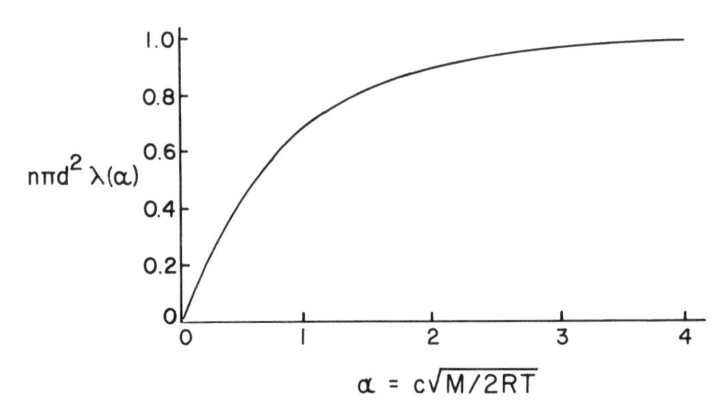

Fig. 2.3. $(n\pi d^2)\lambda(\alpha)$, the ratio of the speed-dependent mean free path to its value at infinite speed, in a one-component system as a function of the scaled speed α.

In a one-component system there is considerable simplification; λ is plotted in Fig. 2.3 as a function of the scaled velocity $c(M/2RT)^{1/2}$. Our expectations are verified; the *relatively slow* molecules have short mean free paths while *relatively fast* ones have values of λ which approach the bound calculated in the previous section. If we use (2.7), λ_A can be determined by numerical or graphical means; the result is

$$\lambda_A = 0.6774/(n_A \pi\, d_{AA}^2) \tag{2.12}$$

about $\frac{2}{3}$ of its limiting value. To determine λ_A in a mixture requires a separate calculation for each composition.

An alternative approach, which handles mixtures as readily as it does pure substances, is to approximate (2.7) by averaging numerator and denominator independently,

$$\lambda_A \approx \bar{c}_A/\bar{z}_A \tag{2.13}$$

an expression which, while often defined as the mean free path, does not properly account for the speed variation illustrated in Fig. 2.3. To determine $\bar{z}_A$ requires the sum of the average frequency of A–A and A–B collisions, the analog to (2.11). The mean collision rate of an A molecule with any B molecule is found by multiplying (2.9) by the probability that an A molecule has its velocity between $\mathbf{c}_1$ and $\mathbf{c}_1 + d\mathbf{c}_1$ and then integrating over all values of $\mathbf{c}_1$ and $\mathbf{c}_2$; the result is

$$\bar{z}_{AB} = n_B \pi\, d_{AB}^2 \iint c_{12} F_A(\mathbf{c}_1) F_B(\mathbf{c}_2)\, d\mathbf{c}_1\, d\mathbf{c}_2 \tag{2.14}$$

The integral in (2.14), which is the mean *relative speed* $\bar{c}_{AB}$, is most easily evaluated by transforming to center-of-mass and relative coordinates,

$$\mathbf{c}_M = (m_A \mathbf{c}_1 + m_B \mathbf{c}_2)/(m_A + m_B), \qquad \mathbf{c} = \mathbf{c}_1 - \mathbf{c}_2 \tag{2.15}$$

To reexpress the product $F_A(\mathbf{c}_1)F_B(\mathbf{c}_2)$ note that the exponential factor is dependent upon the total kinetic energy of the two particles, $(m_A c_1{}^2 + m_B c_2{}^2)/2$. If (2.15) is solved for $\mathbf{c}_1$ and $\mathbf{c}_2$ and these values substituted in the expression for the total kinetic energy, it takes the form $(Mc_M{}^2 + \mu c^2)/2$, where M is $m_A + m_B$, and μ, the reduced mass for an A–B pair, is $m_A m_B/M$. Using these expressions along with (1.22) we find that

$$\begin{aligned} F_A(\mathbf{c}_1)F_B(\mathbf{c}_2) &= \left(\frac{M}{2\pi kT}\right)^{3/2}\left(\frac{\mu}{2\pi kT}\right)^{3/2} \exp[-(Mc_M{}^2 + \mu c^2)/2kT] \\ &= F_M(\mathbf{c}_M)F_R(\mathbf{c}) \end{aligned} \tag{2.16}$$

Note that the transformed representation is a product of Maxwellian distributions of the center-of-mass and relative kinetic energy. To complete the transformation of (2.14) requires expressing $d\mathbf{c}_1\, d\mathbf{c}_2$ in terms of the new coordinates. The result, obtained by the Jacobian method outlined in the Appendix, is $d\mathbf{c}_1\, d\mathbf{c}_2 = d\mathbf{c}_M\, d\mathbf{c}$ so that the integral in (2.14) is the product

$$\int F_M(c_M)\, d\mathbf{c}_M \int F_R(c) c\, d\mathbf{c}$$

which is easily interpreted. The first factor is a normalization integral for particles of mass M and the second factor is the average speed for particles of mass μ so that $\bar{c}_{AB}$ is

$$\bar{c}_{AB} = (8kT/\pi\mu)^{1/2} = [8kT(m_A + m_B)/\pi m_A m_B]^{1/2} \tag{2.17}$$

which when substituted in (2.14) yields

$$\bar{z}_{AB} = n_B \pi\, d_{AB}^2 [8kT(m_A + m_B)/\pi m_A m_B]^{1/2} \tag{2.18}$$

Note that $\bar{c}_{AB}$ is *always* larger than the mean speed of *either* molecule.

The mean collision frequency for an A molecule in a binary mixture is thus

$$z_A = n_A \pi\, d_{AA}^2 2^{1/2} \bar{c}_A + n_B \pi\, d_{AB}^2 \bar{c}_{AB} \tag{2.19}$$

since $\bar{c}_{AA}$ is $2^{1/2}\bar{c}_A$. When combined with (2.13) the mean free path is approximated by

$$\lambda_A \approx [2^{1/2} n_A \pi\, d_{AA}^2 + (1 + m_A/m_B)^{1/2} n_B \pi\, d_{AB}^2]^{-1} \tag{2.20}$$

Compared to the exact one-component result, (2.12), the approximation overestimates λ by 4%. Computer integration of (2.7) allows the same comparison to be made for binary mixtures. The results indicate that the maximum error occurs when $m_A = m_B$; in practice the approximation (2.20) is excellent.

For the example of a dilute mixture of Br_2 in N_2 gas at 1 atm and 300 K considered in Section 2.2, we find that the mean relative speed is 516 m sec^{-1}. Assuming, as before, that $d_{AB} \sim 0.4$ nm, we obtain a collision frequency $\sim 6.4 \times 10^9$ sec^{-1} and a mean free path ~ 31 nm. It is worth noting that λ_{Br_2} is much *larger* than the mean interparticle distance, which is only 3.4 nm at this density.

In addition to quantities already calculated, the total collision rate for all A molecules per unit volume is useful for application to chemical kinetics. It is

$$Z_A = Z_{AA} + Z_{AB} = n_A{}^2 \pi \, d_{AA}^2 \bar{c}_A / 2^{1/2} + n_A n_B \pi \, d_{AB}^2 \bar{c}_{AB} \tag{2.21}$$

which is essentially n_A times the molecular collision frequency (2.19); the factor of $1/2^{1/2}$ in the first term of (2.21) corrects for the fact that this procedure counts each A–A collision twice, i.e., $Z_{AA} = n_A z_{AA}/2$.

2.4. Macroscopic Equations of Transport

If a system is displaced from equilibrium by creating a temperature gradient, heat flow takes place in the opposite direction to reestablish equilibrium. This is a typical *transport* process; energy flows from one region of the system to another. In a one-component system at constant pressure there is a simple relationship between the heat flux and the temperature gradient. As long as the gradient is small, *Fourier's law of heat conduction* is applicable. For a temperature gradient in the z direction the heat flux, J_q, is

$$J_q = -\varkappa \frac{dT}{dz} \tag{2.22}$$

where $\varkappa$ is the thermal conductivity.

For mixtures with concentration (number density) gradients, matter flows are established to restore homogeneity. There can be two contributions to this flux, one due to motion of the system as a whole (center-of-mass motion), the other due to the relative motion of the various components of the mixture. It is the second of these, the diffusive flux J, which serves to homogenize the system. In a binary mixture at constant temperature with a concentration gradient in the z direction, *Fick's first law of diffusion* states that

$$J_A = -D \frac{dn_A}{dz} \tag{2.23}$$

where D is the diffusion coefficient. If the system is also at constant pressure the center of mass is stationary so that the total flux and the diffusive flux are the same, and, since $n_A + n_B = n$, which is constant, $J_A = -J_B$.

In addition to temperature and concentration gradients, transverse velocity gradients may also be established in a fluid. Figure 2.4 illustrates the velocity profile in a system where there is viscous flow. For molecules near the stationary plate the mean x component of velocity, $\bar{u}$, is zero; at the moving plate it is δU. The gradient in $\bar{u}$, $d\bar{u}/dz$, is opposed by the drag force required to keep the plate in motion. The larger the area of the plate, the greater this force. *Poiseuille's law* describes the relation between the force per unit area in the x direction, P_x, and the velocity gradient

$$P_x = -\eta \frac{d\bar{u}}{dz} \tag{2.24}$$

where η is the shear viscosity. Since the viscous force per unit area has the dimensions of pressure it can equally well be interpreted as the momentum flux established to counter a velocity gradient.

The three formulas (2.22)–(2.24) are examples of linear transport equations; they relate the response of a system (the flux) to a small perturbing force (the gradient). The transport coefficients D, η, and $\varkappa$ are the parameters of proportionality, to be determined experimentally. A familiar transport equation is Ohm's law. Here voltage is the force, current the response, and conductivity (the reciprocal of resistance) the transport coefficient. In general, equations of transport are not as simple as these. In a two-component system with a temperature gradient, Fourier's law states that there is only heat flow. However, if the masses of the components

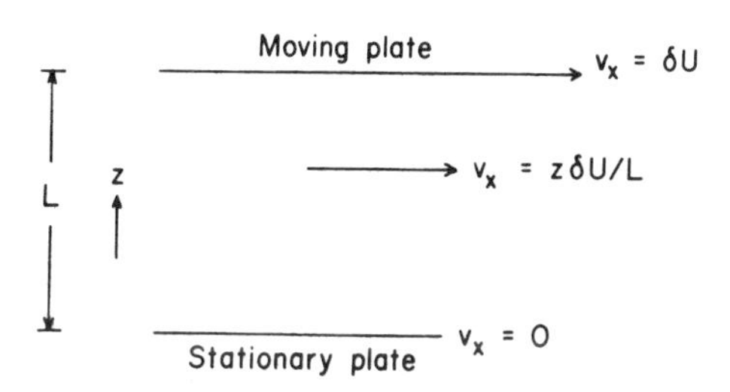

Fig. 2.4. Velocity profile in a gas undergoing viscous flow. The lower plate is stationary while the upper one has a velocity $v_x = \delta U$.

differ significantly, so does the mean speed of each one. There is no reason to expect that the flux of each is the same. In fact, the temperature gradient leads to a flux of matter, the phenomenon of thermal diffusion.[4] Similarly a concentration gradient causes heat flow as well as the matter flow demanded by Fick's law.

Rigorous phenomenological justification of transport equations is the domain of irreversible thermodynamics.[5] By extending the entropy concept to systems slightly displaced from equilibrium and requiring that entropy increases with time, it is possible to generalize flux–force relations like (2.22)–(2.24) to account for coupled phenomena such as thermal diffusion. An important consequence is that there is an inherent correlation between reciprocal processes such as mass flux due to a temperature gradient and heat flux due to a concentration gradient. The transport coefficient for the process and its inverse are not independent. Such constraints on the transport coefficient for a process and its inverse are quite general and are known as Onsager reciprocal relations.[6]

2.5. Solution of the Transport Equations

As presented the transport equations do not explicitly involve time. To include temporal effects consider a system such as the diffusion cell of Fig. 2.5 in which the A molecules are initially confined to the narrow region between 0 and a; when the barriers are removed they diffuse into the (essentially infinite) region containing B molecules. To determine the concentration of A as a function of both position and time consider a part of the cell of cross-sectional area S between z and $z + \Delta z$. The rate of change of the number of A molecules in this volume is the difference between the entering and leaving flows:

$$\left(\frac{\partial N_{\mathrm{A}}}{\partial t}\right)_z = SJ_{\mathrm{A}}(z) - SJ_{\mathrm{A}}(z + \Delta z)$$

Since the volume is $S\,\Delta z$ the number density of A is $N_{\mathrm{A}}/S\,\Delta z$ and because

[4] Thermal diffusion has been used as a basis for isotope separation techniques [K. Clusius and G. Dickel, *Naturwiss.* **26**, 546 (1938)].

[5] For an introduction to this subject see I. Prigogine, *Thermodynamics of Irreversible Processes*, 3rd ed. (New York: Interscience, 1967).

[6] The theorems demonstrating the interconnections were first proved by L. Onsager, *Phys. Rev.* **37**, 405 (1931); **38**, 2265 (1931).

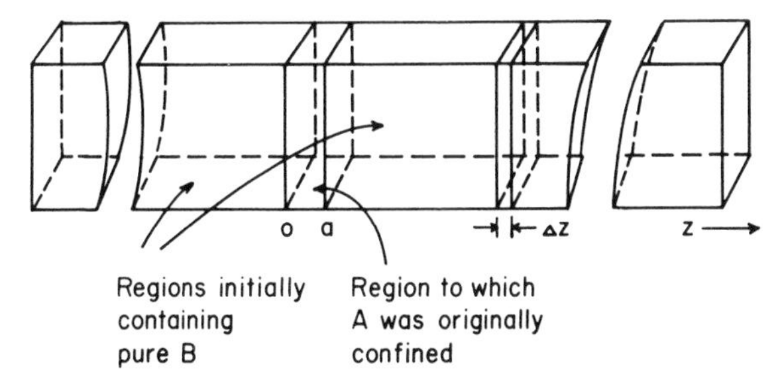

Fig. 2.5. Schematic model of a diffusion cell.

$J(z + \Delta z) = J(z) + \Delta z\, \partial J/\partial z$ we find, with (2.23),

$$\left(\frac{\partial n_A}{\partial t}\right)_z = -\left(\frac{\partial J}{\partial z}\right)_t = \left[\frac{\partial}{\partial z}\left(D\,\frac{\partial n_A}{\partial z}\right)\right]_t$$

or, if D is independent of position,

$$\left(\frac{\partial n_A}{\partial t}\right)_z = D\left(\frac{\partial^2 n_A}{\partial z^2}\right)_t \tag{2.25}$$

which is *Fick's second law of diffusion*. It determines the concentration profile for given boundary conditions.

A similar equation for the temperature can be derived from Fourier's law (2.22). Arguments similar to the ones just given show that

$$\left(\frac{\partial e}{\partial t}\right)_z = \varkappa\left(\frac{\partial^2 T}{\partial z^2}\right)_t \tag{2.26}$$

where e is the energy per unit volume. Since $e = E/V$,

$$\left(\frac{\partial e}{\partial t}\right)_z = \left(\frac{\partial (E/V)}{\partial t}\right)_z = \left(\frac{\partial (E/V)}{\partial T}\right)_v\left(\frac{\partial T}{\partial t}\right)_z = \frac{C_v}{V}\left(\frac{\partial T}{\partial t}\right)$$

and the equation for heat conduction is

$$\left(\frac{\partial T}{\partial t}\right)_z = \frac{\varkappa}{C_v/V}\left(\frac{\partial^2 T}{\partial z^2}\right)_t \tag{2.27}$$

Similarly the velocity profile is determined by the equation

$$\left(\frac{\partial u}{\partial t}\right)_z = \frac{\eta}{\varrho}\left(\frac{\partial^2 u}{\partial z^2}\right)_t \tag{2.28}$$

where ϱ is the mass density.

The three equations (2.25), (2.27), and (2.28) all have the same form. While general solutions for arbitrary boundary conditions present complicated mathematical problems,[7] certain one-dimensional problems are simple and instructive. Consider a cylinder of which one end is held at temperature T^* and the material inside is initially at temperature T_0. This is a model for heat flow in a long metal rod or in a tube containing liquid or gas. Making suitable correspondences it also describes diffusion in a long tube where the solute concentration at one end is kept constant. Ignoring effects due to the surface of the cylinder, the boundary conditions are

$$T = \begin{cases} T^*, & z = 0, \text{ all } t \\ T_0, & z > 0, \ t = 0 \end{cases} \tag{2.29}$$

The solution to (2.29), which may be verified by direct differentiation, is

$$\frac{T - T_0}{T^* - T_0} = \frac{2}{\pi^{1/2}} \int_s^\infty \exp(-y^2)\, dy = 1 - \operatorname{erf}(s) \tag{2.30}$$

where $s = z/[2(D^*t)^{1/2}]$, $D^* = \varkappa V/C_v$, and $\operatorname{erf}(s)$ is the error function discussed in Section 1.4. Since $\operatorname{erf}(0) = 0$ and $\operatorname{erf}(\infty) = 1$, the boundary conditions (2.29) are obviously satisfied.

The temperature profile is a function of the single composite variable s. This observation implies that:

(a) the time required for a point to reach a given temperature is proportional to the square of the distance from the heat source, inversely proportional to the thermal conductivity, and directly proportional to the heat capacity;

(b) the distance to which a given temperature has penetrated is proportional to $t^{1/2}$;

(c) the heat flux required to keep the $z = 0$ surface at constant temperature is proportional to $t^{-1/2}$; this follows by substituting (2.30) in (2.22) and setting $z = 0$.[8]

As an application of (2.30) consider a piece of copper tubing which has one end maintained at 500°C (easily done with a propane torch). Let us estimate the time required for a point 0.3 m (about 10 in.) from the end to reach 100°C (which is too hot to hold) assuming an initial temperature

[7] For general solutions see J. Crank, *The Mathematics of Diffusion* (Oxford: Clarendon Press, 1975).

[8] J. Crank, *The Mathematics of Diffusion* (Oxford: Clarendon Press, 1975), p. 87.

of 20°C. Numerical substitution indicates that s is the solution of the equation $\text{erf}(s) = \frac{5}{6}$; referring to tabulations of the error function we find that $s \sim 0.98$. For copper at room temperature $\varkappa = 420 \text{ J m}^{-1} \text{ sec}^{-1} \text{ K}^{-1}$ and $C_v/V = 3.46 \times 10^6 \text{ J m}^{-3} \text{ K}^{-1}$ so that $D^* = 1.21 \times 10^{-4} \text{ m}^2 \text{ sec}^{-1}$ and thus

$$t = \frac{z^2}{4D^* s^2} = 2.07 \times 10^3 \frac{z^2}{s^2} = 2.15 \times 10^3 z^2$$

A point 0.3 m from the end of the rod reaches 100°C in approximately 193 sec or a little over 3 min.[9]

2.6. Kinetic Theory of Transport and Postulate of Local Equilibrium

The linear relations between forces and fluxes which were discussed in the previous section are empirical. To provide a justification based upon kinetic theory and simultaneously to obtain expressions for the transport coefficients $\varkappa$, η, and D requires developing a molecular model for transport in dilute gases and extending the concept of a velocity distribution to non-equilibrium systems.

For the systems considered the gradients are very small. Thus, in any region of the gas, concentration, temperature, and mean velocity are well-defined experimental quantities. There is no objection, either in principle or practice, to inserting a probe which can measure the local values of these macroscopic parameters (as, for example, one might measure temperature at different points in a room). As the gas is *not* at equilibrium the velocity distribution is positionally dependent but the mean molecular properties change little over distances comparable with the mean free path (estimated in Section 2.2 to be between 30 and 300 nm for a gas at 300 K and 1 atm). In any region, extending over many mean free paths, the properties of the gas can be characterized by the local values of the macroscopic observables.

[9] The solution to the diffusion equation (2.30) implies that the temperature decreases monotonically with z at all times. It is interesting that this uniform behavior is *not* a property of ultrapure materials. Instead, heat is propagated as a wave, in a fashion similar to the propagation of sound. This phenomenon, known as *second sound*, was first discovered in liquid helium at very low temperatures. It has since been observed in solid helium and has been reported to occur as well in the isotopically and chemically pure materials NaF and Bi at very low temperatures. For a discussion of the effect see B. Bertram and D. J. Sandiford, *Scientific American* **222**(5), 92 (1970).

Consider a gas at 1 atm in which the substantial temperature gradient of 1 K mm^{-1} is maintained. In traveling 1 mm a molecule undergoes $\sim 10^4$ collisions. If the *mean* temperature of the gas is ~ 300 K then the mean speed differs by 0.2% for regions 1 mm apart. On the other hand, the relative dispersion in the speed is about 20%. Thus the variation in $\bar{c}$ due to the temperature gradient is far less, in the example considered, than the normal dispersion of the distribution. We can therefore imagine that in each region of the gas the velocity distribution is Maxwellian, but with the temperature characteristic of the particular region. If there are also gradients in concentration and velocity it is natural to extend this assumption and to assert that the positionally dependent velocity distribution is Maxwellian in every region of the gas; however the distribution is characterized by the *local* values of temperature, velocity, and concentration. It must be emphasized that this postulate of *local equilibrium* is only a first approximation. To improve upon it, which is necessary to develop an exact theory, requires elaborate physical and mathematical analysis.

2.7. Simplified Transport Theory: Dilute Hard-Sphere Gas

Depending upon the density, transport in a gas may be either effusive or diffusive in nature. If the gas is so dilute that the mean free path, λ, is greater than the length of the vessel, L, a molecule rarely undergoes a collision while traversing the vessel. It is much more likely to hit the wall. Transport in this effusive domain, known as Knudsen flow,[10] is extremely sensitive to the nature of the interactions between molecules and the walls of the container. While sensitive to gas–surface interactions, Knudsen flow does not depend on the forces between gas molecules.

At high pressures $\lambda < L$ and the gas is in the diffusive domain. To construct a molecular model which accounts for collisional effects, consider Fig. 2.6; the macroscopic gradients are in the z direction. If we assume local equilibrium is established in the volume element $d\tau$, particles in that region have a Maxwellian speed distribution and are moving randomly with mean speed $\bar{c}$.[11] Molecules in $d\tau$ have properties different from those

[10] For consideration of transport when $\lambda > L$ see M. Knudsen, *The Kinetic Theory of Gases* (New York: John Wiley and Sons, 1950).

[11] This is not quite accurate; there are small perturbations to the distribution function induced by the macroscopic gradients. To treat such effects properly requires a far more elaborate treatment than is given here. See R. D. Present, *Kinetic Theory of Gases* (New York: McGraw–Hill, 1958) for details.

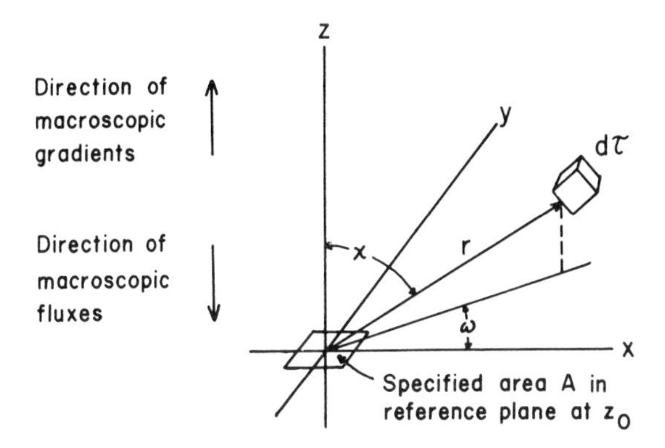

Fig. 2.6. Coordinate geometry for determining the differential molecular flux through a reference plane at z_0.

in the vicinity of a reference plane located at z_0. The further $d\tau$ is from z_0, the greater the differences. If molecules from $d\tau$ reach z_0 *without* undergoing a collision they will on average, contribute to the macroscopic flux through z_0 in a manner determined by the density, momentum, and temperature in $d\tau$. Of course the further $d\tau$ is from z_0 the greater is the chance of a collision which leads to reequilibration in a new region. We assume that such reequilibration is established with each collision, i.e., that molecules are only aware of the last stimulus to which they have been exposed. Thus to compute the flux of any property at z_0 requires determination of the differential flux of molecules originally in $d\tau$ that reach z_0 without an intermediate collision. The differential flux multiplied by the mean local energy or momentum yields the differential energy or momentum flux. Then summing over all regions $d\tau$ yields the corresponding macroscopic flux.

The total number of particles in $d\tau$ is $n(z)\,d\tau$, where $n(z)$ is the number density, which may be position dependent. Of these only the fraction moving in the direction of r can pass through the specified area A in the reference plane z_0. Those molecules originally in $d\tau$ *that have not undergone a collision* spread out uniformly over a sphere of area $4\pi r^2$. Since the projection of A onto this sphere is $A \mid \cos\chi \mid$, the fraction which could pass through A is $A \mid \cos\chi \mid/4\pi r^2$. Collisions reduce the flux through A. In Section 2.2 we found that the fractional change in flux intensity per unit distance traveled is $\lambda^{-1}e^{-r/\lambda}$. Then, since molecules, on the average, travel a distance $\bar{c}\,\Delta t$ during an interval Δt, the total number of molecules initially in $d\tau$ that pass through A in a time t *without undergoing an intermediate collision* is

$$n(z)\,d\tau(A \mid \cos\chi \mid/4\pi r^2)(\lambda^{-1}e^{-r/\lambda})\bar{c}\,\Delta t$$

Since molecules crossing the reference plane from above are moving in the $-z$ direction the differential molecular flux at z_0 is

$$dJ = -n(z)(\cos\chi/4\pi r^2\lambda)\bar{c}e^{-r/\lambda}\,d\tau \tag{2.31}$$

When $0 < \chi < \pi/2$ (molecules arriving from above z_0) the differential flux is negative; when $\pi/2 < \chi < \pi$ it is positive.[12]

Application of (2.31) is straightforward. To obtain Poiseuille's law compute the momentum flux for a system in which the mean value of the x component of velocity is position dependent,

$$P_x = \int m\bar{u}(z)\,dJ \tag{2.32}$$

Expanding $\bar{u}(z)$ in a Taylor's series about z_0 and setting $z_0 = 0$ we find

$$\bar{u}(z) = \bar{u}(z_0) + z\left.\frac{d\bar{u}}{dz}\right|_{z_0} + \frac{z^2}{2}\left.\frac{d^2\bar{u}}{dz^2}\right|_{z_0} + \cdots \tag{2.33}$$

Since viscous flow occurs at constant temperature and density, the momentum flux, with the geometry of Fig. 2.6, is

$$P_x = -\frac{mn\bar{c}}{4\pi\lambda}\int_0^\infty\int_0^\pi\int_0^{2\pi}\left(\frac{\cos\chi}{r^2}\,e^{-r/\lambda}\right)$$
$$\times\left(\bar{u}_0 + r\cos\chi\bar{u}_0' + \frac{r^2\cos^2\chi}{2}\,\bar{u}_0'' + \cdots\right)r^2\sin\chi\,dr\,d\chi\,d\omega$$

where we have substituted $\bar{u}(z_0) = \bar{u}_0$, etc., and $z = r\cos\chi$. As long as gradients are small, higher-order terms in the $\bar{u}(z)$ expansion may be ignored. The integrals are easily done. Because of the dependence on $\cos\chi$ only the term proportional to $\bar{u}_0'$ is nonzero; the result is

$$P_x = -\frac{mn\lambda\bar{c}}{3}\left.\frac{d\bar{u}}{dz}\right|_{z_0} \tag{2.34}$$

precisely Poiseuille's law (2.24). The viscosity is

$$\eta = \frac{mn\lambda\bar{c}}{3} = \frac{2}{3\pi}\,\frac{(\pi mkT)^{1/2}}{\pi d^2} \tag{2.35}$$

[12] It is possible to modify (2.31) to account for the distribution of molecular speeds and thereby to consider the speed dependence of the mean free path. However, such modification does not incorporate the effect of macroscopic gradients on the distribution function. Furthermore, it is still restricted to billiard ball dynamics. Thus, even though it is conceptually more in tune with the previous analysis we do not pursue it here.

where the second expression is determined by introducing $\bar{c}$ from (1.25) and λ from (2.20), where $d = d_{AA}$.[13]

Fourier's law is almost as easily established. The energy flux is

$$J_q = \int \bar{\epsilon}(z)\, dJ \tag{2.36}$$

where $\bar{\epsilon}(z)$ is the z-dependent mean energy. Since heat flow occurs in a system at *constant pressure*, the density is not uniform, $n(z) = p/kT(z)$. Furthermore the mean speed is also position dependent, $\bar{c} = [8kT(z)/\pi m]^{1/2}$. Finally the mean energy of an ideal gas is $\tilde{c}_v T(z)$, where $\tilde{c}_v$ is the constant volume heat capacity *per molecule*. Substituting into the heat flow integral (2.36), expanding $T(z)$ in a Taylor's series about z_0, and integrating, the result is

$$J_q = -\frac{\tilde{c}_v n \lambda \bar{c}}{6} \left.\frac{dT}{dz}\right|_{z_0} \tag{2.37}$$

which is Fourier's law (2.22). The thermal conductivity may be identified as

$$\varkappa = \frac{n \tilde{c}_v \lambda \bar{c}}{6} = \frac{1}{3\pi} \frac{\tilde{c}_v}{m} \frac{(\pi mkT)^{1/2}}{\pi d^2} \tag{2.38}$$

The simple mean free path approach is not adequate for describing mutual diffusion in a binary system at constant temperature. Only in one case is the theory self-consistent—when the molecules are mechanically indistinguishable (self-diffusion). By integrating (2.31) we obtain the total flux of one component, $J_A = \int dJ_A$. Since the only position-dependent quantity is $n_A(z)$, the analysis used in the derivation of Poiseuille's law yields

$$J_A = -\frac{\lambda_A \bar{c}_A}{3} \frac{dn_A}{dz} \tag{2.39}$$

precisely Fick's law (2.23) from which we identify

$$D_A = \frac{\lambda_A \bar{c}_A}{3} \tag{2.40}$$

The same arguments may be applied to component B; the analogs to (2.39) and (2.40) are obtained. From (2.23) we note that there is only a single diffusion coefficient, i.e., $D_A = D_B$. In general $\lambda_A \bar{c}_A \neq \lambda_B \bar{c}_B$ so that the

[13] This expression for λ is used for internal consistency. Its derivation, like this treatment of viscosity, did not consider the variation of mean free path with speed.

Table 2.1. Comparison of Various Theoretical Formulas for the Transport Coefficients of Hard-Sphere Gases

Property	Transport coefficient $x[\pi d^2/(\pi mkT)^{1/2}]$	
	Section 2.6	Exact
mnD	0.159	0.375
η	0.159	0.3125
$m\varkappa$	$0.0796\tilde{c}_v$	$0.3125\tilde{c}_v + 0.7031k$

mean free path technique is internally inconsistent. The inconsistency may be corrected within the framework of this approach; however, the revised formulas are neither simple nor illuminating nor precise.[14] Instead we limit consideration to the one case where the matter flux computed using (2.39) is zero. When the molecules are mechanically indistinguishable, i.e., $m \equiv m_A = m_B$, $d \equiv d_{AA} = d_{AB} = d_{BB}$, (2.40) becomes

$$D = \frac{2}{3\pi} \frac{(\pi mkT)^{1/2}}{nm\pi d^2} \tag{2.41}$$

where $n = n_A + n_B$.

Our treatment is limited to a very simple model, a dilute gas of hard spheres. Nonetheless the results are only approximate. The expressions for η and D, (2.35) and (2.41), differ from the results of an exact theory of hard-sphere transport by a numerical factor. The expression for $\varkappa$, (2.38), is in error in an additional way. The derivation does not account for possible differences in the rate of transport of translational energy (center-of-mass kinetic energy) and of rotational and vibrational energy (internal energy). Exact theory does; the results, assuming internal energy is also equilibrated with each collision, are summarized in Table 2.1.

Both treatments lead to similar qualitative conclusions. The viscosity of a gas is predicted to be independent of the density, an astonishing result when first discovered by Maxwell. Experiment has confirmed the theory, which is readily understood qualitatively. As the density increases the molecular flow increases proportionately. On the other hand, the mean free path decreases and collisions occur more often. The two effects exactly

[14] For a discussion of this point see J. E. Mayer and M. G. Mayer, *Statistical Mechanics* (New York: John Wiley, 1940), p. 30.

balance. Another surprise was the prediction that gas viscosity increases with temperature, a result opposite to the more familiar situation in liquids, where viscosity decreases rapidly as temperature increases.

For the same qualitative reasons $\varkappa$, like η, is density independent. The thermal conductivity is more temperature sensitive than the viscosity because of the dependence on heat capacity. Its behavior is also quite different from that observed in liquids. In gases $\varkappa$ increases with T, while in liquids the opposite behavior is normal.

Unlike η or $\varkappa$, D is density dependent since diffusion, as discussed in the introduction to this chapter, is essentially a direct measure of the mean free path. The temperature dependence is a consequence of the fact that faster molecules diffuse more rapidly. From Table 2.1 we see that the ratio mnD/η is predicted to be constant in dilute gases; this is completely different from Walden's rule in liquids, $D\eta \approx \text{const}$.

2.8. Comparison with Experiment

The predictions of an exact treatment of transport for the dilute gas of hard spheres are summarized in Table 2.2. The correction for internal degrees of freedom is included. The effects of mean speed and collision cross section always occur in the combination $T^{1/2}/\pi d^2$; thus the ratios

$$\frac{mnD}{\eta} = 1.2 \tag{2.42}$$

$$\frac{m\varkappa}{(0.4\tilde{c}_v + 0.9k)\eta} = 2.5 \tag{2.43}$$

are predicted to be constants, independent of temperature or of the particular gas. These ratios are tabulated for a number of dilute gases at

Table 2.2. Exact Theoretical Expressions for Transport Coefficients of Hard Spheres

Flux	Gradient	Transport coefficient, $x[\pi d^2/(\pi mkT)^{1/2}]$
Matter	Concentration	$D = 3/(8mn)$
Momentum	Velocity	$\eta = 5/16$
Energy	Temperature	$\varkappa = 5[(\tilde{c}_v + 2.25k)/16m]$

Table 2.3. The Ratio mnD/η for Gases at Various Temperatures[a]

Temperature (K)	Ne	Ar	N_2	O_2	CH_4
77.7	1.29	1.26	1.33	1.33	1.60 @ 90.2 K
273.2	1.37	1.31	1.34	1.39	1.43
353.2	1.39	1.34	1.39	1.43	1.36

[a] Data from H. H. Landolt and R. Börnstein, *Zahlenwerte und Funktionen*, 6th ed. (Berlin: Springer, 1969), Vol. II, Part 5a, pp. 3, 516.

different temperatures in Tables 2.3, 2.4, and 2.5. From Table 2.3 it is clear that mnD/η does not vary greatly. It is closer to 1.2 for monatomic gases; deviation from the predicted ratio becomes larger as the molecules become more complex. The data indicate that the hard-sphere model accounts for most of the observed phenomena but that it is certainly not precise. Table 2.4 emphasizes the reliability of the simple theory for monatomic gases (for these molecules $0.4\tilde{c}_v + 0.9k = \tilde{c}_v$); there is very little deviation from the predicted value of 2.5. Table 2.5 indicates that the correction for internal energy is extremely important; only if this is made do all dilute gases behave in roughly the same fashion. Had we assumed that kinetic and internal energies were transported in the same way, the invariant ratio would have turned out the same as for monatomic gases, i.e., $m\varkappa/\tilde{c}_v\eta$, a number which varies greatly. The variation of the ratio (2.43) provides a test of the assumption that internal energy is equilibrated with each

Table 2.4. The Ratio $\varkappa m/\eta\tilde{c}_v$ for Monatomic Gases at Various Temperatures[a]

Temperature (K)	He	Ne	Ar	Kr
90	2.44	2.46	2.49	—
195	2.45	2.52	2.51	2.5
273	2.45	2.50	2.48	2.50
373	2.44	2.51	2.53	2.50
491	2.43	2.47	2.47	2.47

[a] Data from H. H. Landolt and R. Börnstein, *Zahlenwerte und Funktionen*, 6th ed. (Berlin: Springer, 1969), Vol. II, Part 4, pp. 398–451; Part 5a, p. 3; Part 5b, pp. 45–53.

Table 2.5. The Ratios $\varkappa m/\eta\tilde{c}_v$ (in Parentheses) and $\varkappa m/\eta(0.4\tilde{c}_v + 0.9k)$ for Polyatomic Gases at Various Temperatures[a]

Temperature (K)	H_2	O_2	CO_2	CH_4	NO
100	2.45	2.37	—	2.42	—
	(2.31)	(1.74)		(1.80)	
200	2.50	2.50	2.28	2.55	2.39
	(2.04)	(1.92)	(1.58)	(1.78)	(1.75)
273	2.50	2.53	2.51	2.60	2.46
	(1.93)	(1.92)	(1.67)	(1.76)	(1.86)
300	2.51	2.58	2.55	2.65	2.50
	(1.94)	(1.96)	(1.69)	(1.79)	(1.89)

[a] Data from H. H. Landolt and R. Börnstein, *Zahlenwerte und Funktionen*, 6th ed. (Berlin: Springer, 1969), Vol. II, Part 4, pp. 398–451; Part 5a, p. 3; Part 5b, pp. 45–53.

collision; its approximate character is apparent since the ratio is much more temperature and species dependent for polyatomic than for monatomic gases.

A better test of the model is given by considering the temperature dependence of the transport coefficients. If molecules were truly hard spheres the values of d, computed using the expressions of Table 2.2 and experimental values of $\varkappa$, η, or D, would be constant. In Figs. 2.7 and 2.8 the apparent temperature dependence of d^{-2} (which is proportional to the transport coefficient) is illustrated. Clearly d^{-2} is not a constant; it increases with temperature. The temperature dependence varies considerably from gas to gas. Furthermore there are also differences in the effect of temperature on d^{-2} depending upon which transport coefficient is used to compute it. The differences are most severe for polyatomic molecules, where the hard-sphere model is least reliable.

The naive model is not adequate. However, a bit more can be extracted from it. Table 2.6 quotes measurements of η and $\varkappa$. Included are three estimates of d, two calculated from transport data using the expression for η and $\varkappa$ of Table 2.2. The third is determined from critical point data via the van der Waals b_0: since $v_c = 3b_0$ and $b_0 = \frac{1}{2}N_0(4\pi d^3/3)$ then $d = (v_c/2\pi N_0)^{1/3}$. In addition the mean free path is evaluated using (2.12) with the value of d determined from viscosity data. The estimates of d from transport and critical point data differ considerably. Nonetheless they are of the same order of magnitude, which suggests that the model is not unreasonable.

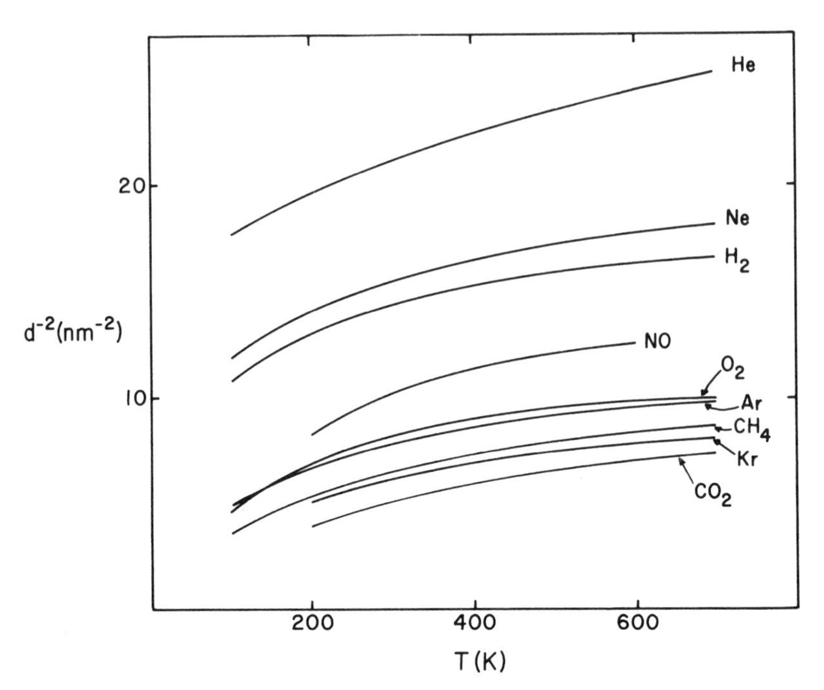

Fig. 2.7. Reciprocal of apparent value of d^2 as determined from thermal conductivity using hard-sphere transport theory (Table 2.2). Transport coefficient data from H. H. Landolt and R. Börnstein, *Zahlenwerte und Funktionen*, 6th ed. (Berlin: Springer, 1969), Vol. II, Part 5b, pp. 45–53.

To improve the description a more realistic intermolecular potential must be used. The following qualitative argument was given by Sutherland.[15] Since molecules are not hard spheres, increasing their relative kinetic energy allows them to approach one another more closely. Thus an increase in temperature leads to a decrease in the mean collision diameter. This may be formulated as

$$d^2(T) = d_\infty^2(1 + A/T) \tag{2.44}$$

where d_∞ and A are constants, with d_∞ the hard-core diameter at infinite temperature. While accounting for deviations from the $T^{1/2}$ prediction the Sutherland modification is inadequate to explain the observed temperature dependence or the differences between the transport coefficients evidenced in Figs. 2.7 and 2.8. To accomplish this a quantitative picture is required.

[15] W. Sutherland, *Phil. Mag.* **36**, 507 (1893).

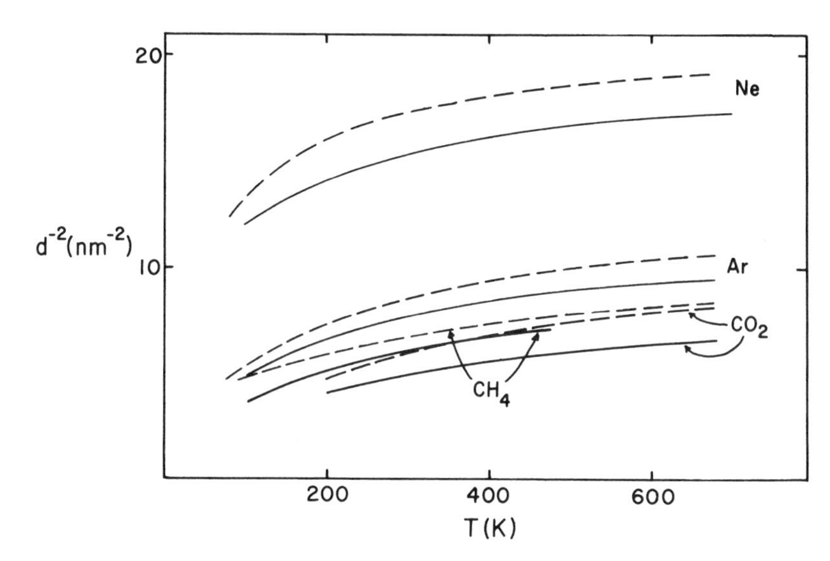

Fig. 2.8. Reciprocal of the apparent value of d^2 as determined from viscosity (—) and self-diffusion coefficients (– –) using hard-sphere transport theory (Table 2.2). Transport coefficient data from H. H. Landolt and R. Börnstein, *Zahlenwerte und Funktionen*, 6th ed. (Berlin: Springer, 1969), Vol. III, Part 5a, pp. 3, 516.

Table 2.6. Molecular Properties of Gases and Related Transport Data Measured at 273.2 K and 1 atm[a]

Gas	Viscosity, $\eta \times 10^5$ ($kg\,m^{-1}\,sec^{-1}$)	Thermal conductivity, $\varkappa \times 10^2$ ($J\,K^{-1}\,m^{-1}\,sec^{-1}$)	Mean free path, λ (nm)	Molecular diameter, d (nm)		
				From η	From $\varkappa$	From critical point data
He	1.85	14.3	168	0.218	0.218	0.250
Ne	2.97	4.60	120	0.258	0.258	0.224
Ar	2.11	1.63	61	0.364	0.365	0.273
H_2	0.845	16.7	108	0.272	0.269	0.260
O_2	1.92	2.42	62	0.360	0.358	0.272
CO_2	1.36	1.48	37	0.464	0.458	0.296
CH_4	1.03	3.04	47	0.414	0.405	0.299

[a] Data from H. H. Landolt and R. Börnstein, *Zahlenwerte und Funktionen*, 6th ed. (Berlin: Springer, 1969), Vol. II, Part 5a, p. 3; Part 5b, pp. 45–53.

Agreement with experiment can be obtained by constructing a rigorous theory for the positionally dependent velocity distribution function; such theories, which provide the tools to determine the deviations from local equilibrium, are based upon analysis of the dynamics of binary collisions. As already pointed out in the derivation of the Maxwell distribution presented in Section 1.2, the net change that occurs in an elastic binary collision is *reorientation of the relative velocity vector*. The reorientation, which rotates the vector $\mathbf{c}_{12}$ through the scattering angle χ, is determined by the intermolecular potential, the collision energy of the two particles, $\frac{1}{2}\mu c_{12}^2$, and the *impact parameter*, b. This latter quantity is simply the distance of closest approach for noninteracting point molecules with the given initial velocities.

To calculate a trajectory requires knowledge of the intermolecular potential, which is not readily measured. To circumvent this difficulty a guess is made as to the form of the potential which is then used to compute the transport coefficients, the second virial coefficient, and a few other properties dependent upon two-particle interaction only. Good agreement between theory and experiment provides *a posteriori* justification for the assumed potential. A simple model potential was suggested by Lennard-Jones.[16] It is

$$u(r) = 4\epsilon\left[\left(\frac{\sigma}{r}\right)^{12} - \left(\frac{\sigma}{r}\right)^{6}\right] \tag{2.45}$$

where ϵ is the depth of the potential well and σ the distance at which $u(r) = 0$. The attractive term, r^{-6}, has some theoretical justification; the choice of r^{-12} dependence to describe the repulsion which is dominant at small separations was made on grounds of computational simplicity. As an illustration of the differences between the hard-sphere model and a more realistic model like the Lennard-Jones potential, consider Fig. 2.9. The hard-sphere diameter is chosen to be the Lennard-Jones σ. In the hard-sphere case the collision is impulsive; the scattering angle is uniquely determined by the impact parameter. At high collision energy the Lennard-Jones potential leads to only slightly different results. However, at low collision energy, the colliding pair is moving relatively slowly as interaction becomes important. Deflections are large and scattering is very different from that of hard spheres. For hard spheres no deflection occurs if $b > \sigma$, a statement that is incorrect for the Lennard-Jones potential. In particular, if the collision energy is small relative to the well depth ϵ, there is significant

[16] J. E. Lennard-Jones, *Proc. Roy. Soc.* **A106**, 463 (1924).

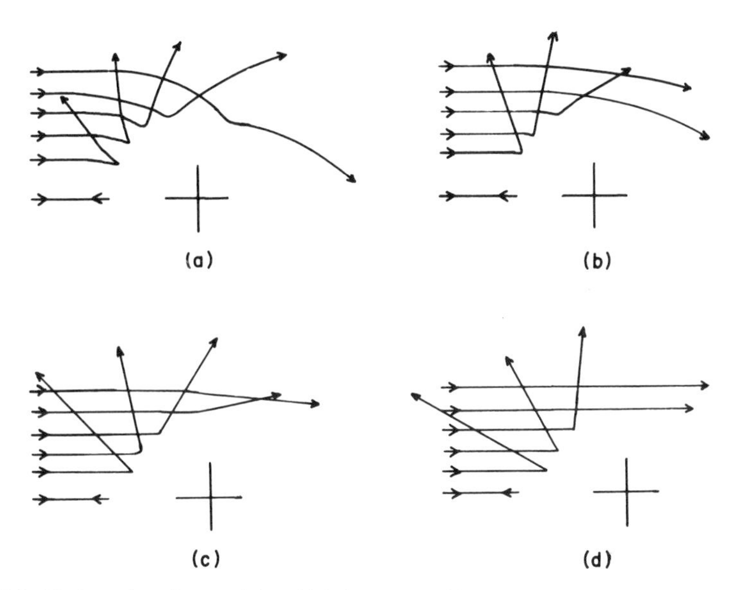

Fig. 2.9. Trajectories of a particle which interacts with a center of force via a Lennard-Jones potential for various reduced impact parameters $b^* = b/\sigma$. As b^* increases from 0, the path changes from a head-on to a glancing collision. Plots are for different reduced collision energies, $T^* = \mu c_{12}^2/2\epsilon$: (a) 0.8, (b) 4, (c) 20, and (d) a hard-sphere potential. For the smaller b^* the trajectories are not greatly affected by changes in T^* although the deflection angle is energy dependent. For the larger b^* there are qualitative differences, most evident in (a), where T^* is the smallest. [Data from J. O. Hirschfelder, C. F. Curtiss, and R. B. Byrd, *Molecular Theory of Gases and Liquids* (New York: John Wiley, 1954), pp. 1132–1146.]

deflection even if b is substantially greater than σ. It is thus apparent that mean free path is an imprecise concept when realistic intermolecular potentials are considered. There can be considerable deflection, even at large b, when the collision energy is small. Thus the collision diameter depends, in effect, upon the collision energy.

Trajectory analysis is used to compute the transport coefficients. Since momentum transfer in collisions accounts for transport, it is apparent that calculations based upon the Lennard-Jones potential (or any other that incorporates intermolecular attraction) lead to rather different predictions than did the hard-sphere model. The differences are naturally most significant at low temperatures, where the hard-sphere model is particularly inappropriate, a phenomenon which is marked by the temperature dependence of the apparent value of d^{-2} (Figs. 2.7 and 2.8). The greater variation of d at low temperature reflects the fact that the mean collision energy

Table 2.7. Parameters for the Lennard-Jones Potential[a]

Gas	From viscosity		From second virial coefficient	
	ϵ/k (K)	σ (nm)	ϵ/k (K)	σ (nm)
Ne	35.7	0.2789	35.60	0.2749
Ar	124	0.3418	119.8	0.3405
Kr	190	0.361	171	0.360
N_2	91.5	0.3681	95.05	0.3698
O_2	113	0.3433	117.5	0.358
CO_2	190	0.3996	189	0.4486
CH_4	137	0.3882	148.2	0.3817

[a] Data from J. O. Hirschfelder, C. F. Curtiss, and R. B. Byrd, *Molecular Theory of Gases and Liquids* (New York: John Wiley, 1954), pp. 1110–1112.

is small so that particle trajectories deviate substantially from those of rigid spheres (Fig. 2.9). By astute choice of ϵ and σ the potential (2.45) can be adjusted to fit either transport or virial data. The corresponding Lennard-Jones parameters are tabulated in Table 2.7 for a few gases. The difference between the parameters best suited to describe the viscosity and those which fit the second virial coefficient is small; the molecular diameters of Table 2.6 show much greater sensitivity to the method of determination. On the other hand, the fact that different parameters are needed indicates that (2.45) does not properly describe the intermolecular interaction. With the development of high-speed digital computers it has become possible to determine a single potential which correlates all two-particle properties. For argon, where the most extensive experimental data are available, such a potential was determined by Dymond and Alder.[17] It is shown in Fig. 2.10; for comparison the Lennard-Jones potential based on viscosity data is also plotted. The curves are significantly different. In the attractive region [$u(r) < 0$] the Dymond–Alder function has a deeper, narrower potential well, while, in the repulsive region, it increases much more slowly. A Lennard-Jones potential using virial parameters would differ very little from the one using viscosity parameters. Neither accurately approximates the Dymond–Alder potential for any value of r. Thus the

[17] J. H. Dymond and. B. J. Alder, *J. Chem. Phys.* **51**, 309 (1969).

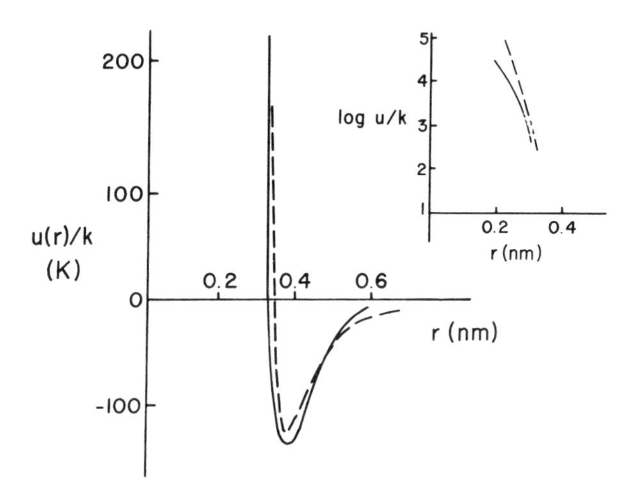

Fig. 2.10. Comparison of Dymond–Alder (—) and Lennard-Jones (– –) potential-energy functions for argon. The insert compares them in the domain where repulsive forces dominate.

agreement of viscosity and virial based parameters in Table 2.7 can be misleading. It would be incorrect to assume that using (2.45) with the two sets of parameters would provide bounds for the true intermolecular potential.

Determining the numerical potential of Fig. 2.10 is a formidable task, even with modern computers. However, once the potential is known, it is relatively simple to use it to calculate thermodynamic and transport properties. It is therefore natural to hope that the argon potential could be expressed in a parametric form and used to compute properties of other gaseous systems. This would be particularly important for determining gas properties in regions of high temperature not easily reached in the laboratory. While the available data are insufficient to provide a completely reliable test, it appears that in a two-parameter form the potential of Fig. 2.10 can be used to correlate the properties of dilute rare gases.[18]

An alternate approach to determination of the pair potential $u(r)$ is to calculate it using quantum mechanics. Perturbation calculations, valid at large intermolecular separations, have been carried out for many atomic and molecular systems. At small interparticle separations the problem is greatly complicated since an atomic pair can no longer be

[18] For an improved, simpler $u(r)$ see R. A. Aziz and H. H. Chen, *J. Chem. Phys.* **67**, 5718 (1977).

described as weakly interacting. Electron indistinguishability, resulting in exchange interactions, must be specifically incorporated; hence, for N-electron atoms, a $2N$-electron wave function, which is not a simple product function, is required. As a result the Ar–Ar potential is still too complicated for accurate calculation to be possible. The He–He and Ne–Ne potentials represent the present limit of accurate theory.[19]

Appendix. Coordinate Transformation by Jacobian Method

To express the differential volume element in terms of new variables is most readily done using Jacobians.[20] Given the coordinate transformation

$$x = x(u, v), \qquad y = y(u, v)$$

the relation between $dx\,dy$ and $du\,dv$ is

$$dx\,dy = |\,J(x, y \mid u, v)\,|\,du\,dv$$

where the Jacobian of the transformation is the determinant

$$J(x, y \mid u, v) = \begin{vmatrix} \left(\dfrac{\partial x}{\partial u}\right)_v & \left(\dfrac{\partial x}{\partial v}\right)_u \\ \left(\dfrac{\partial y}{\partial u}\right)_v & \left(\dfrac{\partial y}{\partial v}\right)_u \end{vmatrix}$$

The formalism may readily be extended to transformations involving more variables by constructing the analogous determinants. As an example consider the transformation from Cartesian to polar coordinates

$$x = r\cos\theta, \qquad y = r\sin\theta$$

The corresponding Jacobian is

$$J(x, y \mid r, \theta) = \begin{vmatrix} \cos\theta & -r\sin\theta \\ \sin\theta & r\cos\theta \end{vmatrix} = r$$

which indicates that $dx\,dy = r\,dr\,d\theta$, the familiar result.

[19] H. Margenau and N. R. Kestner, *Theory of Intermolecular Forces*, 2nd ed. (Oxford: Pergamon Press, 1971), Chapters 2 and 3; J. N. Murrell, in *Rare Gas Solids*, M. L. Klein and J. A. Venables, eds. (London: Academic Press, 1976), Vol. I, Chapter 3.

[20] For proof of the relations to be stated see H. Margenau and G. M. Murphy, *The Mathematics of Physics and Chemistry*, 2nd ed. (Princeton: Van Nostrand, 1956), pp. 192–194.

Problems

2.1. What is the probability that an A atom in air at 300 K and 1 atm travels 1 μm *without* making a collision? 10 μm? (Use the data of Table 2.6 and assume N_2 and O_2 have the same effective molecular diameters.)

2.2. Show that integration of (2.9) yields (2.10). [Choose the polar axis in the direction of $\mathbf{c}_1$ so that $c_{12} = (c_1{}^2 + c_2{}^2 - 2c_1c_2 \cos \theta_2)^{1/2}$.]

2.3. Use the Jacobian method to show that the differential volume element in spherical polar coordinates is $c^2 \sin \theta \, dc \, d\theta \, d\varphi$.

2.4. Use the Jacobian method to prove the statement made in Section 2.4, $d\mathbf{c}_1 \, d\mathbf{c}_2 = d\mathbf{c}_M \, d\mathbf{c}$; the coordinate transformation is given by (2.15).

2.5. Modify the analysis of Section 2.2 and show that in a *two-dimensional* gas the limiting mean free path [the analog to (2.5)] is $0.5[n_A \, d_{AA} + n_B \, d_{AB}]^{-1}$.

2.6. By constructing the two-dimensional analog to (2.18) show that the collision frequency in a two-dimensional gas is

$$2(\pi RT/M_A)^{1/2}[n_A \, d_{AA} + n_B \, d_{AB}(1 + M_A/M_B)^{1/2}]$$

This is the analog to (2.19).

2.7. Imagine that it is possible to construct and maintain a system with the sinusoidal temperature profile $T(z) = T_0 + \delta T \sin(m\pi z/L)$, where L is the width of the container and m is an integer. What is the heat flux at the edges of the container? Show that if the system is allowed to decay to equilibrium the time-dependent temperature profile is

$$T(t, z) = T_0 + \delta T \exp\left[-\left(\frac{m\pi}{L}\right)^2 \frac{\varkappa V t}{C_v}\right] \sin \frac{m\pi z}{L}$$

2.8. Demonstrate that (2.30) satisfies (2.27) and fits the boundary conditions (2.29).

2.9. Another important, and soluble, one-dimensional diffusion problem is the example for which there is constant matter flow into one end of a cylindrical tube. The boundary conditions for this problem are

$$c = c_0, \quad \text{all } z, \quad t = 0$$

$$D \frac{\partial c}{\partial z} = -J = \text{const}, \quad z = 0, \quad \text{all } t$$

Demonstrate by direct differentiation that the solution to (2.25) consistent with these constraints is

$$c = c_0 + \frac{Jz}{D}\left[\frac{\exp(-s^2)}{\pi^{1/2} s} - \frac{2}{\pi^{1/2}} \int_s^\infty \exp(-y^2) \, dy\right]$$

where $s = z/[2(Dt)^{1/2}]$.

2.10. At 15°C and 1 atm the viscosity of NH_3 is 9.6×10^{-6} kg m^{-1} sec^{-1} and the heat capacity is 0.52 cal g^{-1}. Estimate both D and $\varkappa$. The observed value of $\varkappa$ is 0.040 J K^{-1} m^{-1} sec^{-1}; comment on any difference between your estimate and the experimental result.

2.11. The Lennard-Jones potential can be used to compute a temperature-dependent effective hard-sphere d in the following fashion. Assume that d is the value of r at which $u(r)$ is the mean thermal energy, $1.5kT$; this accounts, in a rough way, for the Sutherland effect. Make this assumption and compute $d(T)$ for A at 100, 200, 300, 400, and 500 K. Then compare your results with the data in Fig. 2.8. Why do you think this approach works better at high temperatures?

2.12. At 273 K the viscosity of gaseous Br_2 is 1.46×10^{-5} kg m^{-1} sec^{-1}; estimate the molecular diameter of Br_2 molecules. Using the data in Table 2.6 determine the mean free path of Br_2 molecules in O_2 gas at 273 K and 1 atm.

2.13. The diffusion coefficient in a binary mixture is $D = \frac{3}{8}(1/n\mu\pi d_{AB}^2)(\pi\mu kT/2)^{1/2}$ where μ is the reduced mass of an AB pair. Estimate the diffusion coefficient of Br_2 in air at 273 K and 1 atm assuming that N_2 and O_2 have the same diameter and a mean mass.

2.14. Assume that a tube filled with air at 273 K and 1 atm is placed in contact with a reservoir containing Br_2 at its vapor pressure, 0.0844 atm. How long will it take for Br_2 to become visible at a distance 0.5 m down the tube? This concentration is approximately equivalent to a pressure of 0.006 atm. Use the results of Problem 2.13 to solve this problem.

2.15. The speed-dependent differential flux can be determined by extension of the arguments that led to (2.31). It is

$$dJ = -n(z)F(c; z)[\cos \chi/4\pi r^2\lambda(c)]ce^{-r/\lambda(c)}\, d\tau\, dc$$

where $F(c; z)$ is given by (1.24) with a *position-dependent temperature*. Since the kinetic energy flux is

$$J_q^{(K)} = \int \tfrac{1}{2}mc^2\, dJ$$

show, by expanding $T(z)$ in a Taylor's series around $z - z_0$, that

$$J_q^{(K)} = -\left[\frac{n_0}{3T_0}\int F_0(c)\lambda(c)\frac{1}{2}mc^3\left(\frac{mc^2}{2kT_0} - \frac{5}{2}\right)dc\right]\frac{dT}{dz_0}$$

where $n_0 \equiv n(z_0)$, etc.

2.16. The internal energy of a molecule is $\epsilon - \frac{1}{2}mc^2$; thus the mean internal energy of a molecule in the volume element $d\tau$ is $\bar{\epsilon}(z) - \frac{3}{2}kT(z)$. Assuming that internal energy is equilibrated at each collision and that

$$\bar{\epsilon}_I(z) \equiv \bar{\epsilon}(z) - \tfrac{3}{2}kT(z) \equiv \tilde{c}_I T(z)$$

where $\tilde{c}_I$ is the molecular internal heat capacity show that

$$J_q^{(I)} = -\left[\frac{n_0\tilde{c}_I}{3}\int F_0(c)\lambda(c)c\left(\frac{mc^2}{2kT_0} - \frac{3}{2}\right)dc\right]\frac{dT}{dz_0}$$

Comparison of $J_q^{(K)}$ and $J_q^{(I)}$ indicates that the transport of kinetic and internal energy involves different averaging processes. Thus $\varkappa$ is not simply proportional to the total molecular heat capacity, $\tilde{c}_I + 1.5k$.

2.17. What are the conditions for Knudsen flow in a vessel of length 0.1 m? Assume molecular diameters of ~0.4 nm. To what pressure does this correspond at 298 K?

2.18. Demonstrate that (2.37) follows from (2.36).

General References

Theory of Gas-Phase Transport

L. Boltzmann, *Lectures on Gas Theory*, translation by S. G. Brush (Berkeley: University of California Press, 1964).

S. Chapman and T. G. Cowling, *Mathematical Theory of Non-Uniform Gases*, 3rd ed. (Cambridge: The University Press, 1970), Chapters 5–8.

J. O. Hirschfelder, C. F. Curtiss, and R. B. Byrd, *Molecular Theory of Gases and Liquids* (New York: John Wiley, 1954), Chapters 1, 7 and 8.

M. H. C. Knudsen, *The Kinetic Theory of Gases*, 3rd ed. (New York: John Wiley, 1950).

W. Kauzmann, *Kinetic Theory of Gases* (New York: Benjamin, 1966), Chapter 5.

J. E. Mayer and M. G. Mayer, *Statistical Mechanics* (New York: John Wiley, 1940), pp. 18–30.

R. D. Present, *The Kinetic Theory of Gases* (New York: McGraw–Hill, 1958), Chapters 3, 4, 7, and 8.

Nonequilibrium Thermodynamics

D. D. Fitts, *Nonequilibrium Thermodynamics* (New York: McGraw–Hill, 1962), Chapters 1–3.

S. R. de Groot and P. Mazur, *Nonequilibrium Thermodynamics* (Amsterdam: North-Holland, 1962), Chapters 2–4.

I. Prigogine, *Thermodynamics of Irreversible Processes*, 3rd ed. (New York: Interscience, 1967), Chapters 1–5.

3

Electrolytic Conduction and Diffusion

3.1. Introduction

There are a number of features which distinguish kinetic problems in solutions from those in the gas phase. The most obvious is that the chemical entities being studied are different in the two phases. A sodium ion in the gas phase is a bare ion. In solution it carries with it a sheath of solvent, i.e., it is solvated. This difference is apparent in the equilibrium properties of solutions. In most instances there are both enthalpy changes and volume changes upon solvation. These effects clearly indicate that the solvated species and the pure substance differ significantly, differences which also affect the kinetic properties.

The solvent has another important effect. Unlike the gas phase, in which the medium is a vacuum through which molecules travel in linear paths except for the occasional instance of interaction, collision, and deflection, in solution the molecules of the species of interest are always interacting with solvent. This interaction radically alters the molecular motion. Instead of traveling in straight lines the molecules move randomly through the maze provided by the solvent. One can describe the average motion of molecules in solution by means of the laws of diffusion; the motion of individual molecules is subject to such a variety of influences that precise description is hopeless.

Solvents of high dielectric constant like water stabilize ionic species. Thus much interest in kinetic properties in solution necessarily focuses on the behavior of ions. This significantly complicates the discussion since the coulombic interaction extends over a long range. The requirement of electroneutrality means that ions cannot diffuse through a solution independently of one another even in very dilute solution.

Before we treat the phenomenology of chemical kinetics, assuming we wish to discuss ionic reaction in solution, we must consider the nature of electrical conduction in solutions. To this end we first review some properties of electrolytes and the mechanism for transport of electric current in solution. We shall find that electrical conduction and solute diffusion are processes that are inextricably coupled, a coupling that can be used to relate the transport coefficient describing electrical conduction, the conductivity λ, with that describing diffusion, the diffusion coefficient D. By incorporating some ideas from the general theory of fluid flow we shall be able to relate λ (or D) to molecular properties of the solvated species. Conductance measurements provide a way to determine D, thus establishing limits for the reaction rate in solution. In addition, when carried out in high electric fields, such measurements can be used to imply the existence of ion pairs, species that are different from both solvated molecules and solvated ions. For simplicity, and because it is the case of greatest chemical interest, our discussion will be limited to constant temperature and constant pressure processes.

3.2. Ionic Conduction in Solution—a Review

In metals, current flow is due to electron motion. This is not the case in electrolytic solution; here a potential difference leads to net ionic migration. It is the motion of the ions which accounts for charge transfer in solution.

For consideration of ionic conduction in solution, Ohm's law is most conveniently formulated in a form analogous to the transport equations presented in Section 2.4,

$$\mathbf{j} = -\sigma' \nabla \psi = \sigma' \mathbf{E} \tag{3.1}$$

Here $\mathbf{j}$ is the current density, or equivalently, the charge flux. The electric field $\mathbf{E}$, which is the gradient of the potential ψ, provides the driving force for conduction. The transport coefficient σ' is the specific conductivity, expressed in $\text{ohm}^{-1}\,\text{m}^{-1}$. While (3.1) may appear unfamiliar, it reduces to the common expression $I = V/R$ in the special case of conduction in a wire of uniform cross section (see Problem 3.1).

To relate the transport coefficient σ' to the properties of the solvated species requires recognizing that even pure solvent contains some ions. Thus, the results of conductivity measurements are corrected for the specific conductivity of the cell in the absence of electrolyte, σ_0; the conductivity attributed to electrolyte is then

$$\sigma = \sigma' - \sigma_0 \tag{3.2}$$

Electrolytic conduction is due to the presence of ions; thus σ is a function of the *concentration of equivalents of ions*, $\tilde{C}$.[1] Strong electrolytes like NaCl, HNO_3, etc., are fully dissociated while weak electrolytes like HCN are not. In the former case $\tilde{C}$ is the same as the normality of the electrolyte C; in the latter it is not. As a consequence the conductivity of a strong electrolyte is roughly proportional to C while that of a weak one is not. It is convenient to define the *equivalent conductance* which accounts for the concentration dependence

$$\Lambda \equiv \sigma/\tilde{C} \tag{3.3}$$

Since $\tilde{C}$ is not always easy to measure the *apparent* equivalent conductance

$$\Lambda_{\text{app}} = \sigma/C \tag{3.4}$$

is also useful. The concentration dependence of Λ_{app} is illustrated in Fig. 3.1. For strong electrolytes Λ_{app} decreases slowly with increasing C while for weak electrolytes the decrease is precipitous.

Strong electrolytes at the same concentration obey an approximate additivity rule in Λ_{app}, e.g.,

$$\Lambda_{\text{NaCl}} = \Lambda_{\text{NaNO}_3} + \Lambda_{\text{KCl}} - \Lambda_{\text{KNO}_3} \tag{3.5}$$

a relation which becomes precise in the limit of infinite dilution. If each ionic species contributes independently to the conductivity and if each is completely dissociated it is easy to rationalize (3.5). Thus we decompose Λ as

$$\Lambda = \lambda_+ + \lambda_- \tag{3.6}$$

where the λ are the ion equivalent conductances. Now, since $\tilde{C} = C$ for strong electrolytes (complete dissociation) the data of Fig. 3.1 require that

[1] A 0.1 M solution of NaCl is also 0.1 N. However, a 0.1 M solution of Na_2SO_4 is 0.2 N; it contains 0.2 equivalents of positive (and negative) charge per liter.

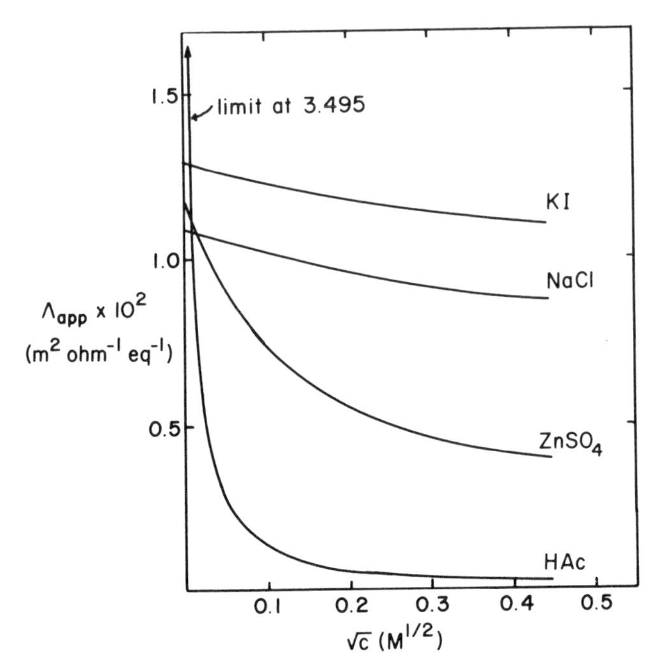

Fig. 3.1. Apparent equivalent conductance, Λ_{app}, of various electrolytes as a function of concentration.

Λ itself be concentration dependent, an observation that is generally consistent with the equation

$$\Lambda = \Lambda^0 - \alpha \tilde{C}^{1/2} \tag{3.7}$$

where Λ^0 is the equivalent conduction at infinite dilution. Here α is a species-dependent constant. A model accounting for the concentration dependence is discussed in Section 3.7.

From (3.6) and (3.7) it is apparent that a quantity of basic interest in electrolytic conductance is the ion equivalent conductance at infinite dilution. Ordinary conductance measurements cannot separate the contribution of individual ions. By designing experiments that distinguish between ion migration toward positive and negative electrodes in a current-carrying cell it is possible to measure λ_+ and λ_- separately. A tabulation of the results of such measurements is given in Table 3.1.[2]

[2] A variety of techniques is available for determining electrolyte and ion conductivity. For a description of the methods and their drawbacks see R. A. Robinson and R. H. Stokes, *Electrolyte Solutions* (London: Butterworths, 1955), Chapter 5.

Table 3.1. Ion Equivalent Conductivities at Infinite Dilution, λ^0, in Aqueous Solution at 25°C[a]

Cation	$\lambda^0 \times 10^2$ ($m^2\ ohm^{-1}\ eq^{-1}$)	Anion	$\lambda^0 \times 10^2$ ($m^2\ ohm^{-1}\ eq^{-1}$)
H^+	3.498	OH^-	1.978
Li^+	0.3866	F^-	0.544[b]
Na^+	0.5011	Cl^-	0.7635
K^+	0.7352	Br^-	0.7820
Rb^+	0.778	I^-	0.769
Cs^+	0.773	NO_3^-	0.7144
Ag^+	0.6192	ClO_3^-	0.646
NH_4^+	0.734	BrO_3^-	0.558
$\frac{1}{2}Mg^{2+}$	0.5306	IO_3^-	0.405
$\frac{1}{2}Ca^{2+}$	0.5950	$Acetate^-$	0.409
$\frac{1}{2}Sr^{2+}$	0.5946	HSO_4^-	0.50[b]
$\frac{1}{2}Ba^{2+}$	0.6364	HCO_3^-	0.445[b]
$\frac{1}{2}Cu^{2+}$	0.54	CN^-	0.78[b]
$\frac{1}{2}Zn^{2+}$	0.53	$\frac{1}{2}C_2O_4^{2-}$	0.240
$\frac{1}{2}Ni^{2+}$	0.49[b]	$\frac{1}{2}SO_4^{2-}$	0.800
$\frac{1}{2}Fe^{2+}$	0.54[b]	$\frac{1}{3}Fe(CN)_6^{3-}$	1.00
$\frac{1}{2}Pb^{2+}$	0.70[b]	$\frac{1}{4}Fe(CN)_6^{4-}$	1.11
$\frac{1}{3}Fe^{3+}$	0.684[b]		
$\frac{1}{3}Cr^{3+}$	0.67[b]		
$\frac{1}{3}La^{3+}$	0.695		

[a] Except where noted data from H. S. Harned and B. B. Owen, *The Physical Chemistry of Electrolyte Solutions*, 3rd ed. (New York: Reinhold, 1958), p. 231.
[b] H. H. Landolt and R. Börnstein, *Zahlenwerte und Funktionen*, 6th ed. (Berlin: Springer, 1969), Vol. II, Part 7, pp. 258–267.

3.3. Nernst–Einstein Relation

According to Ohm's law, applying an electric field to an ionic solution leads to an ion current which creates a concentration gradient. This gradient immediately sets up a diffusive flow to attempt to restore homogeneity. Both processes involve ion flow; the diffusive flux and the charge flux are inextricably coupled. The consequence, as we shall see, is that the ion equivalent conductance is related to the ion diffusion coefficient.

Consider a single *cationic* species of equivalent conductance λ_+, ion concentration n_+ and ionic charge $+z_+$. From (3.3) and (3.6) such a species, at low concentration, contributes $z_+ n_+ \lambda_+$ to the specific conductance of the system, and, from (3.1), the cationic contribution to the current density is

$$\mathbf{j}_+ = z_+ n_+ \lambda_+ \nabla\psi \tag{3.8}$$

The ion flux required to produce this current density is $(\mathbf{j}_+/z_+F)$, where F, Faraday's constant, is the number of coulombs per equivalent of charge. Thus the total cation flux in a potential field, found by combining (3.8) with Fick's law (2.23) is[3]

$$\mathbf{J}_+ = \mathbf{J}_+(\text{diff}) + \frac{\mathbf{j}_+}{z_+F} = -D_+ \nabla n_+ - \frac{n_+\lambda_+}{F} \nabla\psi \tag{3.9}$$

The corresponding anion flux is

$$\mathbf{J}_- = -D_- \nabla n_- + \frac{n_-\lambda_-}{F} \nabla\psi \tag{3.10}$$

where the sign reversal in the second term reflects the fact that a potential gradient drives positive and negative ions in opposite directions.

To establish the relationship between D and λ assume, for simplicity, that the anionic species is essentially immobile (as would be the case were it very large) so that $\mathbf{J}_- = 0$. Now assume further that the potential gradient can be adjusted, if only in principle, to balance the concentration gradient so that $\mathbf{J}_+ = 0$. The necessary condition, from (3.9), is

$$\nabla n_+ = -\frac{n_+\lambda_+}{FD_+} \nabla\psi \tag{3.11}$$

With these constraints the system remains electrically neutral throughout *as there is no net ionic flow.*

[3] This formulation is based upon the simple form of Fick's law (2.23). In fact the flux of a species depends not only on its own concentration gradient but also on the concentration gradients of *all* other species in solution. A more complete theory would have to account for such phenomena. While the precise expressions may be affected, the general conclusions are not significantly different from those based upon (3.9) where *cross-diffusion* terms are ignored. This and other coupled phenomena are discussed in S. R. de Groot, *Thermodynamics of Irreversible Processes* (Amsterdam: North-Holland, 1963).

Consider the potential energy of the cations due to the electric potential; it is simply

$$U_+ = z_+ e\psi$$

where e is the electronic charge. As ψ is position dependent, so is U_+. For the example being treated—no ionic flow, constant temperature, constant pressure—the cation properties may be treated from an equilibrium viewpoint. Thus the concentration profile of the positive species is related to its potential energy by a Boltzmann distribution

$$n_+ = n_+^0 \exp(-\beta U_+) = n_+^0 \exp(-\beta z_+ e\psi) \tag{3.12}$$

where n_+^0 is a reference state concentration which is not position dependent. Taking the gradient and substituting in (3.11) immediately yields

$$(-\beta z_+ e \nabla\psi) n_+ = -(n_+ \lambda_+ / D_+ F) \nabla\psi$$

from which we obtain the *Nernst–Einstein* relation[4]

$$\lambda_+ = \frac{eFD_+ z_+}{kT} = \frac{F^2 D_+ z_+}{RT} \tag{3.13}$$

Had the cations been assumed immobile a similar formula would be found relating λ_- to $D_- z_-$.

The derivation of (3.13) was based on the assumption that only cations are current carriers. However, the result is general. Ion equivalent conductance is a *property of the individual ionic species*, independent of the nature of the counterion (at least in dilute solution). Choosing a counterion for which $D = \lambda = 0$ simplifies the argument; it does not affect the result. Limiting consideration to dilute solution, (3.6) and (3.13) can be combined and we find[5]

$$\Lambda^0 = \frac{F^2}{RT} (z_+ D_+ + z_- D_-) \tag{3.14}$$

[4] We can now recognize why only the cation is viewed from an equilibrium standpoint. Its electrochemical potential is $\tilde{\mu}_+ = \mu_+^0 + RT \ln n_+ + z_+ F\psi$ which, with (3.12), is $\tilde{\mu}_+ = \mu_+^0 + RT \ln n_+^0$. Since both μ_+^0 and n_+^0 are position independent so is $\tilde{\mu}_+$, as it must be in equilibrium. The electrochemical potential for the anion is $\tilde{\mu}_- = \mu_-^0 + RT \ln n_- - z_- F\psi$. Since electroneutrality demands that $z_+ n_+ = z_- n_-$ we find, using (3.12), that $\tilde{\mu}_- = \mu_-^0 + RT \ln n_-^0 - (z_+ + z_-)F\psi$ which, since ψ is position dependent, is not constant. Thus the anions are not at equilibrium and we see the importance of the assumption of anionic immobility for establishing (3.13).

[5] W. Nernst, *Z. Physik. Chem.* **2**, 613 (1888).

3.4. Electrolyte Diffusion

According to the Nernst–Einstein relation (3.13), the conductance and the diffusion coefficient are proportional and the data of Table 3.1 can be used to calculate single-ion diffusion coefficients. Thus $D_{Cl^-} \sim 2D_{Li^+}$, and, were a concentration gradient established in a solution of LiCl, Cl^- would initially diffuse twice as fast as Li^+. Such a situation cannot last; charge separation creates an electric field accelerating Li^+ and braking Cl^-.

To determine the electrolyte diffusion coefficient combine (3.9) and (3.10) with (3.13),

$$\mathbf{J}_\pm = -D_\pm\left(\nabla n_\pm \pm \frac{Fz_\pm}{RT} n_\pm \nabla\psi\right) \tag{3.15}$$

The plus signs refer to cations and the minus signs to anions. With a binary electrolyte $(A^{+z_+})_{\nu_+}(B^{-z_-})_{\nu_-}$ at concentration n we have

$$n_+ = \nu_+ n, \qquad n_- = \nu_- n \tag{3.16}$$

and

$$\nu_+ z_+ = \nu_- z_- \tag{3.17}$$

To maintain electroneutrality the ion fluxes must be coupled,

$$z_+\mathbf{J}_+ = z_-\mathbf{J}_- \tag{3.18}$$

Combining (3.15)–(3.18) we obtain the local electric field due to the coupled-ion flow

$$\nabla\psi = -\left(\frac{D_+ - D_-}{z_+D_+ + z_-D_-}\right)\frac{RT}{F}\,\frac{\nabla n}{n}$$

which when substituted in (3.15) leads, with (3.16), (3.17), and the observation that $\mathbf{J} = \mathbf{J}_+/\nu_+$, to alternative forms

$$\mathbf{J} = -\left[\frac{(z_+ + z_-)D_+D_-}{(z_+D_+ + z_-D_-)}\right]\nabla n = -\frac{(\nu_+ + \nu_-)D_+D_-}{(\nu_-D_+ + \nu_+D_-)}\nabla n = -D\nabla n \tag{3.19}$$

a result also obtained by Nernst.[6] The electrolyte diffusion coefficient is an average, always intermediate between D_+ and D_-, which is reasonable since the local field induced by the difference in ion mobility accelerates the less mobile ion and slows up the faster one.

[6] W. Nernst, *Z. Physik. Chem.* **2**, 613 (1888).

3.5. Stokes' Law and the Microscopic Interpretation of λ^0

One of the early results of continuum fluid mechanics was the determination of the frictional force exerted on a spherical particle as it moved through a viscous fluid. *Stokes' law* relates the force on the particle $\mathbf{f}$, to the particle radius r, its drift velocity $\mathbf{v}$, and the viscosity of the medium η,

$$\mathbf{f} = 6\pi\eta r\mathbf{v} \tag{3.20}$$

a result derived assuming that the fluid close to the particle is moving with the same velocity as the particle, i.e., the fluid *sticks* to the particle. A great body of experimental evidence demonstrates that (3.20) accurately accounts for the behavior of macroscopic particles. How and why (3.20) can be used to describe the motion of ions in a solvent is unclear. From the standpoint of the ion the solvent is certainly *not* a continuous medium; it has a molecular structure.[7] Nonetheless we shall assume that (3.20) describes the average force acting on ions migrating in a solution. A partial justification is found in the molecular insights that will be obtained. However, the application of (3.20) to ion migration must be considered to be a postulate, not a consequence of fluid mechanics.

In an electric field $\mathbf{E} = -\nabla\psi$, the force on a representative ion of charge z is $ze\mathbf{E}$; if the ion is not being accelerated this force is balanced by friction in the same direction so that

$$ze\mathbf{E} = 6\pi\eta r\mathbf{v} \tag{3.21}$$

The ion mobility u, defined as velocity per unit electric field strength,[8] is therefore

$$u = \mathbf{v}/\mathbf{E} = ze/6\pi\eta r \tag{3.22}$$

To relate u to λ^0 consider an ionic species with mean velocity $\mathbf{v}$; the ion

[7] It is a question of some interest whether the solvent can be presumed to stick to the solvated ion as it diffuses through the solution. If instead it is assumed that the solvent slips by the solute the factor 6π in (3.20) is reduced to 4π. This has no great qualitative effect on the picture to be developed. For a derivation of Stokes' law see R. M. Fuoss and F. Accascina, *Electrolyte Conductance* (New York: Interscience, 1959), pp. 53–59. Theoretical studies on hard-sphere liquids indicate that Stokes' law applies if it is assumed that solvent slips by the molecules [B. J. Alder, D. M. Gass, and T. E. Wainwright, *J. Chem. Phys.* **53**, 3813 (1970)].

[8] Since $\mathbf{E}$, $\mathbf{v}$, and $\mathbf{j}$ are oriented in the same direction, they are proportional to the same unit vector $\mathbf{n} = \mathbf{v}/|\mathbf{v}| = \mathbf{E}/|\mathbf{E}| = \mathbf{j}/|\mathbf{j}|$. It is thus permissible to "divide" the vectors since their $\mathbf{n}$ dependence cancels.

Table 3.2. Comparison of Crystal Radii, r_c, with Ionic Radii in Aqueous Solution, r_λ, as Estimated from (3.23)

Species	r_λ (nm)	r_c^a (nm)	Species	r_λ (nm)	r_c^a (nm)
H^+	0.026		Ca^{2+}	0.31	0.10
Li^+	0.23	0.06	Sr^{2+}	0.31	0.11
Na^+	0.18	0.10	Ba^{2+}	0.29	0.14
K^+	0.12	0.13	OH^-	0.046	
Rb^+	0.12	0.15	Cl^-	0.12	0.18
Cs^+	0.12	0.17	Br^-	0.12	0.20
Mg^{2+}	0.35	0.07	I^-	0.12	0.21

[a] L. Pauling, *The Nature of the Chemical Bond*, 3rd ed. (Ithaca: Cornell, 1960), p. 518.

flux is $\mathbf{J} = n\mathbf{v}$, where n is the ion concentration. Recognizing that the current density is

$$\mathbf{j} = zF\mathbf{J} = zFn\mathbf{v}$$

we find, using (3.22) and Ohm's law (3.1),

$$\sigma = \mathbf{j}/\mathbf{E} = nzF\mathbf{v}/\mathbf{E} = nzFu = nz^2eF/6\pi\eta r$$

from which, using (3.3), the ion equivalent conductance at infinite dilution can be calculated,

$$\lambda^0 = \sigma/\tilde{C} = \sigma/nz = Fu = zeF/6\pi\eta r \tag{3.23}$$

Thus λ^0 can be used to determine ionic radii in solution. Some values of r are given in Table 3.2; for comparison, crystal radii of the ions are also included.(9) Except for H^+ and OH^-, for which r appears to be absurdly small, the Stokes' law values are reasonable and comparable to the crystal radii. However, the trends in both the alkali and alkaline earth ion series are opposite. In part this may be rationalized by recognizing that a small ion, like Li^+, binds its water of solvation more strongly than a larger ion, like Cs^+; this suggests that the effective ionic radius of solvated Li^+ could be larger than that of Cs^+. However, the major source of the discrepancy is the use of Stokes' law itself. Applying a formula designed to describe the forces on macroscopic objects in continuum fluids to a molecular

(9) Using the slip condition [see footnote (7) in this Chapter] the values of r estimated from (3.23) increase by 50%.

problem cannot possibly be correct; the fact that it works as well as it does is somewhat surprising.

An important consequence of (3.23) is *Walden's rule*

$$\eta\lambda^0 = (\text{const})/r \tag{3.24}$$

Therefore, if the radius of the solvated ion is an invariant, the product $\eta\lambda^0$ is constant. Assuming that r is constant for both anion and cation, a more general form is

$$\eta\Lambda^0 = \text{const} \tag{3.25}$$

Figure 3.2 illustrates that Walden's rule accounts for most of the temperature dependence of ion conductivity in water solution; over the range considered the viscosity of water decreases tenfold. Again H^+ and OH^- are anomalous. The more highly charged ions deviate the least from Walden's rule; a possible interpretation is that such ions bind their water of solvation more strongly and are less affected by temperature variation.

Assuming no change in the nature of solvation, Walden's rule can be used for crudely estimating conductivity in different solvents. One example, for NaCl and KCl in water–methanol solution, is given in Table 3.3. Here, (3.25) appears accurate as long as the solution is rather dilute (25 wt. % methanol is 16 mole %), but, as the methanol concentration increases, deviations are substantial suggesting major changes in solvation. A second example, for KI in a variety of solvents, is given in Table 3.4. The Walden's rule product is roughly constant for the organic solvents but water and SO_2 differ greatly.

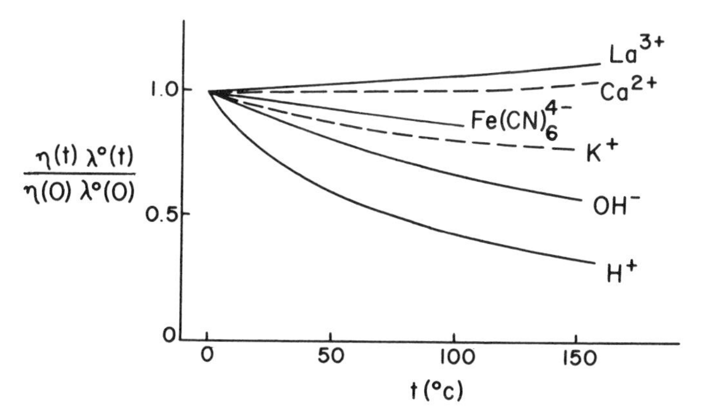

Fig. 3.2. Test of Walden's rule. The ratio $[\eta(t)\,\lambda^\circ(t)/\eta(0)\,\lambda^\circ(0)]$ for a series of ions as a function of temperature.

Table 3.3. Tests of Walden's Rule. The Product $\Lambda^0\eta$ for KCl and NaCl in Methanol–Water Mixtures at 25°C[a]

Methanol (wt. %)	$\eta \times 10^3$ ($kg\ m^{-1}\ sec^{-1}$)	$\Lambda^0\eta \times 10^5$ ($ohm^{-1}\ m\ kg\ sec^{-1}\ eq^{-1}$)	
		KCl	NaCl
0	0.8949	1.341	1.131
25	1.475	1.334	1.150
50	1.54	1.15	1.01
75	1.15	0.896	0.808
100	0.541	0.565	0.527

[a] H. S. Harned and B. B. Owen, *The Physical Chemistry of Electrolyte Solutions*, 3rd ed. (New York: Reinhold, 1958), p. 706.

Table 3.4. Tests of Walden's Rule. The Product $\Lambda\eta$ for KI in Various Solvents at 25°C[a]

Solvent	$\Lambda \times 10^2$ ($ohm^{-1}\ m^2\ eq^{-1}$)	$\eta \times 10^3$ ($kg\ m^{-1}\ sec^{-1}$)	$\Lambda\eta \times 10^5$
Acetonitrile	1.982	0.345	0.684
Acetone	1.855	0.316	0.586
Nitromethane	1.240	0.611	0.758
Methanol	1.148	0.546	0.627
Ethanol	0.509	1.096	0.560
Furfural	0.431	1.490	0.642
Acetophenone	0.398	1.620	0.644
Water[b]	1.5038	0.8937	1.344
SO_2[c]	2.65	0.394	1.044

[a] J. O'M. Bockris and A. K. N. Reddy, *Modern Electrochemistry* (New York: Plenum Press, 1970), Vol. 1, p. 386.
[b] H. S. Harned and B. B. Owen, *The Physical Chemistry of Electrolyte Solutions*, 3rd ed. (New York: Reinhold, 1958), p. 697.
[c] At 0°C.

3.6. The Mobility of H^+ and OH^- in Water

In the previous section we saw that Stokes' law could be used to estimate reasonable values of ionic solvation radii and that $\eta\lambda^0$ is only slightly dependent upon temperature for a variety of ions in aqueous solution. There are two glaring exceptions, H^+ and OH^-; their solvation radii, 0.026 and 0.046 nm, respectively, are ridiculously small. The temperature dependence of $\eta\lambda^0$ is very great.

A possible explanation for these large conductivities (small radii) is given if we consider the structure of water as shown schematically in Fig. 3.3. Proton diffusion in water can proceed by a cooperative method that does *not* require ion migration over large distances. First a hydrogen bond is broken and reformed as in Fig. 3.3a; the distance traveled is $\lesssim 0.05$ nm. This then is repeated as in Fig. 3.3b. The original water molecule may then rotate as in Fig. 3.3c and be prepared to accept another proton. The net effect is the displacement of a proton by $\sim$0.3 nm without any particular H^+ actually moving that distance. This mechanism is unique to H^+ (or OH^-) in water. Ionic migration of any other species requires significant solvent motion to permit the physical displacement of the ion; the net effect of such motion must be that a solvent molecule and a solute molecule exchange positions in the fluid.

The proton-jump model for diffusion is based upon the premise that the structure of liquid water derives from an icelike hydrogen-bonded network; in addition to the two normal OH bonds in water each oxygen atom is linked to neighboring ones via hydrogen bonds. The linkage is imperfect and there are defects in the network. These imperfections, known

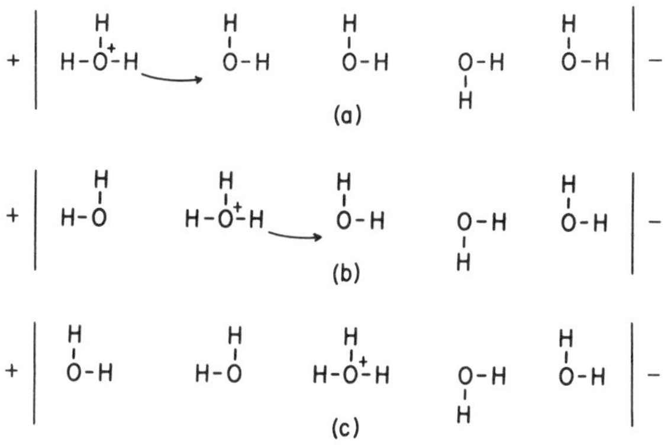

Fig. 3.3. Proposed mechanism for proton diffusion in water. (a) Proton jump. (b) Proton jump. (c) Rotation of water molecule.

as Bjerrum faults,[10] are of two types: a pair of oxygen atoms with two hydrogen atoms between them or a pair of oxygen atoms with no hydrogen atom between them. In some steps of the diffusional process such faults must be formed as shown in Fig. 3.3c, where molecular rotation has broken the hydrogen-bond network by creating an (O—H H—O) domain. While no theory of fault dynamics in water has been shown to correlate its electrical properties, the Bjerrum model is known to account for the anomalous electrical properties of ice.[11]

3.7. *The Concentration Dependence of* Λ

From the data presented in Fig. 3.1 it is clear that, at low concentration, the equivalent conductance decreases with increasing concentration

$$\Lambda = \Lambda^0 - \alpha C^{1/2} \tag{3.26}$$

which can be accounted for if the effect of the ion atmosphere is taken into consideration. At equilibrium an ion in solution is surrounded by a spherical distribution of charge, just balancing the ionic charge itself, as shown in Fig. 3.4a. When an electric field $\mathbf{E}$ is applied the ion migrates in one direction while the ionic atmosphere, being oppositely charged, moves in the opposite direction. Since the ion cloud is large it may be treated as if it were a charged colloidal particle moving under the influence of an electric field in a viscous fluid. In colloid chemistry this is known as *electrophoresis.* As the ion atmosphere migrates it naturally tries to carry its central ion with it; thus due to the electrophoretic effect there is an effective field $\mathbf{E}_E$ in the opposite direction to the applied field which decelerates the central ion.

In addition to the electrophoretic drag force, there is an additional effect because a moving ion deforms its own ion atmosphere. As shown in Fig. 3.4b an ion tends to pile up charge in front of it as it moves and tends to leave a diminishing wake behind as the atmosphere tries to readjust and reform the spherical distribution about the ion. The effect is to *separate charge* which produces an electric field called the *relaxation field* $\mathbf{E}_R$, in the direction of, but opposing the applied field. Considering both electrophoresis and relaxation the local field at the ion is $\mathbf{E} - \mathbf{E}_E - \mathbf{E}_R$. Since the mean

[10] N. Bjerrum, *Science* **115**, 385 (1952).

[11] L. Onsager and M. Dupuis, in *Electrolytes*, B. Pesce, ed. (New York: Pergamon, 1962), pp. 27–46.

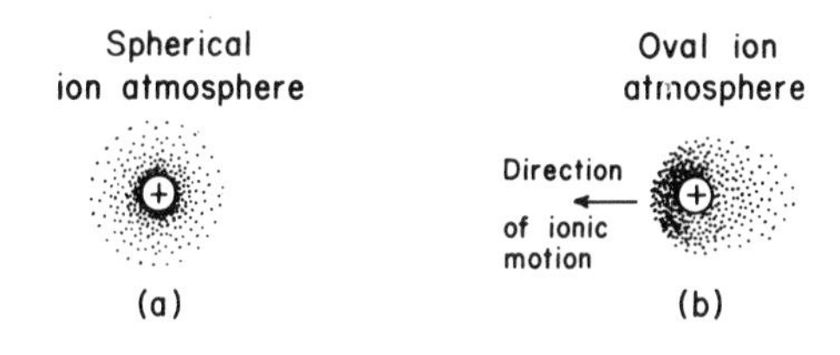

Fig. 3.4. Deformation of ion atmosphere due to ionic motion. (a) Spherical ion cloud surrounding representative ion at equilibrium. (b) Oval ion cloud surrounding representative migrating ion.

ionic velocity, from (3.21), is proportional to the field we have

$$\mathbf{v} \propto \mathbf{E} - \mathbf{E}_E - \mathbf{E}_R$$

Thus, using (3.22), the ion mobility is

$$u = u^0[1 - (\mathbf{E}_E + \mathbf{E}_R)/\mathbf{E}] = u^0 - u_E - u_R$$

where u^0 is the mobility at infinite dilution.[12] Using the same arguments which led to (3.23) yields an expression for λ which accounts for the effect of the ion atmosphere,

$$\lambda = \lambda^0[1 - (\mathbf{E}_E + \mathbf{E}_R)/\mathbf{E}] = \lambda^0 - \lambda_E - \lambda_R \tag{3.27}$$

At very low electrolyte concentration the ion atmosphere is very diffuse and both $\mathbf{E}_E$ and $\mathbf{E}_R$ are weak. As the concentration increases $\mathbf{E}_E$ and $\mathbf{E}_R$ will also increase. Thus the ratio $(\mathbf{E}_E + \mathbf{E}_R)/\mathbf{E}$ increases as concentration increases, accounting qualitatively for (3.26).

Estimates of $\mathbf{E}_E$ and $\mathbf{E}_R$ in dilute solution can be obtained using Debye–Hückel theory in its simplest form. Assume the electrolyte to be point charges interacting via coulomb forces only. Ignore the molecular structure of both solvent and ions. Since the electrophoretic effect is due to the migration of the ion cloud through the fluid, the reduction in ion mobility can be calculated using Stokes' law (3.21)

$$u_E = \frac{\mathbf{v}}{\mathbf{E}} = \frac{ze}{6\pi\eta r} \tag{3.28}$$

where u_E is the *mobility of the ion atmosphere*; r is the radius of the ion atmosphere. From the Debye–Hückel theory of ion activity coefficients,[13]

(12) Since the vectors $\mathbf{E}$, $\mathbf{E}_E$, and $\mathbf{E}_R$ are parallel they can be treated as scalars in division. See footnote (8) in this chapter for discussion of a similar problem.

(13) See, e.g., T. L. Hill, *Introduction to Statistical Thermodynamics* (Reading, Mass.: Addison-Wesley, 1960), p. 321*ff.*

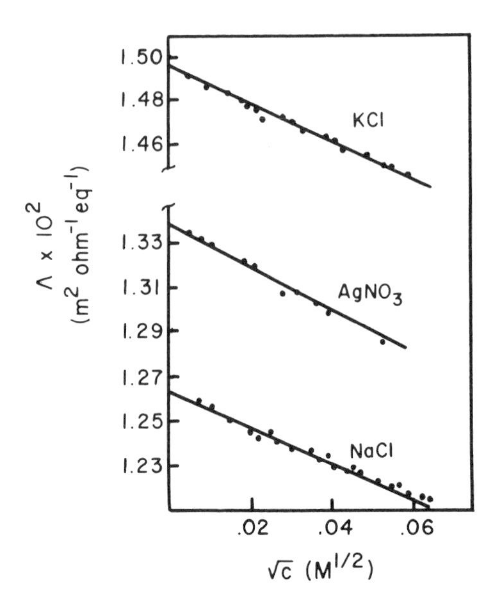

Fig. 3.5. Comparison of equivalent conductance, Λ, of various electrolytes as predicted by Debye–Hückel–Onsager equation (3.30) with those observed experimentally. [Adapted from J. O'M. Bockris and A. K. N. Reddy, *Modern Electrochemistry* (New York: Plenum Press, 1970), Vol. 1, p. 436.]

r can be identified as

$$r = \varkappa^{-1} = \left(\frac{e^2}{\epsilon kT} \sum_i n_i z_i^2\right)^{-1/2} \tag{3.29}$$

where ϵ is the dielectric constant of the solvent and n_i are the ion concentrations. Combining (3.28) and (3.29) shows immediately that $u_E \propto C^{1/2}$. The effect of the relaxation field is far harder to compute and we shall not attempt it here. Qualitatively, this field is produced by distortion of the ion atmosphere. The corresponding ionic displacements depend upon *individual* ion mobilities. However, as in the problem of electrolyte diffusion (see Section 3.4), ionic motions are coupled. As a result the precise dependence of $\mathbf{E}_R$ upon the ion equivalent conductance is rather involved. Onsager[14] developed a quantitative description of the relaxation field; he demonstrated that, like $\mathbf{E}_E$, $\mathbf{E}_R$ is also proportional to $\varkappa$ and thus to $C^{1/2}$.

Onsager's final expression for the equivalent conductance has the form

$$\Lambda = \Lambda^0 - (A + B\Lambda^0)C^{1/2} \tag{3.30}$$

where A depends on η, ϵ, and T while B is a function of λ_+, λ_-, ϵ, and T. A comparison of (3.30) with experiment for various electrolytes is shown in Fig. 3.5. A more sensitive test is to compare the Onsager coefficient

[14] L. Onsager, *Z. Physik* **28**, 277 (1927).

Table 3.5. Observed and Calculated Values of $A + B\Lambda^0$ from (3.30) in Aqueous Solution at 25°C[a]

Electrolyte	Observed slope × 10^4 ($ohm^{-1}\ m^{7/2}\ eq^{-3/2}$)	Calculated slope × 10^4 ($ohm^{-1}\ m^{7/2}\ eq^{-3/2}$)
LiCl	2.56	2.30
$NaNO_3$	2.60	2.35
KBr	2.78	2.54
KCNS	2.42	2.46
CsCl	2.40	2.54
$MgCl_2$	4.56	4.60
$Ba(NO_3)_2$	5.08	4.76
K_2SO_4	4.43	5.04

[a] J. O'M. Bockris and A. K. N. Reddy, *Modern Electrochemistry* (New York: Plenum Press, 1970), Vol. 1, p. 437.

$A + B\Lambda^0$ with that determined experimentally from the slope of the equivalent conductance; some values are tabulated in Table 3.5. It is clear that the Debye–Hückel–Onsager theory accounts satisfactorily for the behavior of Λ at low concentrations. As the ion concentration increases into regions where Debye–Hückel theory no longer accurately describes electrolyte activity, there are also severe deviations from (3.30).

The theory of conductivity has been improved by eliminating some of the assumptions implicit in its derivation.[15] The coupling of the relaxation field and the electrophoretic field has been accounted for. The effect of finite ion size and of ion-pair formation has also been taken into consideration.

3.8. The Wien Effects

Ohm's law is a nearly ideal example of the linear relation between force (the electric field) and flux (the current). However, in aqueous solution at field strengths $\gtrsim 10^5\ V\ m^{-1}$, Ohm's law no longer holds; the conductance becomes noticeably field dependent.[16] In solutions of strong electrolytes

[15] R. M. Fuoss and L. Onsager, *J. Phys. Chem.* **66**, 1722 (1962); **67**, 621 (1963); **68**, 1 (1964).

[16] In nonpolar solvents Ohm's law breaks down at much lower field strengths; measurable deviation occurs at $\mathbf{E} \sim 10^3\ V\ m^{-1}$ as ϵ (the dielectric constant) → 1.

the conductance increases 5–10% in high fields and approaches a limiting value Λ_∞ for fields $\gtrsim 2\times 10^7$ V m^{-1} (first Wien effect).[17] In solutions of weak electrolytes the conductance at high fields is proportional to the field strength. No saturation occurs; in some instances there is even a quadratic dependence of conductance on field strength (second Wien effect).[18]

The first effect arises because the ion atmosphere is greatly modified at high fields. The ion velocity is very large and the ion traverses distances comparable to that of the radius of the ion atmosphere in times so short that the atmosphere cannot reform. As a consequence the relaxation field is destroyed and the electrophoretic field significantly altered. There is less retardation; λ increases, approaching a high field limit.

The second effect is altogether different. Since the conductance is proportional to field strength, no high field saturation is found. It appears that the *number* of charge carriers increases at high fields. This may be understood if we postulate a second nonconducting species, the ion pair. For a weak electrolyte AB there are then two equilibria[19]:

$$\mathrm{AB} \overset{K_0}{=} (\mathrm{A}^+, \mathrm{B}^-) \overset{K_1}{=} \mathrm{A}^+ + \mathrm{B}^- \tag{3.31}$$

In the ion pair (A^+, B^-) the ions are not yet individually solvated but are partially separated by interpolated solvent. The equilibrium constant K_0 is *not* field dependent. However, high fields naturally favor separation of the ion pair; more ions are produced, i.e., K_1 is an increasing function of the field. For this reason the field dependence of conductance in weak electrolytes is also known as the *dissociation field effect*. Arguing from (3.31), the apparent equilibrium constant is

$$\frac{[\mathrm{A}^+][\mathrm{B}^-]}{[\mathrm{AB}]_{\mathrm{tot}}} = \frac{K_1}{1 + K_0^{-1}} \tag{3.32}$$

where $[AB]_{tot} = [AB] + [(A^+, B^-)]$. A quantitative theory describing the effect of field variation has been developed by Onsager.[20]

The Wien effects are more than just curiosities. They significantly affect the kinetics of electrode processes, particularly in solvents of low

[17] M. Wien, *Phys. Z.* **29**, 751 (1928).
[18] M. Wien and L. Schieb, *Phys. Z.* **32**, 545 (1931).
[19] For a discussion of the statistical mechanics of ion-pair formation see N. Davidson, *Statistical Mechanics* (New York: McGraw–Hill, 1962), pp. 507–512. A kinetic derivation of K_1 is given in Section 9.13.
[20] L. Onsager, *J. Chem. Phys.* **2**, 599 (1934).

dielectric constant. Since there is always a potential gradient near an electrode due to the formation of an electrical double layer, there may be significant non-Ohmic behavior in the vicinity of an electrode.[21] Such phenomena must be considered when interpreting the kinetics of electrode reactions.

Another nontrivial application of the Wien effect is in discriminating between CO_2 (aq) and H_2CO_3 in order to accurately measure the true equilibrium constant for the aqueous ionization[22]

$$H_2CO_3 = H^+ + HCO_3^-, \qquad K_I$$

The equilibrium constant for the hydration reaction

$$H_2O + CO_2 = H_2CO_3, \qquad K_H$$

is not field dependent whereas K_I is. The apparent ionization constant is given by (3.32) with $K_1 = K_I$ and $K_0 = K_H$; high-field conductance measurements permit distinction between the two forms of dissolved CO_2.

Problems

3.1. Consider a wire of cross section A and length l in which there is a potential drop V and a current I. Show that (3.1) reduces to $V = IR$, where $R = l/A\sigma'$.

3.2. A weak electrolyte is only partially dissociated. The equilibrium $AB \rightleftharpoons A^+ + B^-$ has equilibrium constant K. Show that

$$K = \tilde{C}^2/(C - \tilde{C}) = \Lambda_{app}^2 C/[\Lambda(\Lambda - \Lambda_{app})]$$

In an approximate form, valid when $\tilde{C}$ is very small, Λ is set equal to Λ^0; the resulting expression for K is the *Ostwald dilution law*.

3.3. The electrolyte diffusion coefficient, (3.19), may be related to ion equivalent conductance using (3.13). Show that an alternative form for D is

$$D = \frac{RT}{F^2} \frac{z_+ + z_-}{z_+ z_-} \frac{\lambda_+ \lambda_-}{\Lambda}$$

[21] There is an enormous literature on electrode kinetics; see P. Delahay, *Double Layer and Electrode Kinetics* (New York: Interscience, 1965).

[22] D. Berg and A. Patterson, Jr., *J. Am. Chem. Soc.* **75**, 5197 (1953); K. F. Wissbrun, D. M. French, and A. Patterson, Jr., *J. Phys. Chem.* **58**, 693 (1954).

3.4. Tracer diffusion studies on NaCl in water at 25°C show that $D(Na^+) = 1.32 \times 10^{-9}$ m² sec⁻¹ and $D(Cl^-) = 1.98 \times 10^{-9}$ m² sec⁻¹. Using (3.19) compute $D(NaCl)$; the measured value is 1.53×10^{-9} m² sec⁻¹. Compare the three values of D with those calculated using (3.13) and the result of Problem 3.3.

3.5. Determine the local electric field in a cell of length 0.1 m containing an aqueous solution of NaCl at 25°C. Assume a linear concentration profile with C maintained at 0.01 M at one end of the cell and 0.015 M at the other end. Estimate the drift velocity of the ions due to this field. Make sure you handle units properly.

3.6. Consider a membrane of thickness d across which there is a potential difference $\Delta\Phi$. Show that the current density across the membrane is

$$\mathbf{j} = \frac{\sigma}{d}\,\Delta\Phi = G\,\Delta\Phi$$

where G is the conductivity per unit area of membrane surface. Now show that

$$G = F^2(z_+D_+ + z_-D_-)C^*/RTd$$

where C^* is the concentration of charge carriers in the membrane.

3.7. Use the data of Table 3.4 to estimate the ionic radius of K^+ in the various solvents listed. You may assume, at this level of approximation, that K^+ and I^- have the same ionic radii. Can you suggest any explanation for the changes observed?

3.8. Liquid ammonia is an ionizing solvent with a high dielectric constant. Its structure is probably like that of water, governed by a hydrogen-bonded network, in this case with one hydrogen bond and three normal bonds per ammonia molecule. Unlike the proton mobility in water, the proton mobility in NH_3 is *not* anomalously high. The value of λ^0 is typical of univalent cations in this solvent. Can you suggest any reasons which might account for this difference?

3.9. For 1–1 electrolytes in water solution at 25°C the values of A and B in the Onsager equation are $A = 1.90 \times 10^{-4}$ ohm⁻¹ m² eq⁻¹ (m³ eq⁻¹)$^{1/2}$ and $B = 7.24 \times 10^{-3}$ (m³ eq⁻¹)$^{1/2}$. Using these values and data from Table 3.1 calculate:

(a) the specific conductivity of a 0.1 M solution of KCl; the experimental value is 1.2896 ohm⁻¹ m⁻¹;

(b) the specific conductivity of a 0.1 M solution of HAc ($K_A = 1.8 \times 10^{-5}$ M).

3.10. For a *binary electrolyte* the expressions for A and B in the Onsager equation are

$$A = 9.218 \times 10^{-6}(z_+ + z_-)^{3/2}/\eta(\epsilon T)^{1/2}$$

$$B = 3.13 \times 10^{-4}(z_+ + z_-)^{1/2}w/(\epsilon T)^{3/2}$$

where all quantities are in SI units and w is a dimensionless function of the ion equivalent conductances:

$$w = 2qz_+z_-/(1 + q^{1/2})$$
$$q = z_+z_-(\lambda_+ + \lambda_-)/[(z_+ + z_-)(z_+\lambda_- + z_-\lambda_+)]$$

By using mixed solvents the dielectric constant of the medium can be varied continuously. The values of ϵ in H_2O–CH_3OH mixtures at 25°C are

CH_3OH (wt. %):	0	25	50	75	100
ϵ:	78.54	67.8	56.3	44.9	32.66

Using these data and those of Table 3.3 calculate Λ for 10^{-3} M and 0.1 M solutions of KCl for the various solvents. Do you think the estimates of Λ at the higher concentrations are reliable?

3.11. Show that (3.32) follows from (3.31).

General References

J. O'M. Bockris and A. K. N. Reddy, *Modern Electrochemistry* (New York: Plenum Press, 1970), Vol. 1, Chapters 4 and 5.

H. S. Harned and B. B. Owen, *The Physical Chemistry of Electrolyte Solutions*, 3rd ed. (New York: Reinhold, 1958), Chapters 4, 6, and 7.

R. A. Robinson and R. H. Stokes, *Electrolyte Solutions* (London: Butterworths, 1955), Chapters 5, 6, 7, 10, and 11.

4

Determination of Rate Laws

4.1. Introduction

Altering the constraints on a chemical system poses two distinct questions. What is the new equilibrium configuration? How rapidly does the system approach this new state? The first is a problem of applied thermodynamics; the second is the central problem of chemical kinetics.

The transport phenomena discussed in Chapters 2 and 3 are the simplest examples of kinetic phenomena, if not the most familiar. For the cases considered, a single macroscopic property (number, charge, momentum, or energy density) was displaced from its equilibrium value. This perturbation caused a flux in the opposite direction, proportional to the displacement. The proportionality constant is the transport coefficient. For simple gaseous systems the relations between displacement and flux and the transport coefficient were determined using a hard-sphere model to describe molecular interaction.

Reacting systems are much more complex. Corresponding to the linear phenomenological transport equations (Fick's, Fourier's, Ohm's, or Poiseuille's laws) are the nonlinear *rate laws* for chemical reactions. The rate constant is the analog of the transport coefficient. The goal of theory is to develop a molecular model which accounts for the rate law and which provides an expression for the rate constant. These problems are, in many ways, still poorly understood; a unified molecular theory such as was developed for treating transport in gases does not yet exist. Instead, reaction mechanisms, which are proposed as models for the underlying molecular events, are constructed to be consistent with the observed rate laws. Then very approximate theories for the rate constants are formulated in terms of the elementary steps which comprise the mechanism. General theories, based upon molecular distributions, are either approximations of uncertain validity (if computationally tractable) or fairly rigorous (but applicable to only the simplest systems).

In practical terms the study of chemical kinetics involves three separate problems. The first is to establish, experimentally, the rate law which describes the kinetic observations. The second is to determine a mechanism (molecular model) which is consistent with the rate law.[1] The third is to develop a theory for the rate constant in terms of the specific mechanism.

In this chapter we discuss three general problems. First, we consider the rate law's relationship with and distinction from both reaction stoichiometry and reaction mechanism. Second, we describe some of the experimental techniques used in establishing rate laws. Finally, we analyze some specific examples.

4.2. *Stoichiometry, Rate Law, and Mechanism*

The reactions between hydrogen and the halogens have the stoichiometry

$$H_2 + X_2 = 2HX \tag{4.1}$$

However, when Bodenstein[2] studied the kinetics of the iodine and bromine reactions he found that the experimental data led to completely different formulas to describe the rate of production of HX

$$\frac{d[\mathrm{HI}]}{dt} = k[\mathrm{H_2}][\mathrm{I_2}] \tag{4.2}$$

$$\frac{d[\mathrm{HBr}]}{dt} = \frac{k'[\mathrm{H_2}][\mathrm{Br_2}]^{0.5}}{1 + k''[\mathrm{HBr}]/[\mathrm{Br_2}]} \tag{4.3}$$

These equations, discovered after repeated experimentation, and the way in which the parameters k, k', k'' depend upon the external conditions (temperature, pressure, etc.) provide the data which are used to construct a mechanism. A simple bimolecular mechanism, closely related to the

[1] As we shall see in the next chapter there are always many mechanisms which are consistent with the data and a choice between them is made on various bases.

[2] M. Bodenstein, *Z. Phys. Chem.* **29**, 295 (1899); M. Bodenstein and S. C. Lind, *Z. Phys. Chem.* **57**, 168 (1907).

stoichiometry, was proposed to account for (4.2)[3]

$$H_2 + I_2 \rightharpoonup 2HI \tag{4.4}$$

For many years this was believed to be an accurate molecular description. No simple mechanism accounted for (4.3). The following complex series of steps, which could not have been guessed from the stoichiometry (4.1),

$$\begin{array}{ll} Br_2 \rightleftharpoons 2Br & \text{(equil. dissoc.)} \\ Br + H_2 \rightharpoonup HBr + H & \text{(slow)} \\ H + Br_2 \rightharpoonup HBr + Br & \text{(fast)} \\ H + HBr \rightharpoonup H_2 + Br & \text{(fast)} \end{array} \tag{4.5}$$

was finally shown to be consistent with (4.3). Later research suggested that (4.2) and (4.4) were deceptively simple; the first step in the formation of HI is very likely the dissociation of the halogen. The subsequent steps are different from (4.5) and there are still numerous unanswered questions.

The fact that seemingly simple kinetics like those in the formation of HI could still be an active subject for controversy for three-quarters of a century is indicative of a major problem in mechanistic assignment. *It is impossible to prove a mechanism correct.* Rather one shows that various plausible mechanisms are inconsistent with the data. By amassing sufficient chemical evidence one hopes to show that all but one chemically reasonable mechanism is impossible. However, this can never really be done since there are always more complex mechanisms which can account for the data. Thus, even given compelling evidence, there remains a chance that further experimentation will require revising the model which we call a mechanism.

4.3. Elementary Rate Laws and the Principle of Mass Action

A general expression for the stoichiometry of a chemical reaction is

$$\sum_i \nu_i Y_i = 0 \tag{4.6}$$

where ν_i are the stoichiometric coefficients and Y_i represent the species which react. The ν_i are *positive for products* and *negative for reactants*;

[3] We use the notation A + B = C to describe the stoichiometry of a reaction. The notation A + B ⇀ C is used to mean a specific molecular encounter.

in these terms the hydrohalogenation reaction (4.1) becomes

$$-H_2 - X_2 + 2HX = 0$$

where $\nu_1 = \nu_2 = -1$ and $\nu_3 = +2$. If the reaction represented by (4.6) proceeds in a single step, the concentration changes of the species involved are simply related at any time,

$$\{[Y_i] - [Y_i]_0\}/\nu_i = \{[Y_j] - [Y_j]_0\}/\nu_j = \Delta\xi(t) \tag{4.7}$$

where $[Y_i]$ is the concentration of species i at time t and $\xi(t)$ is the *progress variable* for the reaction.[4] A kinetic formulation in terms of a single progress variable is only correct if the reaction involves just one step. This is clear if we refer to the mechanism (4.5); the important reaction intermediates, Br and H, would not even appear in (4.7) since they do not occur in the overall stoichiometry.

Because the rates of change of *all the components* are simply related *at all times* in *single-step reactions*, the rate of reaction may be defined as

$$R = \frac{d\xi}{dt} = \frac{1}{\nu_i}\frac{d[Y_i]}{dt} \tag{4.8}$$

At equilibrium R is zero, which we interpret to mean that the rate of reaction and its inverse precisely balance. To determine the rate of the forward process in a single-step reaction note that a reaction occurs in one step only if all reactants are close together at the same time. Stated differently the reacting molecules must collide essentially simultaneously. Since collision probability is proportional to the product of the number density of the molecules which participate in a kinetic event, the equilibrium forward and reverse reaction rates are

$$\begin{aligned} R_f(\text{eq}) &= k_f(T, p) \prod_{\text{react.}} [Y_i]_{\text{eq}}^{-\nu_i} \\ R_r(\text{eq}) &= k_r(T, p) \prod_{\text{prod.}} [Y_i]_{\text{eq}}^{\nu_i} \end{aligned} \tag{4.9}$$

where the k's are the *rate constants*.[5] Since the two rates are equal at

[4] For a discussion of progress variables and chemical equilibria see, e.g., F. T. Wall, *Chemical Thermodynamics*, 3rd ed. (San Francisco: Freeman, 1974), Chapter 10.

[5] The pressure dependence of k in *single-step* reactions is usually small; this problem is treated by C. A. Eckart, *Ann. Rev. Phys. Chem.* **23**, 239 (1972). Gas-phase unimolecular reaction, for which k is a sensitive function of p, is *not* a single-step process (Section 5.4).

equilibrium, (4.9) can be rearranged to yield

$$\frac{k_f(T, p)}{k_r(T, p)} = \prod_i [Y_i]_{\text{eq}}^{\nu_i} \tag{4.10}$$

In ideal gases or ideal solutions the right-hand side of (4.10) is the equilibrium constant for reaction $K_{\text{eq}}(T, p)$. More generally it is activity, not concentration, which specifies a concentration-independent equilibrium constant. Defining the activity as

$$a_i \equiv \gamma_i [Y_i]_{\text{eq}}$$

where γ_i is the activity coefficient,[6] we find from (4.10)

$$K_{\text{eq}}(T, p) = \prod a_i^{\nu_i} = \frac{k_f(T, p)}{k_r(T, p)} \prod \gamma_i^{\nu_i} \tag{4.11}$$

The ratio of forward and reverse rate constants can be identified as the equilibrium constant whenever $\prod \gamma_i^{\nu_i} = 1$, a condition satisfied in ideal systems such as dilute gases. In addition, by redefining the standard state of the reacting species, the constraint may be satisfied if the γ_i do not vary significantly with reactant concentration (see Problem 4.2). It is clearly to the kineticist's advantage to study systems under conditions where the activity coefficient product is unity. The activity correction requires that considerable care be exercised when comparing kinetic parameters determined under grossly different conditions.

Assuming that away from equilibrium the reaction rates of the forward and reverse steps are still given by (4.9),[7] we expect that $R_f \neq R_r$ and that the overall rate of reaction is

$$\frac{d\xi}{dt} \equiv R = R_f - R_r = k_f \prod_{\text{react.}} [Y_i]^{-\nu_i} - k_r \prod_{\text{prod.}} [Y_i]^{\nu_i} \tag{4.12}$$

which is precisely of the form (4.2) as long as k_r or [HI] are small. From (4.11) we note that if K_{eq} varies with temperature or pressure one (or both) of the rate constants must do so too. Furthermore, the activity dependence indicates that the rate constants may be concentration dependent. The

[6] For a discussion of activity, activity coefficient, and their relationship to the equilibrium constant see F. T. Wall, *Chemical Thermodynamics*, 3rd ed. (San Francisco: Freeman, 1974), pp. 365–366, 419–420.

[7] This is equivalent to assuming that the collision frequency (2.18) is unaltered in a system displaced from equilibrium.

rate law (4.12) is consistent with the single-step molecular model

$$\sum_{\text{react.}} -\nu_i Y_i \rightleftharpoons \sum_{\text{prod.}} \nu_i Y_i$$

Complicated sequential mechanisms such as (4.5) require a collection of progress variables, one for each elementary reaction. With such models, and assumptions about the magnitude of the rate constants, it becomes possible to provide molecular interpretations of rate expressions such as (4.3). In multistep sequences often only the forward rate law is established experimentally. However, as the concentration of reactants and products are related via the equilibrium constant and as $R_f(\text{eq}) = R_r(\text{eq})$, knowledge of the forward rate law can be used to limit the possible rate laws for reverse reaction (Problem 4.20).

The rate law (4.12) is unlike any of the linear transport equations discussed in Chapters 2 and 3. The response is the reaction rate $\dot{\xi}$. However, the right-hand side of (4.12) bears little apparent resemblance to the thermodynamic driving force for reaction, the affinity, A, defined as

$$A \equiv -\sum_i \nu_i \mu_i$$

where μ_i is the chemical potential.[8] The condition for chemical equilibrium is that A be zero; the chemical transport equation is thus

$$\dot{\xi} = LA \tag{4.13}$$

with L the corresponding transport coefficient. The two expressions for $\dot{\xi}$, (4.12) and (4.13), are not obviously equivalent and L cannot be immediately related to the rate constants. In fact the equations are not the same except in the vicinity of equilibrium.[9] Further from equilibrium the linear relation (4.13) is inadequate and it must be augmented to include *all* powers of A. As a result the standard form (4.12) is by far the more practical in application. The relationship between L and the rate constants is usually complicated by activity effects. However, if these may be ignored, the result is simple; L is proportional to the equilibrium reaction rates, (4.9),

$$L = R_f(\text{eq})/RT = R_r(\text{eq})/RT$$

[8] The affinity is discussed by F. T. Wall, *Chemical Thermodynamics*, 3rd ed. (San Francisco: Freeman, 1974), Chapter 10.

[9] For a simplified demonstration of this limited equivalence see A. Katchalsky and P. F. Curran, *Nonequilibrium Thermodynamics in Biophysics* (Cambridge, Mass.: Harvard University Press, 1965), pp. 91–97.

4.4. Reaction Order and Molecularity

To distinguish between a rate law determined experimentally and one proposed on the basis of an assumed mechanism we use *decimal* numbers as exponents in *experimental* rate laws and *integers or fractions* as exponents in *mechanistic* rate laws. In the presence of argon the rate law for the reaction

$$H_2 + D_2 = 2HD$$

determined experimentally is

$$\frac{d[\mathrm{HD}]}{dt} = 2k(T)[\mathrm{H_2}]^{0.38}[\mathrm{D_2}]^{0.66}[\mathrm{A}]^{0.98} \tag{4.14}$$

On the other hand, an assumed mechanism for the aqueous reaction

$$I^- + ClO^- = IO^- + Cl^-$$

predicts a rate law[10]

$$\frac{d[\mathrm{IO^-}]}{dt} \propto [\mathrm{I^-}]^1[\mathrm{ClO^-}]^1[\mathrm{OH^-}]^{-1}$$

Equally important is the molecularity of an elementary reaction. One of the fast steps in the hydrogen bromine kinetic sequence (4.5) is

$$H + Br_2 \rightharpoonup HBr + Br \tag{4.15}$$

which is a statement about the molecular processes that lead to reaction. When an atom of hydrogen interacts with a bromine molecule, there is a probability of the transformation to HBr and Br. According to (4.15) the rate of HBr production is predicted to be

$$\frac{d[\mathrm{HBr}]}{dt} \propto [\mathrm{H}]^1[\mathrm{Br_2}]^1$$

The process is unimolecular in both H and Br_2. Since elementary reactions such as (4.15) represent molecular collision processes the exponents in the predicted rate expressions must *always* be *positive integers*. Thus experimental rate laws like those for the $H_2 + Br_2$ reaction (4.5) or the $H_2 + D_2$ reaction (4.14) indicate mechanisms which involve more than one elementary step.

[10] This reaction is discussed in Problem 5.10.

4.5. The Principle of Detailed Balance

We have seen that dynamic equilibrium requires that rate constants be related to the equilibrium constant in a single-step reaction. For sequential reactions the overall equilibrium constant is the product of the equilibrium constants for the individual reactions, and, in an n-step process, the extension of (4.11) is

$$K_{\text{eq}} = \prod_{\alpha=1}^{n} K_{\text{eq}}^{(\alpha)} = \left(\prod_{\alpha=1}^{n} \frac{k_f^{(\alpha)}}{k_r^{(\alpha)}}\right)\left(\prod_{\alpha=1}^{n} \prod_{i} \gamma_i^{\nu_i^{(\alpha)}}\right) \tag{4.16}$$

A more interesting situation arises in systems of which the following is a well-known example:

$$\begin{array}{ll} (A \rightleftharpoons B) & H_2CO_3 \rightleftharpoons CO_2(aq) + H_2O \\ (B \rightleftharpoons C) & CO_2(aq) + H_2O \rightleftharpoons H^+ + HCO_3^- \\ (C \rightleftharpoons A) & H^+ + HCO_3^- \rightleftharpoons H_2CO_3 \end{array} \tag{4.17}$$

It would appear possible that equilibrium could be established by means of the cycle $A \rightarrow B \rightarrow C \rightarrow A$ even if the forward and reverse steps in each elementary step did not proceed at the same rate. An expression analogous to (4.16) could be derived for an equilibrium constant even though (4.11) might not be true for the individual steps. It is precisely such a possibility that is forbidden by microscopic reversibility.[11]

Newtonian mechanics and quantum mechanics have in common the property of time-reversal invariance. The Newtonian equations of motion for a system of N interacting particles are

$$m_i\ddot{x}_i = -\partial U(x_1, \ldots, z_N)/\partial x_i \qquad \text{for all } x, y, z, \quad i = 1, \ldots, N \tag{4.18}$$

Changing $t \rightarrow -t$ does not affect the accelerations $\ddot{x}_i$; on the other hand the velocities $\dot{x}_i$ do change sign. As a consequence, for every solution of (4.18) there is an identical one in which all particles have their velocities reversed.[12] For classical point particles the dynamics of a colliding pair

[11] The following treatment is adapted from N. Davidson, *Statistical Mechanics* (New York: McGraw–Hill, 1962), pp. 230–235.

[12] The dynamics are more complicated when magnetic fields are present. Since the field couples with the velocity by means of terms like $\mathbf{v} \times \mathbf{B}$ it is also necessary to reverse the field direction when the velocities are reversed. In this way time-reversal invariance can be generalized.

are specified by their velocities, $\mathbf{c}_1$ and $\mathbf{c}_2$, and impact parameter $\mathbf{b}$. Due to time-reversal invariance the transition probability for the change of state $(\mathbf{c}_1, \mathbf{c}_2, \mathbf{b}) \rightarrow (\mathbf{c}_1', \mathbf{c}_2', \mathbf{b}')$ must equal that for the change $(-\mathbf{c}_1', -\mathbf{c}_2', -\mathbf{b}') \rightarrow (-\mathbf{c}_1, -\mathbf{c}_2, -\mathbf{b})$.[13]

Quantum mechanics describes the change of state which occurs in an A–B collision in terms of the wave functions of the interacting particles; the wave function for the pair changes from $\Psi_1(\mathrm{A, B}) \rightarrow \Psi_2(\mathrm{A, B})$. Ignoring magnetic fields, the equations of motion can be used to demonstrate that the time-reversed collision is one involving particles with reversed momenta and angular momenta. Denoting that change of state as $\Psi_2^-(\mathrm{A, B}) \rightarrow \Psi_1^-(\mathrm{A, B})$ and defining the transition probability for the change from state 1 to state 2 as $w(1 \rightarrow 2)$, time-reversal invariance requires that

$$w(1 \rightarrow 2) = w(2^- \rightarrow 1^-)$$

Since neither collision nor time reversal affects energy, the states 1, 2, 2^-, and 1^- are isoenergetic and therefore equally probable in an equilibrium system. As a consequence the *number* of pairs making the transition $1 \rightarrow 2$ exactly equals the number which go from $2^- \rightarrow 1^-$. This is the quantum version of detailed balance.[14]

We can now show that each reaction in the system (4.17) equilibrates individually. On the molecular level in an *equilibrium system* the number of $\mathrm{A} \rightarrow \mathrm{B}$ transitions, each molecule being in a specified state, is exactly equal to the number of back transitions for the time-reversed molecules. Since a molecular state and its time-reversed partner are chemically indistinguishable (the two states are degenerate) the overall effect, when averaged over a Boltzmann distribution of energies, is that the rate of forward and backward reaction in the *individual* steps of (4.17) must be equal at equilibrium. In addition detailed balance implies a similar dynamic equilibrium among molecules undergoing transitions between any subset of states.

[13] For particles which interact via spherical potentials a more restricted principle can be demonstrated. Velocity reversal is *not* required for equivalence of transition probability so that the transition $(\mathbf{c}_1', \mathbf{c}_2', \mathbf{b}') \rightarrow (\mathbf{c}_1, \mathbf{c}_2, \mathbf{b})$ and its inverse $(\mathbf{c}_1, \mathbf{c}_2, \mathbf{b}) \rightarrow (\mathbf{c}_1', \mathbf{c}_2', \mathbf{b}')$ are equally probable, a result demonstrated in the derivation of the Maxwell distribution, Section 1.2.

[14] Microscopic reversibility is the transition probability constraint. Detailed balance is its consequence in an equilibrium system. See B. W. Morrissey, *J. Chem. Ed.* **52**, 296 (1975).

4.6. Experimental Methods for Determining Rate Laws

A rate law is established by monitoring the time dependence of the concentrations of the species that take part in the reaction and by determining how this dependence is affected by altering the constraints on the reacting system. Since temperature usually has a considerable effect upon the reaction rate, the reaction vessel is always thermostatted. While temperature control is rarely difficult unless the reaction is highly exothermic and simultaneously very rapid, the accurate measurement of time may pose a significant problem, particularly for fast reactions. In most chemically interesting systems there are two major, closely related, experimental problems. The first is to find a property of the system that is tightly coupled to the concentration of one (or more) reactant and which is easy to monitor *continuously*. The second, more directly related to the problem of mechanism, is to be able to detect, identify, and monitor the trace species that can be intermediates in the reaction sequence.

Many different techniques have been used to detect trace intermediates. These include:

(a) observation of characteristic emission spectra, either ultraviolet, visible, or infrared;
(b) titration of the intermediate with a specific reagent followed by observation of the product;
(c) observation of a free radical by means of electron paramagnetic resonance;
(d) observation of gaseous atomic intermediates by atomic resonance spectroscopy;
(e) observation of an unusual oxidation state by polarographic methods.

Each of these techniques has drawbacks. For example, if a reactant has an emission band in the same region of the spectrum as the intermediate, spectroscopic detection will fail. Other difficulties arise because the speed of the reaction may limit the time available for observation.

An almost unlimited number of ways have been proposed to monitor reaction. Among the more common are:

(1) pressure (in some gas-phase reactions);
(2) dielectric constant or index of refraction;
(3) light absorption at a specific frequency;
(4) EMF with respect to an ion-specific half-cell;
(5) volume changes (dilation);

(6) conductance;
(7) optical rotation;
(8) concentration (by removing an aliquot, quenching the reaction, and titrating);
(9) vapor chromatography;
(10) mass spectrometry;
(11) electron paramagnetic resonance spectroscopy.

It is not always simple to determine the concentrations from the measured property. Consider light absorption in a multicomponent system. The frequency-dependent absorbance $A(t, \nu)$ is related to the concentrations by the Beer–Lambert law

$$A(t, \nu) = l \sum_i \epsilon_i(\nu) C_i(t) \tag{4.19}$$

where l is the path length, $\epsilon_i(\nu)$ the extinction coefficient of species i, and $C_i(t)$ its concentration. If only one species absorbs, A is directly proportional to the concentration of that species. If more than one species absorbs, measurements are required over a range of frequencies to determine the $C_i(t)$. Similar difficulties exist no matter what monitoring procedure is used. The general time-dependent response of an evolving system, $R(t)$, is

$$R(t) = \sum_i a_i C_i(t) \tag{4.20}$$

where the a_i are determined by the properties of species i and by the measurement procedure used[15] (Problems 4.21 and 4.22).

In all cases the time required for a measurement (the response time of a particular technique) *must be short* compared with the time during which significant composition changes occur. For example, method 8 is well suited to reactions with half-times on the order of hours but requires great ingenuity when applied to reactions with short half-times.[16] The practical limits of various techniques are summarized in Fig. 4.1.

For reactions which are essentially complete between periods ranging from 1 msec to 1 sec *rapid mixing* techniques must be used. Two such

[15] I thank Professor L. Parkhurst for pointing out to me the practical significance of this problem.

[16] Quenching, by spraying a reacting mixture into a cold liquid or onto a cold plate, can be completed within milliseconds. Combining this approach with flow methods allows method 8 to be applied to reactions with half-times $\lesssim$100 msec. See B. Chance, Q. H. Gibson, R. H. Eisenhardt, and K. K. Lonberg-Holm, eds., *Rapid Mixing and Sampling Techniques in Biochemistry* (New York: Academic Press, 1966), Chapter 6.

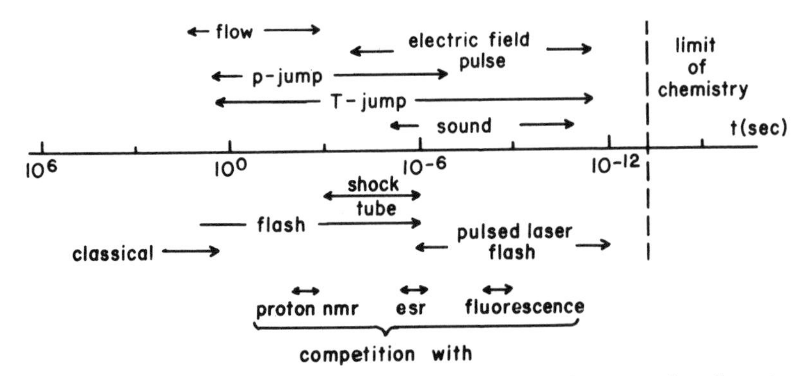

Fig. 4.1. Practical range of applicability of various methods for measuring the rates of chemical reaction.

procedures are the stopped flow and the continuous flow methods. In stopped flow[17] the reactants are driven into a small reaction chamber in such a way that mixing is essentially complete within 0.1 msec. The reaction is then monitored automatically using sensitive recording techniques synchronized with the *arrest of fluid flow*. In continuous flow[18] experiments reactants continuously enter a reaction chamber, mix, and are forced into a flow tube where the reaction occurs. Depending upon the rate of flow, each point in the tube corresponds to a different time after the onset of reaction. By monitoring at different points the degree of reaction at different times can be determined.

When the reaction time is less than 1 msec, flow methods are too slow to insure homogeneity. Shock tube[19] methods are useful for studying high-temperature gas-phase reactions which take place in time spans of 1 msec to 1 μsec. Here a *driver gas* (usually He or H_2 at a few atmospheres pressure) is constrained by a diaphragm at one end of a long tube; the remainder of the tube is filled with reactant mixture at low pressure (typically 10^{-3} atm). When the diaphragm is ruptured the driver gas expands abruptly,

(17) The basic idea for stopped flow was introduced by B. Chance, *J. Franklin Inst.* **229**, 455, 613, 737 (1940). For a modern apparatus see Q. H. Gibson and L. Milnes, *Biochem. J.* **91**, 161 (1964).

(18) This method was pioneered by Rutherford for studying gas-phase reactions [E. Rutherford, *Phil. Mag.* **44**, 422 (1897)]. The mixing methods that are now in common use were designed by H. Hartridge and F. J. W. Roughton, *Proc. Roy. Soc.* (*London*) **A104**, 376 (1923), and by G. A. Millikin, *Proc. Roy. Soc.* (*London*) **A155**, 277 (1936).

(19) Shock tubes were invented by M. P. Vieille, *Compt. Rend.* **129**, 1228 (1889). For a discussion of modern design see H. T. Nagamatsu, *Fundamental Data Obtained from Shock Tube Experiments*, R. Ferri, ed. (New York: Pergamon, 1961), pp. 86–136.

producing a shock wave which travels down the tube at supersonic speed. The shock front is very sharp. As it passes any point in the reactant mixture the reactant gases are compressed, heating them to temperatures $\sim 10^3$–10^4 K; this compression, which takes ~ 1 μsec, provides one limit for the technique's applicability. Reaction begins as the gases are heated with passage of the shock front. To study the system one relies on the fact that the driver gas lags behind the shock front, taking a few milliseconds to flow down the tube. An observation point is established near the end of the tube. The first changes to be seen take place when the shock front passes. The gas behind the front has been reacting from the instant the shock wave passed by. It is pushed past the monitoring station by the driver gas. The later the reactants pass the point, the longer they will have had to react. Thus a time profile can be established.

Shock tubes are of limited utility. A more general approach to the study of reactions which are complete in the range 1 msec–1 nsec is to use *fast reaction* methods.[20] An equilibrium system is perturbed by an external stimulus applied for a very short time (always less than the half-time for reestablishing equilibrium). A common approach is to effect a temperature jump in the system by a brief burst of heating. If the equilibrium is temperature sensitive the concentration of reactants must readjust; by synchronizing an automatic recording technique with the onset or termination of the heating pulse the relaxation to the new equilibrium state can be followed. There are many other stimuli that can be used to perturb the system. These include dilation (pressure jump), electric field (Wien effect), etc. Any method that can perturb the system very rapidly is potentially useful for such an experiment.

Perturbation methods displace a chemical equilibrium but no new species are formed. In flash photolysis novel, unstable chemical species are produced by irradiation. Their reactions with the species initially present are the domain of photochemistry (treated in Chapter 6). Such reactions

[20] The original fast reaction techniques were based upon flash photolysis [G. Porter, *Proc. Roy. Soc.* (*London*) **A200**, 284 (1950)] and the Wien effect [M. Eigen and J. Schoen, *Z. Elektrochem.* **59**, 483 (1955)]. An enormous literature exists on the subject. For recent discussions of some techniques see literature on: (1) temperature jump, G. W. Hoffman, *Rev. Sci. Instr.* **42**, 1643 (1971), and D. H. Turner, G. W. Flynn, N. Sutin, and J. V. Beitz, *J. Am. Chem. Soc.* **94**, 1554 (1972); (2) pressure jump, A. Jost, *Ber. Bunsenges. Phys. Chem.* **70**, 1057 (1966); (3) electric field methods, L. C. M. de Maeyer, in *Methods in Enzymology*, Vol. 16, K. Kustin, ed. (New York: Academic Press, 1969), p. 80; (4) ultrasonic methods, K. G. Plass, *Acoustica* **19**, 236 (1967/68); (5) flash photolysis, G. Porter and M. R. Topp, *Proc. Roy. Soc.* (*London*) **A315**, 163 (1970).

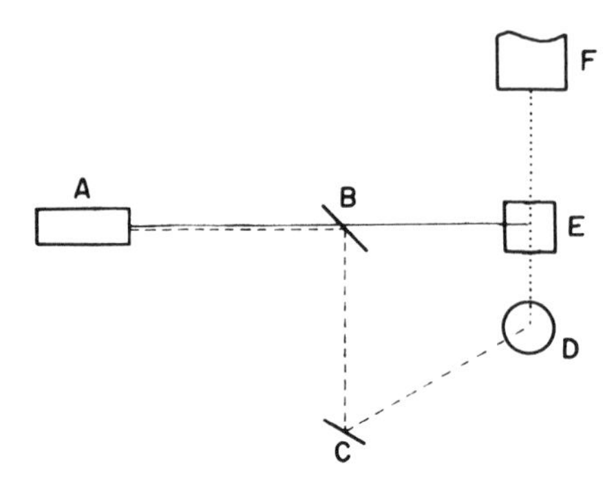

Fig. 4.2. Schematic diagram of apparatus for nanosecond flash photolysis. The components are: (A) pulsed laser source; (B) beam splitter; (C) movable mirror; (D) scintillation solution; (E) reaction vessel; (F) spectrograph. (—) Path of exciting pulse. (– –) Path of pulse to scintillation solution (varied by moving C). (· · ·) Path of analyzing pulse.

are completely different from those studied using perturbation methods. Flash photolysis is designed to investigate fast *photochemical* reaction.

An alternate approach to the study of fast reactions is by means of *competition methods.* Here the *rate* of a process is measured separately in systems where chemical reaction can and cannot occur. If the process is perturbed by the reaction, the *difference* in the rates is a measure of the rate of reaction. For example, an unstable species may be formed photochemically by irradiation and its lifetime determined. The experiment is then altered by adding a substance which can react with the unstable species; it will thus have a shorter lifetime from which the rate of the chemical reaction can be deduced. To be a useful technique the chemical and physical processes must proceed at roughly similar rates which limits the applicability of this approach, as is emphasized in Fig. 4.1. Other competition experiments (and the corresponding observables) have been made using polarography (current), nuclear magnetic resonance (line width), electron spin resonance (line width), and diffusion (diffusion coefficient).[21]

Ultrafast flash photolysis ($\sim$ 1 nsec) can be studied using pulsed lasers.[22] Here a pulse of radiation of duration $\sim$ 1 nsec is formed and passed through a beam splitter, as illustrated in Fig. 4.2. Half of the beam is sent through the reaction vessel as an initiating flash. The other half of the beam is reflected from a mirror into a solution containing a scintillation material which fluoresces almost instantly producing a nanosecond pulse of continuous radiation which passes through the reactant vessel. Concentrations are determined by ordinary spectroscopic means; the scintillation radiation provides the exciting light. The time at which the concentration

[21] For an introduction to these methods see D. N. Hague, *Fast Reactions* (London: Wiley-Interscience, 1971), pp. 47–75. Photochemical competition (quenching) is discussed in Chapter 6.

[22] G. Porter and M. R. Topp, *Proc. Roy. Soc.* (*London*) **A315**, 163 (1970).

is monitored is the lag between the times when the initiating flash and the scintillation radiation pass through the reactant vessel. The time lag is specified by the paths of the light beams and is, therefore, fixed by the geometry of the optical system. By varying the position of the mirror that sends light to the scintillation solution, different time lags are produced (0.3 m $\equiv$ 1 nsec). The technique has only one serious limitation: pulsed lasers cannot produce exciting radiation at every frequency of interest. Thus not all systems are amenable to study. Still faster reactions have been studied using a variation of the pulsed-laser technique; the picosecond range is now accessible.[23]

4.7. Integrated Rate Laws—First Order

Most kinetic measurements monitor the time dependence of the *concentration* of the reacting species, not the reaction rate itself. Thus the rate law is derived from the observed time dependence of the concentrations. Simple rate laws have characteristic integrated forms. For a first-order rate law,

$$\frac{d[\mathrm{A}]}{dt} = -k[\mathrm{A}]^{1.0} \tag{4.21}$$

The integrated expression is

$$[\mathrm{A}] = A_0 e^{-kt}$$

where A_0 is the concentration of A at $t = 0$. Exponential time dependence always implies first-order kinetics.[24] If the experimentally determined decay constant is *independent* of the concentration of any other species present, it is a true rate constant and the data provide strong evidence for a first-order reaction, even though (4.21) may not describe the kinetics. To see why caution is needed consider the possible experimental rate law

$$\frac{d[\mathrm{A}]}{dt} = -\frac{d[\mathrm{B}]}{dt} = -k_+[\mathrm{A}]^{1.0} + k_-[\mathrm{B}]^{1.0} \tag{4.22}$$

[23] P. M. Rentzepis and C. J. Michele, *Anal. Chem.* **42**(14), 20A (1970); P. M. Rentzepis, *Science* **169**, 239 (1970); T. L. Netzel and P. M. Rentzepis, *Chem. Phys. Lett.* **29**, 327 (1974).

[24] As we shall see exponential decay is characteristic kinetic behavior in many cases. By itself it is not adequate evidence for the rate law (4.21). The observed decay constant is seldom a true rate constant; it is generally a complicated combination of the rate constants for individual steps.

which is characteristic of a reversible reaction. The product B may itself react to reproduce A. In terms of the progress variable ξ,

$$[A] = A_0 - \xi, \qquad [B] = B_0 + \xi$$

which when substituted in (4.22) and integrated, yields

$$\xi = (k_+A_0 - k_-B_0)\{1 - \exp[-(k_+ + k_-)t]\}/(k_+ + k_-)$$

Here the time dependence is also exponential but the decay constant is the *sum of the forward and reverse* rate constants. Thus, for exponential relaxation, either of the rate laws, (4.21) or (4.22), is consistent with a *concentration-independent* decay constant. The individual rate constants may be evaluated if the equilibrium constant is also known. Ignoring activity effects, $k_- = k_+K_{eq}^{-1}$ and we find

$$k_+ = \varkappa/(1 + K_{eq}^{-1}) \tag{4.23}$$

where $\varkappa$ is the decay constant. If $K_{eq} \gg 1$, k_+ is essentially equal to the decay constant; the reverse step in (4.22) is unimportant and the two rate laws are kinetically indistinguishable.

First-order processes are very common. The most familiar is radioactive decay, e.g., $^{14}C \rightharpoonup {}^{14}N + e$, the nuclear reaction used in carbon dating for which $k = 1.24 \times 10^{-4}\ yr^{-1}$. An interesting chemical example is the gas-phase decomposition of N_2O_5 which obeys the stoichiometry

$$-2N_2O_5 + 4NO_2 + O_2 = 0$$

but follows the rate law

$$\frac{d[N_2O_5]}{dt} = -k[N_2O_5]$$

with $k = 3.38 \times 10^{-5}\ sec^{-1}$ at 298 K. This suggests a first step of fragmentation of an N_2O_5 molecule; it provides *no information* about the fragments produced or the subsequent steps in the reaction. To detail the further course of reaction more experiments are needed. Examples of the reversible first-order rate law (4.21) are found in many reactions involving rearrangement, isomerization, or conformational change. An interesting case is the chair$_1 \rightharpoonup$ chair$_2$ rearrangement in cyclohexane shown in Fig. 4.3. The rate law is

$$\frac{d[\text{chair}]_1}{dt} = -k[\text{chair}]_1 + k[\text{chair}]_2$$

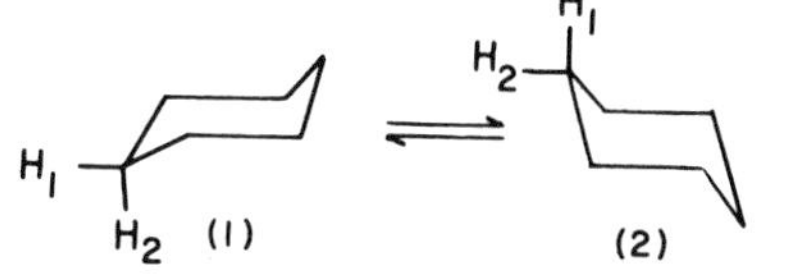

Fig. 4.3. Chair$_1 \rightleftharpoons$ chair$_2$ interconversion in cyclohexane showing exchange of axial and equatorial hydrogen atoms.

since the specific rate constants for both forward and backward steps must be equal; at 206 K the rate constant is 52.5 sec^{-1}.[25]

4.8. Isolation Methods

While integrated rate expressions can be found for second- and higher-order rate laws, in practice this is rarely useful. To see why, consider the five second-order rate laws in Table 4.1. If one monitors the concentration of A, all five rate laws lead to the same general time dependence. In terms of the progress variable (4.7), these rate laws all have the form

$$-\dot{\xi} = a_2\xi^2 + a_1\xi + a_0$$

where the coefficients a_i depend upon the initial concentrations and stoichiometric coefficients of the reacting species and upon the rate constants in a variety of complicated ways. Integration of the rate expression always yields a result that involves three parameters,

$$\frac{[\mathrm{A}]}{A_0} = 1 + \frac{\alpha\varrho(1 - e^{-\lambda t})}{\varrho e^{-\lambda t} - 1} \tag{4.24}$$

where α, ϱ, and λ are determined by the a_i.[26] To derive and use such expressions for eliciting a rate law is a time-consuming, demanding, and error-prone task.[27]

[25] F. R. Jensen, D. S. Noyce, C. H. Sederholm, and A. J. Berlin, *J. Amer. Chem. Soc.* **84**, 386 (1962). The experiment is particularly interesting. Reactants and products are chemically indistinguishable but it is still possible to investigate their rate of interconversion. The reaction rate is deduced from nuclear magnetic resonance line-width measurements.

[26] The same form is deduced if the rate laws of Table 4.1 describe nonelementary systems for which autocatalysis (A or B as products) or autoinhibition (C or D as reactants) occur. However, in some such instances λ may be negative or imaginary; as a consequence unchecked growth or oscillatory behavior may be predicted. In Chapter 7 we shall see that such behavior cannot describe a reacting system at all times although it may be approximately correct for some periods in systems far from equilibrium.

[27] For a compendium of integrated rate laws see A. A. Frost and R. G. Pearson, *Kinetics and Mechanism*, 2nd ed. (New York: John Wiley, 1961), Chapters 2 and 8.

Table 4.1. Possible Elementary Reversible Second-Order Rate Laws

$$\frac{-d[\mathrm{A}]}{dt} = \begin{cases} k_+[\mathrm{A}]^{2.0} - k_-[\mathrm{C}]^{1.0} \\ k_+[\mathrm{A}]^{1.0}[\mathrm{B}]^{1.0} - k_-[\mathrm{C}]^{1.0} \\ k_+[\mathrm{A}]^{2.0} - k_-[\mathrm{C}]^{2.0} \\ k_+[\mathrm{A}]^{1.0}[\mathrm{B}]^{1.0} - k_-[\mathrm{C}]^{2.0} \\ k_+[\mathrm{A}]^{1.0}[\mathrm{B}]^{1.0} - k_-[\mathrm{C}]^{1.0}[\mathrm{D}]^{1.0} \end{cases}$$

Rather than determining a rate law under conditions where all concentrations vary simultaneously, it is easier to *isolate* one species and hold the concentration of the others constant. Consider a reaction which goes to completion and follows the rate law

$$\frac{d[\mathrm{A}]}{dt} = -k[\mathrm{A}]^a[\mathrm{B}]^b[\mathrm{C}]^c \tag{4.25}$$

where a, b, and c are the unknown exponents. If the reaction mixture is *flooded* with B and C so that during reaction their concentration is invariant, the apparent rate law is

$$\frac{d[\mathrm{A}]}{dt} = -kB_0{}^bC_0{}^c[\mathrm{A}]^a = -\varkappa[\mathrm{A}]^a \tag{4.26}$$

which may be easily integrated. If $a = 1.0$, the kinetics are pseudo-first-order; the dependence of $\varkappa$ on the initial concentrations B_0 and C_0 then establishes the exponents b and c. When $a \neq 1.0$, decay is not exponential; the integrated form of (4.26) is

$$[\mathrm{A}]^{1-a} = (a-1)\varkappa t + A_0{}^{1-a} \tag{4.27}$$

Plots of $[\mathrm{A}]^{1-a}$ as a function of t then determine a by trial and error; a is varied until a straight-line plot is obtained. As seen in Fig. 4.4 if a is chosen too large, curvature is positive and vice versa if a is too small. Once a is known, $\varkappa$ is evaluated from the slope of the line, and, by varying B_0 and C_0, the other exponents can be determined.

Flooding is useful even if the rate law is more complicated. Consider two examples

$$\frac{d[\mathrm{A}]}{dt} = -k_1[\mathrm{A}]^a[\mathrm{B}]^b - k_2[\mathrm{A}]^{a'}[\mathrm{C}]^c \tag{4.28}$$

$$\frac{d[\mathrm{A}]}{dt} = -k_+[\mathrm{A}]^a[\mathrm{B}]^b + k_-[\mathrm{C}]^c[\mathrm{D}]^d \tag{4.29}$$

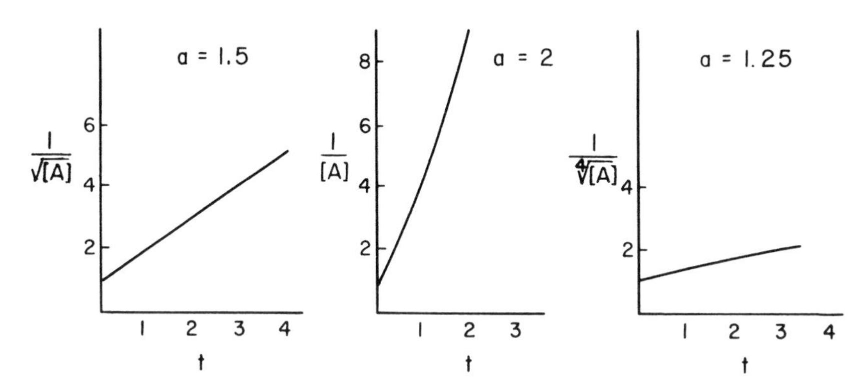

Fig. 4.4. Plots of $[A]^{1-a}$ vs. time for various choices of a in (4.27). A linear relation is assumed when $a = 1.5$.

In (4.28) there are two *parallel* reactions which consume A. If conditions can be adjusted so that little B is present while C is in great excess only one path is significant; the rate law is approximately

$$\frac{d[A]}{dt} = -k_2 C_0^c [A]^{a'}$$

and normal isolation experiments can be carried out. The rate law (4.29) includes a step in which A is *produced.* If initially there is a large excess of B the reaction, in its early stages, appears to be of order a; as reaction progresses it is affected by the reverse step and isolation conditions no longer apply. To maintain isolation one of the reaction products must be removed, as is possible if it precipitates or is volatile.

A practical indication that isolation conditions have not been established is when the trial and error method for determining the exponent a outlined in Fig. 4.4 fails. Then concentrations other than that of A vary significantly during the course of reaction. If no changes in the experimental conditions are adequate to circumvent this problem a different approach must be used.

4.9. Relaxation Methods

The observation central to the use of relaxation methods is simply that, sufficiently close to equilibrium, *all reactions follow pseudo-first-order kinetics.* Consider the rate law (4.29); near equilibrium

$$(-[A] + A_{eq})/\nu_A = \xi, \quad (-[B] + B_{eq})/\nu_B = \xi, \quad \text{etc.} \tag{4.30}$$

where ξ is a progress variable which *equals zero* at equilibrium. As long as the reaction is sufficiently close to equilibrium the only terms which significantly contribute to the reaction rate are those linear in ξ. Introducing (4.30) into (4.29), expanding, and keeping only linear terms we find

$$\begin{aligned} \frac{d\xi}{dt} &= -k_+A_{\mathrm{eq}}^a B_{\mathrm{eq}}^b + k_- C_{\mathrm{eq}}^c D_{\mathrm{eq}}^d - \xi[k_+(-a\nu_A B_{\mathrm{eq}} - b\nu_B A_{\mathrm{eq}})A_{\mathrm{eq}}^{a-1}B_{\mathrm{eq}}^{b-1} \\ &\quad + k_-(c\nu_c D_{\mathrm{eq}} + d\nu_D C_{\mathrm{eq}})C_{\mathrm{eq}}^{c-1}D_{\mathrm{eq}}^{d-1}] \\ &= -\varkappa\xi \end{aligned} \tag{4.31}$$

The first term in (4.31) vanishes since, at equilibrium, the rate of reaction is zero. When integrated (4.31) is

$$\xi = \xi_0 e^{-\varkappa t} \tag{4.32}$$

a result only valid when ξ_0 is small. The decay constant $\varkappa$ is the sum of two terms, characteristic of the forward and of the reverse step in the reaction. In general we can write

$$\varkappa = \varkappa_f + \varkappa_r \tag{4.33}$$

where $\varkappa_f$ is the contribution to the decay constant from the forward step in the reaction, etc. In Table 4.2 expressions for $\varkappa_f$ are given for a number of rate laws; with these results it is possible to construct the decay constant that describes *any* relaxation experiment close to equilibrium. The relation between the order of a particular step in the reaction and the concentration dependence of the corresponding term in $\varkappa$ is particularly simple. For a first-order step $\varkappa$ is concentration independent, for a second-order step $\varkappa$

Table 4.2. Contribution of Forward Step in Reaction to Overall Decay Constant for Different Rate Laws

Forward rate expression	$\varkappa_f$ (defined in text)
$k_+[A]^{1.0}$	k_+
$k_+[A]^{2.0}$	$4k_+A_{\mathrm{eq}}$
$k_+[A]^{1.0}[B]^{1.0}$	$k_+(A_{\mathrm{eq}} + B_{\mathrm{eq}})$
$k_+[A]^{3.0}$	$9k_+A_{\mathrm{eq}}^2$
$k_+[A]^{2.0}[B]^{1.0}$	$k_+(4A_{\mathrm{eq}}B_{\mathrm{eq}} + A_{\mathrm{eq}}^2)$
$k_+[A]^{1.0}[B]^{1.0}[C]^{1.0}$	$k_+(A_{\mathrm{eq}}B_{\mathrm{eq}} + A_{\mathrm{eq}}C_{\mathrm{eq}} + B_{\mathrm{eq}}C_{\mathrm{eq}})$

is linear in concentration, etc. By measuring $\varkappa$ as a function of concentration the order of the reaction and the specific rate constants can be established.

Systems for which there is only a single kinetic process are the exception, not the rule. In general there is a spectrum of relaxation times. As long as these are well separated the individual decay constants are easily measured. However, deducing the coupled rate laws and the rate constants is more difficult.[28] If the kinetic processes proceed at similar rates considerable sophistication in data analysis is needed to obtain reliable values of the decay constants.[29]

Relaxation methods were pioneered by Eigen and by Norrish and Porter[30] as a method of studying fast reactions in systems where the difficulty of obtaining homogeneity forbids the use of ordinary rapid mixing techniques. However, by using sophisticated electronic monitoring devices they can be applied more generally. Any system can be perturbed by adding a reactant, then stirred continuously, thermostatted, and monitored in the region near equilibrium. The concentration changes are small but accurately measurable using modern analytical methods.

Perturbation approaches such as temperature jump, pressure jump, etc., are based on a simple principle. For a system initially at equilibrium the forward and reverse rates of reaction are equal. In a temperature-jump experiment the system is heated so rapidly that the temperature increases faster than the concentrations can respond. The initial reaction rate is determined by the rate constants at the new temperature T and concentrations are set by the equilibrium at the original temperature T_0; the initial reaction rate is then

$$\left(\frac{d[\mathrm{A}]}{dt}\right)_{t=0} = -k_+(T)A_0{}^aB_0{}^b + k_-(T)C_0{}^dD_0{}^d$$

for the rate law (4.29). Unless $[k_+(T)/k_-(T)] = [k_+(T_0)/k_-(T_0)]$ the rates of forward and reverse steps are no longer equal and the system must readjust to a new equilibrium state. Analysis in terms of the progress variable leads to the previous result, (4.32). Naturally there must be a

[28] For a general treatment of the concentration dependence of decay constants for a system of sequential reactions see G. W. Castellan, *Ber. Buns. Gesell.* **67**, 898 (1963).

[29] A useful method for separating many overlapping exponentials is given by R. D. Dyson and I. Isenberg, *Biochemistry* **10**, 3233 (1971). Where only two processes overlap a simple subtraction procedure can be used; see D. S. Honig and K. Kustin, *Inorg. Chem.* **11**, 65 (1972) for an example.

[30] M. Eigen, *Disc. Faraday Soc.* **17**, 194 (1954); R. G. W. Norrish and G. Porter, *Disc. Faraday Soc.* **17**, 40 (1954).

detectable change. For essentially irreversible reactions there are measurable concentrations of either reactants or products but not both. Perturbation does not alter this situation so the technique is not applicable.[31] An instance where perturbation is not practical is in the study of cyanide–myoglobin binding; cyanide binds too strongly with myoglobin.[32] However the azide–myoglobin system, in which the binding is much weaker, can be studied using temperature jump.[33]

Inhomogeneity problems are avoided using this method; the system is initially uniform and a sharp increase in temperature will not introduce concentration gradients. However, care must be taken to avoid temperature gradients. In addition, the perturbation must be rapid enough so that heating is completed before reaction commences. The heating pulse must be synchronized with the monitoring device so that a record of the relaxation may be obtained. If reaction is so rapid that significant changes occur before heating is complete it is no longer permissible to consider the heating pulse instantaneous. The observed response is a function of both the heating curve and the chemical relaxation.[34]

Relaxation methods have wide applicability. The temperature-jump study of the carbonyl addition reaction[35]

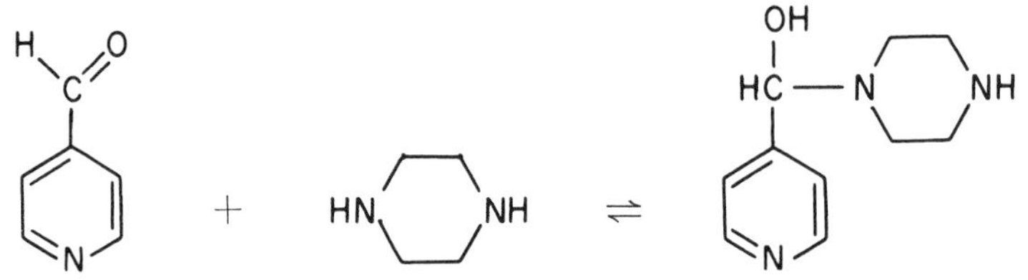

Pyridine-4-aldehyde (P) + Piperazine (A) $\rightleftharpoons$ Carbinolamine (C)

monitored spectrophotometrically, was shown to proceed via parallel pathways

$$P + A \rightleftharpoons C \quad \text{and} \quad P + AH^+ \rightleftharpoons CH^+$$

[31] The general constraints which limit the application of perturbation methods are derived by M. Eigen and L. de Maeyer, in *Techniques of Organic Chemistry*, 2nd ed., A. Weissberger, ed. (New York: Interscience, 1963), Vol. 8, Part 2, pp. 929–941.

[32] P. George and G. I. H. Hanina, *Disc. Faraday Soc.* **20**, 216 (1955).

[33] D. E. Goldsack, W. S. Eberlein, and R. A. Alberty, *J. Biol. Chem.* **240**, 2312 (1965).

[34] For analysis of the consequences of different heating pulses see M. Eigen and L. de Maeyer, in *Techniques of Organic Chemistry*, 2nd ed., A. Weissberger, ed. (New York: Interscience, 1963), Vol. 8, Part 2, pp. 917–928.

[35] H. Diebler and R. N. F. Thornley, *J. Am. Chem. Soc.* **95**, 896 (1971).

where AH^+ and CH^+ are the protonated forms of piperazine and carbinolamine. Furthermore, the possible mechanism

$$PH^+ + A \rightleftharpoons CH^+$$

involving the protonated species PH^+ could be ruled out.

Another example is the hemoglobin $\rightleftharpoons$ oxyhemoglobin equilibrium

$$HbO_2 \rightleftharpoons Hb + O_2$$

A dilution-jump technique, in which oxyhemoglobin is injected into an anaerobic buffer,[36] requires the establishment of a new equilibrium via deoxygenation of HbO_2; the rate of reaction is monitored spectrophotometrically.

It is even possible to consider using perturbation methods to study biological processes *in vivo* by means of concentration jump (injection of a reactant into the cellular material) and then monitoring the response of the system. Such an experiment could be done nondestructively and would be more representative of the biological process than any *in vitro* experiment. By changing the perturbation one would determine which reactants affect the process of interest as well as the rate of reaction.

4.10. Determination of Rate Law from Data—Two Examples

Many sources of error complicate the deduction of a rate law from kinetic data. In each experiment the time dependence of the concentration has a specific functional form which is usually not self-evident from the data unless isolation or relaxation methods are used. In addition no data ever precisely fit a trial function. However, if all errors in the experiment are random, probabilistic methods can be used to determine whether the trial function is reasonable and to estimate the parameters of the function. As long as only a single chemical process is significant, isolation and relaxation data are most readily treated using linear least-squares analysis, described in the Appendix. This procedure provides the most reliable estimate of the decay constant. Then, by varying experimental conditions the concentration dependence of the decay constant can be obtained. With such information probabilistic methods are again useful. A presumed rate

[36] G. G. Hammes, ed., *Investigations of Rates and Mechanisms of Reactions*, Part. II (New York: Wiley-Interscience, 1974), p. 57.

law can be tested for consistency with the kinetic measurements and the rate constants can be determined.

In general the kineticist is faced with the problem of deducing rate laws that describe consecutive reactions involving many steps. The corresponding rate equations are coupled, and, in most cases, the functions that are to be fit cannot be determined using linear least-squares analysis. A vast literature of nonlinear function-fitting methods exists to treat these problems.[37]

4.10.1. A Concentration-Dependent Rate Constant

The electron-transfer reaction

$$\mathrm{Eu(II)} + \mathrm{Fe(III)} = \mathrm{Eu(III)} + \mathrm{Fe(II)}$$

is representative of systems for which the kinetics depend upon species other than those indicated by stoichiometry. Studies carried out between pH 1 and 3 were consistent with the binary rate law

$$-\frac{d[\mathrm{Eu(II)}]}{dt} = k'[\mathrm{Eu(II)}][\mathrm{Fe(III)}] \tag{4.34}$$

However the apparent rate constant k' is pH dependent.[38] Values of k' determined at three temperatures are given in Table 4.3. Since k' decreases as $[H^+]$ increases, a possibility worth testing is that k' might be a linear function of $[H^+]^{-1}$. Figure 4.5 shows that the data are well represented by such a plot suggesting that

$$k' = k_0 + k_{-1}/[\mathrm{H}^+] \tag{4.35}$$

While the fit is gratifying, it is hardly conclusive since many other functions are qualitatively similar, e.g.,

$$k' = a/(b + [\mathrm{H}^+]) \tag{4.36}$$

However, when this hypothesis is tested (see Problem 4.10), it is not in accord with the data. The parameters k_0 and k_{-1} of (4.35) can be estimated

[37] For a brief discussion, with examples from enzyme kinetics, see J. T. Wong, *Kinetics of Enzyme Mechanisms* (London: Academic Press, 1975), Chapter 11. A more extensive development is given by Y. Bard, *Nonlinear Parameter Estimation* (New York: Academic Press, 1974).

[38] D. W. Carlyle and J. H. Espenson, *J. Am. Chem. Soc.* **90**, 2272 (1968).

Table 4.3. Apparent Binary Rate Constant, k', for the Reaction of Eu(II) and Fe(III) as a Function of [H⁺] and Temperature[a]

$T = 1.4°C$		$T = 15.8°C$		$T = 25.0°C$	
$[H^+]$ (M)	$k' \times 10^{-4}$ (M^{-1} sec^{-1})	$[H^+]$ (M)	$k' \times 10^{-4}$ (M^{-1} sec^{-1})	$[H^+]$ (M)	$k' \times 10^{-4}$ (M^{-1} sec^{-1})
0.031	8.14	0.031	22.9	0.03	51.7
0.04	6.92	0.036	17.9	0.04	38.9
0.05	4.92	0.05	13.4	0.05	28.8
0.1	2.70	0.07	8.54	0.0667	24.6
0.2	1.33	0.1	6.86	0.08	18.1
0.5	0.729	0.13	5.91	0.111	13.3
0.879	0.616	0.2	3.30	0.167	9.14
0.939	0.613	0.876	1.51	0.2	7.65
		1.00	1.25	0.4	4.08
				0.953	2.11

[a] D. W. Carlyle and J. H. Espenson, *J. Am. Chem. Soc.* **90**, 2272 (1968).

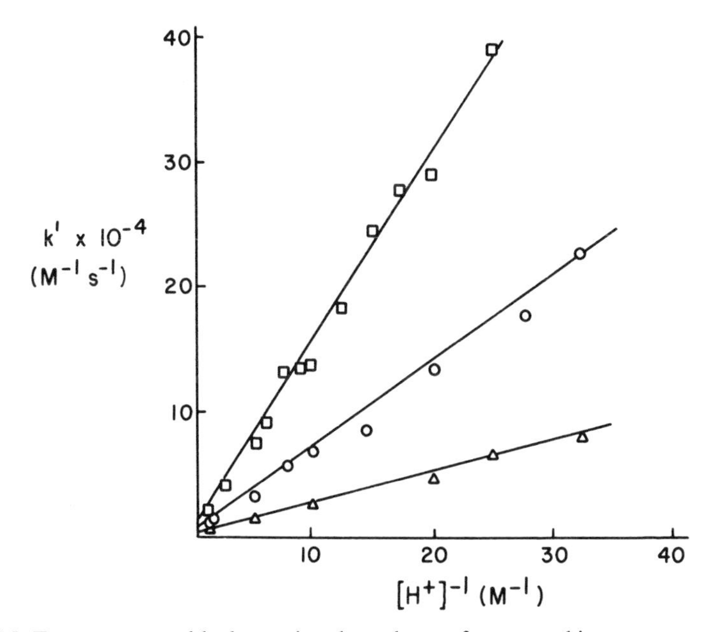

Fig. 4.5. Temperature and hydrogen-ion dependence of apparent binary rate constant in Fe(III) + Eu(II) system. □, 25.0°C. ○, 14.8°C. △, 1.4°C. [Adapted from D. W. Carlyle and J. H. Espenson, *J. Am. Chem. Soc.* **90**, 2272 (1968).]

by visually fitting the straight line or by carrying out a linear least-squares analysis (see Appendix). Using the latter procedure, the values of the parameters and their corresponding standard deviations at 1.4°C are $k_0 = (2.49 \pm 0.07)\times 10^3\ M^{-1}\ \text{sec}^{-1}$ and $k_{-1} = (2.45 \pm 1.1)\times 10^3\ \text{sec}^{-1}$. The correlation coefficient is 0.997, a result, given eight data, that would occur by chance less than 1 in 10^6. One may wonder, given such apparent significance of the correlation, why the standard deviation in k_{-1} is so large. First, it must be remembered that comparison is being made with *chance* occurrence, not with another proposed model.[39] Second, Fig. 4.5 indicates that values of k determined for large values of $[H^+]^{-1}$ are not sufficiently consistent to precisely establish the intercept of the line. Even with the numerical uncertainties it is clear that (4.35) represents the data quite well and that, with (4.34), the rate law is similar in form to (4.28). There are two kinetically significant pathways; the complete rate law is

$$-\frac{d[\text{Eu(II)}]}{dt} = k_0[\text{Eu(II)}][\text{Fe(III)}] + k_{-1}\frac{[\text{Eu(II)}][\text{Fe(III)}]}{[\text{H}^+]} \tag{4.37}$$

4.10.2. A Relaxation Experiment

The complexation reaction between $Rh_2(OAc)_4 \cdot 2H_2O$ and 5′-AMP (adenosine monophosphate) has been investigated using temperature jump.[40] It provides an interesting example of relaxation kinetics because two equilibria are maintained. Denoting the dirhodium complex as M and the 5′-AMP as L, the important equilibria are

$$\begin{aligned} \text{M}\cdot 2\text{H}_2\text{O} + \text{L} &\underset{k_{-1}}{\overset{k_1}{\rightleftharpoons}} \text{M}\cdot \text{L(H}_2\text{O)} + \text{H}_2\text{O}, \qquad K_1 = 1893\ M^{-1} \\ \text{M}\cdot \text{L(H}_2\text{O)} + \text{L} &\underset{k_{-2}}{\overset{k_2}{\rightleftharpoons}} \text{M}\cdot 2\text{L} + \text{H}_2\text{O}, \qquad K_2 = 202\ M^{-1} \end{aligned} \tag{4.38}$$

where the equilibrium constants were determined at 25°C.[41] The approach to equilibrium was followed by a spectrophotometric technique which was sensitive to the concentration of the dirhodium complex $M \cdot 2H_2O$. The decay constant was determined at 25°C for a variety of *total* concentrations of both M and L, c_M, and c_L; the data are given in Table 4.4.

[39] When (4.36) is assumed to represent k, the correlation coefficient is 0.964, a result that would occur by chance only once in 10^4; of course, the hypothesis (4.35) correlates substantially better. The inappropriateness of (4.36) is only demonstrated when k^{-1} is plotted vs. $[H^+]$ (see Problem 4.10).

[40] K. Das, E. L. Simmons, and J. L. Bear, *Inorg. Chem.* **16**, 1268 (1977).

[41] L. Rainen, R. A. Howard, A. P. Kimball, and J. L. Bear, *Inorg. Chem.* **14**, 2752 (1975).

Table 4.4. Concentration Dependence of the Decay Constant in the $Rh_2(OAc)_4 \cdot 2H_2O$-(5'-AMP) System[a]

c_M (mM)	c_L (mM)	$\varkappa \times 10^{-3}$ (sec^{-1})	[M] (mM)	[L] (mM)
1.94	1.96	5.88	0.79	0.67
1.92	2.91	6.25	0.50	1.21
1.92	3.38	7.14	0.41	1.51
1.91	3.85	7.69	0.33	1.84
1.90	4.31	8.70	0.27	2.19
1.89	4.76	9.09	0.23	2.54

[a] K. Das, E. L. Simmons, and J. L. Bear, *Inorg. Chem.* **16**, 1268 (1977).

To establish the rate law we must determine whether the observed relaxation is a consequence of reestablishing the first or second equilibrium in (4.38). If we assume that the experiment is monitoring the reversible elementary reaction $M + L \rightleftharpoons ML$, the decay constant can be found using Table 4.2. The result is

$$\varkappa_1 = k_1([M] + [L]) + k_{-1} = k_1([M] + [L] + 1/K_1) \qquad (4.39)$$

A similar expression can be found for $\varkappa_2$. The kinetically significant variables for $\varkappa_1$ are the concentrations of unreacted M and L which are determined by the equilibrium relations

$$\frac{[ML]}{[M][L]} = K_1, \qquad \frac{[ML_2]}{[ML][L]} = K_2 \qquad (4.40)$$

and the stoichiometric constraints

$$\begin{aligned} c_M &= [M] + [ML] + [ML_2] \\ c_L &= [L] + [ML] + 2[ML_2] \end{aligned} \qquad (4.41)$$

The values of [M] and [L] are also included in Table 4.4. To test whether (4.39) adequately accounts for the data, $\varkappa$ is plotted against $[M] + [L] + 1/K_1$ in Fig. 4.6. According to (4.39) we expect a straight line *passing through the origin* in good agreement with the data. If it is *assumed* that the line passes through the origin, linear least-squares analysis (see Appendix)

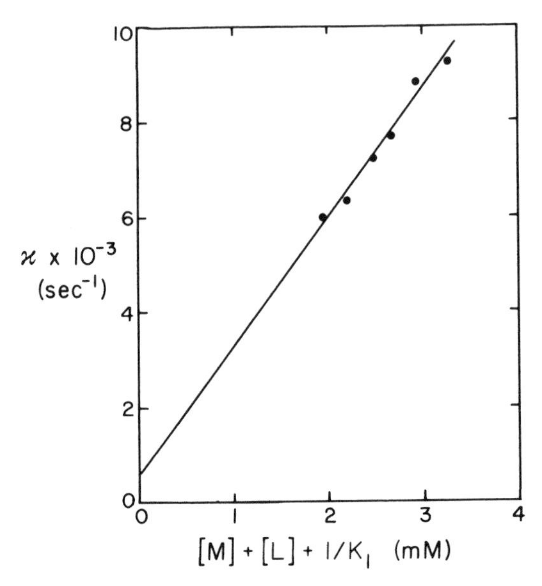

Fig. 4.6. Dependence of the decay constant on concentration for the $Rh_2(OAc)_4$-5′-AMP complexation reaction [K. Das, E. L. Simmons, and J. L. Bear, *Inorg. Chem.* **16**, 1268 (1977)].

yields the rate constants $k_1 = (2.85 \pm 0.03) \times 10^6\ M^{-1}\ \text{sec}^{-1}$, $k_{-1} = (1.51 \pm 0.02) \times 10^3\ \text{sec}^{-1}$. Had we assumed that the $ML + L \rightleftharpoons ML_2$ reaction was responsible for the observed relaxation no correlation would be found (Problem 4.11).

4.11. Temperature Dependence of Rate Constants

The experimental characterization of a chemical reaction is not completed by deducing the rate law. In addition the way in which the rate constants depend upon temperature, pressure, etc., must be determined. The variation of k with temperature is striking; repeated experimentation in the late nineteenth century showed that it could be represented in the form

$$k(T) = A \exp(-E_a/RT) \tag{4.42}$$

Arrhenius[42] was the first to interpret E_a as an activation energy and present, as qualitative justification for this point of view, the rationalization based

[42] S. Arrhenius, *Z. Phys. Chem.* **1**, 110 (1887).

upon kinetic theory discussed in Section 1.4. To determine A and E_a linear least-squares analysis is used; T^{-1} is the independent variable and $\ln k(T)$ is the dependent one.

In some cases, either on experimental grounds or to effect comparison with a theoretical model for the rate constant, one wishes to express $k(T)$ in the form

$$k(T) = A'T^m e^{-\theta/T} \tag{4.43}$$

If m is known on theoretical grounds, A' and θ may be estimated by the least-squares method if $\ln k - m(\ln T)$ is treated as the dependent variable. A direct extension of the least-squares approach is not possible if m is unknown because $\ln T$ and $1/T$ are not independent quantities. However, by determining E_a for both high- and low-temperature data, m may be estimated. Since (4.42) defines E_a as

$$E_a = RT^2 \frac{d(\ln k)}{dT} = -R \frac{d(\ln k)}{d(1/T)}$$

it is possible to relate E_a and θ via (4.43)

$$E_a = R\theta + mRT \tag{4.44}$$

from which we note that

$$m = \frac{E_a(\text{high } T) - E_a(\text{low } T)}{R(\text{high } T - \text{low } T)} \tag{4.45}$$

With this and nearby estimates of m it is possible, using the least-squares procedure for *known m*, to determine both A' and θ as functions of m. The optimal set of parameters may be defined as that which leads to the largest correlation coefficient. An example for which there is definite curvature in a plot of $\ln k$ vs. $1/T$ is the reaction $Cl(^2P) + CH_4 \rightharpoonup HCl + CH_3$, illustrated in Fig. 4.7.[43]

An alternate way to determine the parameters A', m, and θ is to fit the data using a nonlinear least-squares procedure. However, all such methods require reasonable initial choices for the parameters so the method outlined in the previous paragraph is a necessary preliminary.[44]

[43] Temperature-dependent activation energies are discussed by W. C. Gardiner, Jr., *Acc. Chem. Res.* **10**, 326 (1977).

[44] For details see N. R. Draper and H. Smith, *Applied Regression Analysis* (New York: John Wiley, 1966), Chapter 10.

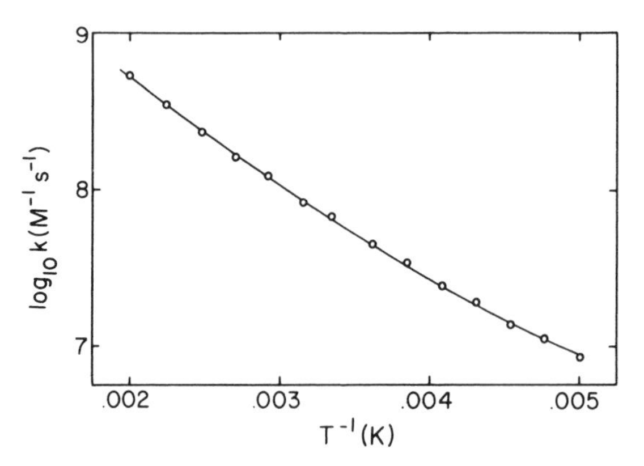

Fig. 4.7. Arrhenius plot of the rate constant for the reaction $Cl(^2P) + CH_4 \rightarrow HCl + CH_3$. Note the definite curvature [D. A. Whytock, J. H. Lee, J. V. Michael, W. A. Payne, and L. J. Stief, *J. Chem. Phys.* **66**, 2690 (1977)].

Comparing (4.11) and (4.42) it is clear, when activity effects are properly taken into account, that the change in internal energy during reaction, ΔE_{rxn}, is

$$\Delta E_{rxn} = E_a^{(f)} - E_a^{(r)} \tag{4.46}$$

and therefore that E_a is a measure of the energy required to effect a chemical transformation. It may never be less than ΔE_{rxn}. For an exothermic reaction ($\Delta E_{rxn} < 0$), the reactants have sufficient energy to form products. One might then expect that $E_a \approx 0$ for such systems. However, in many cases E_a is substantial, which indicates that the reactants must go through a highly energetic intermediate in order to form products. Similarly, for endothermic reactions ($\Delta E_{rxn} > 0$), E_a need only be as large as ΔE_{rxn} but in many cases it is substantially greater, again indicating a highly energetic intermediate. The relative energies of the reactants, products, and intermediate are shown schematically in Fig. 4.8.

Care is required when comparing thermodynamic reaction energies with kinetic activation energies. Rate constants, as they refer to *concentrations*, presume measurements at constant volume. Equilibrium constants

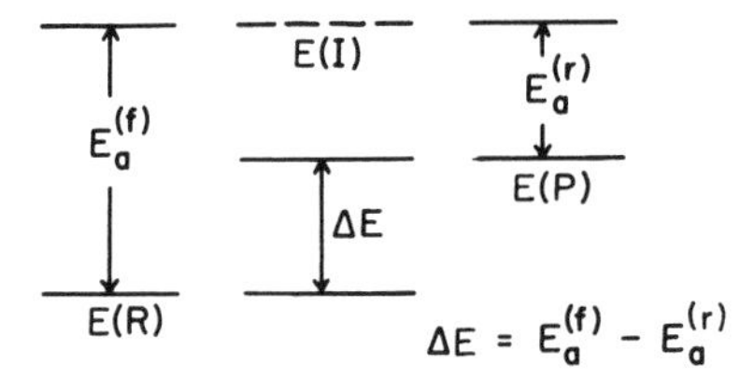

Fig. 4.8. Schematic diagram of the activation barrier for the reaction R(eactants) $\rightleftharpoons$ P(roducts) via an I(ntermediate) which is more energetic than either R or P.

are generally determined at constant pressure; as a result thermodynamic data provide the enthalpy of reaction, not the energy of reaction. Transformation is required to obtain ΔE_{rxn}.[45]

In rare instances $E_a < 0$. Almost invariably this indicates that the rate law determined experimentally does *not* describe an elementary step. However, if $0 \lesssim E_a \lesssim -10$ kJ mol^{-1} it is possible that the rate law could describe an elementary step. Small negative E_a may also be due to experimental error.

Appendix. Linear Least-Squares Analysis

A major problem in interpreting data is to determine whether experimental quantities are correlated on the basis of an assumed theoretical model. Since there are always errors of measurement, correlation is never exact and a method for judging whether correlation is significant is required. From probability theory random errors have a Gaussian distribution.[46] If *all errors are random*, the best procedure for fitting data to a model is the method of least squares.

As an example, consider the data presented in Fig. 4.6; *experimental* values of the decay constant $\varkappa$ are plotted against *experimental* values of the quantity $[M] + [L] + 1/K_1$. If there are N data with ordinates y_i and abscissas x_i the problem is to test whether the functional form

$$y = kx \tag{4A.1}$$

is consistent with these data, to determine k, and to estimate the error in k. For any value of k the difference

$$d_i \equiv kx_i - y_i \tag{4A.2}$$

need not be zero. If the functional form (4A.1) is correct and all experimental errors are random the best estimate of k is found by minimizing the quantity

$$\sum_i d_i^2 = \sum y_i^2 - 2k \sum x_i y_i + k^2 \sum x_i^2 \tag{4A.3}$$

[45] See F. T. Wall, *Chemical Thermodynamics*, 3rd ed. (San Francisco: Freeman, 1974), Chapter 3.

[46] For an introduction to data analysis see H. D. Young, *Statistical Treatment of Experimental Data* (New York: McGraw–Hill, 1962), in particular, pp. 101–132. A comprehensive treatment is given by N. R. Draper and H. Smith, *Applied Regression Analysis* (New York: John Wiley, 1966), Chapter 10.

with respect to k. The result is

$$\frac{d}{dk}(\sum d_i^2) = 2k \sum x_i^2 - 2 \sum x_i y_i = 0$$

and we obtain

$$k = \sum x_i y_i / \sum x_i^2 \tag{4A.4}$$

More generally we have no reason to expect that the assumed function passes through a fixed point and we search for a fit to the function

$$y = kx + q \tag{4A.5}$$

Proceeding as in the previous paragraph but minimizing $\sum d_i^2$ with respect to k and q yields

$$\begin{aligned} k &= (N \sum x_i y_i - \sum x_i \sum y_i)/\Delta x \\ q &= (\sum x_i^2 \sum y_i - \sum x_i y_i \sum x_i)/\Delta x \end{aligned} \tag{4A.6}$$

where

$$\Delta x = N \sum x_i^2 - (\sum x_i)^2 \tag{4A.7}$$

The questions of the significance of correlation and of the *statistical* error in k and q may be answered using probabilistic arguments.[47] Significance is tested via the correlation coefficient, with Δy analogous to (4A.7),

$$r \equiv (N \sum x_i y_i - \sum x_i \sum y_i)/(\Delta x \, \Delta y)^{1/2} = k(\Delta x/\Delta y)^{1/2} \tag{4A.8}$$

When $r = 1$ there is perfect correlation, i.e., (4A.5) is *exact* for each datum; when $r = 0$ there is no correlation. Since $1 > r > 0$ in any series of experiments we may refer to tables of correlation coefficients[48] for the answer to this question: What is the probability that the same value of r would be found by *chance* if the variables were unrelated? Finally the standard deviation in k and q can be calculated[49]:

$$\begin{aligned} \sigma_k &= \sigma(N/\Delta x)^{1/2}, \qquad \sigma_q = \sigma(\sum x_i^2/\Delta x)^{1/2} \\ \sigma &= [\sum (y_i - kx_i - q)^2/N]^{1/2} \end{aligned} \tag{4A.9}$$

[47] See H. D. Young, *Statistical Treatment of Experimental Data* (New York: McGraw–Hill, 1962) for details.

[48] See H. D. Young, *Statistical Treatment of Experimental Data* (New York: McGraw–Hill, 1962), p. 164 for such a table.

[49] If $q = 0$ then correlation is via (4A.1). In this case $\sigma_q \equiv 0$, $r = \sum x_i y_i/(\sum x_i^2 \sum y_i^2)^{1/2}$, and $\sigma_k = [\sum (y_i - kx_i)^2/\sum x_i^2]^{1/2}$.

Problems

4.1. Recent research [J. H. Sullivan, *J. Chem. Phys.* **46**, 73 (1967); **39**, 3001 (1963)] has indicated that the $H_2 + I_2$ reaction may proceed by one of the two following sets of elementary steps:

$$I_2 = 2I \quad \text{(equilibrium dissociation)}$$

$$\text{(i)} \quad H_2 + 2I \xrightarrow{k} 2HI$$

$$\text{(ii)} \quad H_2 + I = H_2I \quad \text{(equilibrium addition)}$$

$$H_2I + I \xrightarrow{k'} 2HI$$

(a) Show that either pathway accounts for the overall stoichiometry.
(b) Write the rate law for each pathway and show that each is consistent with Bodenstein's experimental results (4.2).
(c) Relate the overall equilibrium constant to the rate constants of the individual steps. You may need to include some other reactions.

4.2. Kinetics of ionic reactions are studied in solutions of *constant ionic strength* (usually 0.1 M or greater).

(a) Refer to Debye–Hückel theory and show that, if the reactants are in mM concentration, their activity coefficients are independent of reactant concentration.
(b) If μ_i^* is the reference chemical potential of Y_i at infinite dilution and $\gamma_i(I)$ is the activity coefficient at ionic strength I show that an equilibrium constant at ionic strength I is

$$(k_f/k_r) = K_{\text{eq}}(T, p, I) = \prod \exp(-\nu_i \tilde{\mu}_i/RT)$$

where $\tilde{\mu}_i = \mu_i^* + RT \ln \gamma_i(I)$.

4.3. At 25°C the forward rate constant for the aqueous reaction

$$Co(NH_3)_5Br^{2+} + Hg^{2+} + H_2O = Co(NH_3)_5(H_2O)^{3+} + HgBr^+$$

has the approximate ionic strength dependence, $\log k_f = \log k_f^0 + 4.1(I^{1/2})$. [J. N. Brönsted and R. Livingston, *J. Am. Chem. Soc.* **49**, 435 (1927)]. Use Debye–Hückel theory to estimate the activities and thus obtain the ionic strength dependence of k_r.

4.4. The kinetics of the hydrogen peroxide oxidation of sodium dithionite were studied by monitoring the disappearance of dithionite ion spectrophotometrically under different initial concentrations of H_2O_2 [C. Creutz and N. Sutin, *Inorg. Chem.* **13**, 2041 (1974)]. The quantities tabulated are the *times* after which $[S_2O_4^{2-}]$ has dropped to specific values. At a pH of 6.5 the data given in Table 4.5 are representative:

(a) Determine the dependence of the rate law on $[S_2O_4^{2-}]$ for each value of $[H_2O_2]$ and find the decay constants.

Table 4.5

$[S_2O_4^{2-}]$ (mM)	t (sec)			
	0.0450[a]	0.0899	0.135	0.180
0.5	0	0	0	0
0.45	0.55	0.25	0.15	0.15
0.4	1.15	0.55	0.35	0.25
0.35	1.80	0.85	0.55	0.45
0.3	2.50	1.20	0.75	0.60
0.25	3.20	1.55	0.95	0.80
0.2	4.05	1.95	1.20	1.00

[a] Initial concentration of $[H_2O_2]_0$ (M).

(b) From the decay constants determine the dependence of the rate law on $[H_2O_2]$.

4.5. Show that T (temperature) jump is a useful kinetic probe only if $\Delta H_{rxn} \neq 0$. Show that P (pressure) jump requires $\Delta V_{rxn} \neq 0$. (Consider the effect of the perturbation on the equilibrium constant.)

4.6. An "old chemists' tale" is that, near 300 K, the reaction rate doubles with each 10 K increase in temperature. What is the "universal" E_a?

4.7. Show that in a tracer exchange reaction of the type

$$AX^* + BX \rightleftharpoons AX + BX^*$$

the *rate of exchange* of X* is first order regardless of the details of the reaction mechanism. Assume that $[AX^*] \ll [AX]$ and that there is *no* isotope effect, i.e., that labeled and unlabeled molecules react at the same rate.

4.8. Many reactions can lead to a variety of products. The simplest case is

$$A \xrightarrow{k_1} B \qquad A \xrightarrow{k_2} C$$

(a) What is the rate law for the disappearance of A?
(b) If B(0) = C(0) = 0, what are B(t) and C(t)?
(c) How can both k_1 and k_2 be determined experimentally if there is *no way* to monitor B or C; if B and/or C can be monitored?

4.9. The reaction of pyridoxal phosphate (PLP) with glutamate (Glu) in the presence of Cu^{2+} has been studied by measuring the initial rate of reaction

Table 4.6

$[PLP]_0$ (mM)	$[Glu]_0$ (mM)	$[Cu^{2+}]_0$ (mM)	Initial rate (μM sec^{-1})
0.1	8.0	0.2	0.12
0.1	8.0	1.0	0.11
0.2	8.0	0.2	0.24
0.3	8.0	1.0	0.33
0.5	8.0	1.0	0.56
0.5	8.0	0.2	0.59
0.2	0.7	0.4	0.023
0.2	1.9	0.4	0.054
0.2	3.2	0.4	0.094
0.2	8.0	0.4	0.23
0.2	11.0	0.4	0.31

[M. E. Farago and T. Mathews, *J. Chem. Soc. A.*, 609 (1969)]. Some results are tabulated in Table 4.6 for 25°C and pH 4.0.

(a) Deduce the rate law.

(b) What is the rate constant?

(c) What is the standard deviation in the rate constant?

4.10. Fe(III) exists in two forms in acid solution; these are readily interconverted according to the equilibrium

$$Fe^{3+}_{(aq)} + H_2O = FeOH^{2+}_{(aq)} + H^+, \qquad K_a$$

(a) If the only species that reacts with Eu(II) (see Section 4.10.1) is $FeOH^{2+}_{(aq)}$ show that the apparent rate constant for the electron-transfer reaction would be

$$k' = k^* K_a/([H^+] + K_a)$$

(b) Test this possibility by plotting $1/k'$ vs. $[H^+]$ using the data in Table 4.3 for $T = 1.4$°C.

(c) Use a least-squares procedure to determine k^* and K_a.

4.11. (a) Show that if the reaction $ML + L \rightleftharpoons ML_2$ were being monitored for the system discussed in Section 4.10.2 the decay constant would be

$$\varkappa = k_2([ML] + [L] + 1/K_2)$$

(b) Use the data in Table 4.4 and the values of K_1 and K_2 to compute [ML].

(c) Test the hypothesis of part (a) both graphically and via a least-squares procedure.

4.12. (a) Use the data of Table 4.3 to compute k_0 and k_{-1} for the Eu(II) + Fe(III) reaction at 15.8°C and 25°C.
(b) What is E_a for the two pathways?

4.13. (a) Show that if for all data, $y_i = kx_i + q$, the correlation coefficient r (4A.8) is unity.
(b) When variables are uncorrelated the value of x has no influence on that of y. Construct an argument to show that, when N is large, for such variables,

$$\sum x_i y_i = \frac{1}{N} \sum x_i \sum y_i$$

and thus that $r = 0$.

4.14. The electron-transfer reaction

$$FeR_3^{2+} + IrCl_6^{2-} \rightleftharpoons FeR_3^{3+} + IrCl_6^{3-}$$

where R is 4,7-dimethyl-1,10-phenanthioline, has been studied using temperature jump [J. Halpern, R. J. Legare, and R. Lumry, *J. Am. Chem. Soc.* **85**, 680 (1963)]. The experiments were carried out under conditions where the initial concentrations of the reactants were equal, i.e., $[FeR_3^{2+}]_0 = [IrCl_6^{2-}]_0 = C_0$. The results suggested that the decay constant is proportional to C_0.
(a) Assume a simple bimolecular reversible mechanism and show that

$$\varkappa = k_+\{[FeR_3^{2+}]_{eq} + [IrCl_6^{2-}]_{eq}\} + k_-\{[FeR_3^{3+}]_{eq} + [IrCl_6^{3-}]_{eq}\}$$

(b) The equilibrium constant for the reaction is K. Show that

$$\varkappa = 2C_0k_+/K^{1/2} = 2C_0k_-/K^{1/2} = 2C_0(k_+k_-)^{1/2}$$

Table 4.7

t (°C)	K	$C_0 \times 10^6$ (M)	$\tau \times 10^6$ (sec)
10	0.55	10	33
		7.5	55
		5	66
18	0.40	10	32
		7.5	42
		5	59
30	0.28	10	23
		7.5	31
		5	44

(c) The data were obtained ($\tau \equiv \varkappa^{-1}$) as shown in Table 4.7. Determine k_+ and k_- at the three temperatures.
(d) Determine A and E_a for both forward and reverse steps.

4.15. (a) For the system of Problem 4.14 write the second-order rate law.
(b) In an ordinary kinetic experiment with initially equal reactant concentrations C_0, show that the rate of product formation is

$$\dot{X} = k_+C_0^2 - 2k_+C_0X + (k_+ - k_-)X^2$$

(X is the concentration of either product).
(c) *Refer to a table of integrals* and show that

$$\frac{(1-K)y + K + K^{1/2}}{(K-1)y - K + K^{1/2}} \, \frac{K^{1/2} - K}{K^{1/2} + K} = e^{\varkappa t}$$

with $y \equiv X/C_0$ and $\varkappa$ given by one of the formulas in problem 4.14b. This expression can be rewritten to yield an equation for y,

$$y(t) = \left(1 + \frac{1}{K^{1/2}} \coth \frac{\varkappa t}{2}\right)^{-1}$$

(d) If the experiment is to be carried out using a stopped-flow apparatus which can measure decay times $\gtrsim 1$ msec, what initial concentrations are required at 4°C? at 50°C?
(e) Would it be possible to determine the rate constants using techniques which only measure decay times $\gtrsim 1$ sec?
(f) Plot $y(t)$ for the initial conditions $C_0 = 10^{-7}\ M$, $t = 4$°C.

4.16. The reaction of Cr(VI) with I^- has been studied by measuring the initial rate of disappearance of $[I^-]$ [K. E. Howlett and S. Sarsfield, *J. Chem. Soc.*, 683 (1968)]. The reactive form of Cr(VI) is the $HCrO_4^-$ ion. The data in Table 4.8 describe the initial conditions at 25°C and an ionic strength of 0.1 M.[50]
The analysis of the data is somewhat complicated since Cr(VI) may exist in a number of forms; the following equilibria are important:

$$H_2O + Cr_2O_7^{2-} = 2HCrO_4^-, \qquad K_1 = 0.018\ M$$
$$H^+ + HCrO_4^- = H_2CrO_4, \qquad K_2 = 5\ M^{-1}$$

(a) Show that, after equilibration of Cr(VI) and H^+, but before reaction with I^-, the hydrogen-ion concentration, h, is

$$h = [H^+]_0/\{1 + K_2[HCrO_4^-]_0\}$$

[50] After R. G. Wilkins, *The Study of Kinetics and Mechanism of Transition Metal Compounds* (Boston: Allyn and Bacon, 1974), p. 56.

Table 4.8

$[H^+]_0$ (mM)	$[Cr_2O_7^{=}]_0$ (mM)	$[I^-]$ (mM)	$-(d[I]/dt)_0$ (μM sec^{-1})
20.0	20.0	10.0	0.14
20.0	20.0	20.0	0.55
20.0	20.0	30.0	1.27
20.0	20.0	40.0	2.13
4	20.0	30.0	0.052
16	20.0	30.0	0.76
40	20.0	30.0	4.97
49.5	20.0	30.0	7.10
12.0	19.8	30.0	0.45
12.0	40.0	30.0	0.80
12.0	60.0	30.0	1.02
12.0	80.0	30.0	1.20
12.0	139.7	30.0	1.65

(b) Set $[HCrO_4^-]_0 = x$ and show that $[HCrO_4^-]_0$ is found by solving the equation

$$2x^2 + K_1(1 + K_2[H^+]_0)x - 2K_1[Cr(VI)]_0 = \frac{K_1[H^+]_0(K_2x)^2}{1 + K_2x}$$

(c) Since the right-hand side of the equation in (b) can be neglected show that

$$[HCrO_4]_0^- = x = \frac{K_1}{4}\left[\left\{(1 + K_2[H^+]_0)^2 + \frac{16[Cr(VI)]_0}{K_1}\right\}^{1/2} - 1 - K_2[H^+]_0\right]$$

(d) Use the results from (c) and (a) and show that the rate law is $d[I^-]/dt = -k[HCrO_4^-][I^-]^2[H^+]^2$.

(e) What is the rate constant, k?

4.17. The stoichiometry for the reaction in Problem 4.16 is

$$2HCrO_4^- + 6I^- + 14H^+ = 2Cr^{3+} + 3I_2 + 8H_2O$$

and a more detailed study of the reaction kinetics leads to the rate law

$$\frac{d[I^-]}{dt} = -\frac{kk_2[HCrO_4^-]^2[I^-]^2[H^+]^2}{k_{-1}[I_2] + k_2[HCrO_4^-]}$$

(a) What experimental approach can you devise for confirming this result?
(b) Could it be established by monitoring the initial rate of the back reaction?

4.18. The reaction

$$Fe^{3+} + HCrO_4^- = FeCrO_4^+ + H^+$$

was studied using excess Fe^{3+} and H^+ [J. H. Espenson and S. R. Helzer, *Inorg. Chem.* **8**, 1051 (1969)]. Small changes in $[FeCrO_4^+]$ were monitored with the result that

$$\frac{d \ln[FeCrO_4^+]}{dt} = k$$

where k was found to be proportional to $[Fe^{3+}]/[H^+]$. What is the rate law?(51)

4.19. For the reaction $Cl(^2P) + CH_4 \rightarrow CH_3 + HCl$ some of the values of k_2 are [D. A. Whytock, J. H. Lee, J. V. Michael, W. A. Payne, and L. J. Stief, *J. Chem. Phys.* **66**, 2690 (1977)]:

$k_2 \times 10^{14}$ (cm^3 molecule^{-1} sec^{-1}):	90.6	57.1	37.7	2.34	1.91	1.46
T (K):	500	447	404	220	210	200

Determine E_a in the two temperature regions. Then estimate m. Use your value of m to estimate θ and the correlation coefficient. If you have access to a computer vary m and refine your results. Statistical analysis of the 14 data reported in the paper led to $m = 2.5$, $\theta = 608$.

4.20. Sometimes the law of mass action can be used to deduce *possible* forms for the reverse rate law if the forward rate law is known. Consider the reaction

$$2Hg^{2+} + 2Fe^{2+} = Hg_2^{2+} + 2Fe^{3+}$$

for which the forward rate law is $R_f = k_f[Hg^{2+}][Fe^{2+}]$. At equilibrium forward and backward rates are equal. Formulate the equilibrium constant, determine some different expressions for the quantity $[Hg^{2+}][Fe^{2+}]$, and show that R_r may have the form (among others)

$$R_r \propto [Fe^{3+}][Hg_2^{2+}]^{1/2} \quad \text{(i)}$$
$$\propto [Fe^{3+}]^2[Hg_2^{2+}]/[Fe^{2+}][Hg^{2+}] \quad \text{(ii)}$$
$$\propto [Fe^{3+}]^3[Hg_2^{2+}]^{3/2}/[Fe^{2+}]^2[Hg^{2+}]^2 \quad \text{(iii)}$$

4.21. Show that absorbance measurements made at a single frequency determine the decay constants when relaxation methods are used, even if more than one constituent absorbs at the monitoring frequency.

4.22. A system contains two components, A and B, which absorb in the same frequency range. The differential absorbance $\Delta A(\nu) = A(\nu, t) - A(\nu, \infty)$ was measured at two frequencies. The molar extinction coefficients are

(51) After R. G. Wilkins, *The Study of Kinetics and Mechanism of Transition Metal Compounds* (Boston: Allyn and Bacon, 1974), p. 57.

Table 4.9

Frequency	ϵ_A	ϵ_B
ν_1	1000	2000
ν_2	1500	500

Table 4.10

t (sec):	0	0.02	0.04	0.08	0.12	0.16	0.20	0.24	0.28
$\Delta A(\nu_1)$:	7.5	8.9	9.5	9.5	8.7	7.8	6.3	5.3	4.3
$\Delta A(\nu_2)$:	13.9	12.3	10.8	8.5	6.8	5.8	4.3	3.5	2.8

given in Table 4.9, and the measured differential absorbances are given in Table 4.10.

(a) Determine $[A(t)]$, $[B(t)]$; the path length of the cell is 0.1 m.
(b) Are the data consistent with a single relaxation process?

4.23. The rate of halogen hydrolysis $X_2 + H_2O \rightleftharpoons X^- + H^+ + XOH$, where $X = Cl$, Br, or I, could not be determined until the advent of fast reaction techniques, especially temperature-jump [M. Eigen and K. Kustin, *J. Amer. Chem. Soc.* **84**, 1355 (1962)].

(a) Derive the expression relating the reciprocal relaxation time $(1/\tau)$ to the forward and reverse rate constants and concentrations in the halogen hydrolysis reaction.
(b) Calculate (within 10% accuracy) the relaxation time, τ, for bromine hydrolysis when total bromine concentration is $3 \times 10^{-3}\ M$; at 293 K and ionic strength 1.0 M, $K_{eq} = [BrOH][Br^-][H]/[Br_2] = 6.9 \times 10^{-9}\ M^2$ and $k_+ = 110\ sec^{-1}$.

General References

Microscopic Reversibility and the Principle of Detailed Balance

R. K. Boyd, *Chem. Rev.* **77**, 93 (1977).

S. R. deGroot and P. Mazur, *Non-Equilibrium Thermodynamics* (Amsterdam: North-Holland, 1962), pp. 92–100.

D. ter Haar, *Elements of Statistical Mechanics* (New York: Rinehart, 1954), pp. 381–382.

A. Messiah, *Quantum Mechanics*, translation by J. Potter (Amsterdam: North-Holland, 1966), pp. 664–675.

Experimental Methods in Chemical Kinetics

C. F. Bernasconi, *Relaxation Kinetics* (New York: Academic Press, 1976), Chapters 11–16.
J. N. Bradley, *Shock Waves in Chemistry and Physics* (London: Methuen, 1962), Chapters 4 and 10.
A. Ferri, ed., *Fundamental Data Obtained from Shock Tube Experiments* (New York: Pergamon, 1961), Chapters 3 and 4.
G. G. Hammes, ed., *Investigations of Rates and Mechanisms of Reactions*, Part II (New York: John Wiley, 1974).
K. Kustin, ed., *Methods in Enzymology*, Vol. 16 (New York: Academic Press, 1969).
P. M. Rentzepis, *Adv. Chem. Phys.* **23**, 189 (1973).

Relaxation Theory

C. F. Bernasconi, *Relaxation Kinetics* (New York: Academic Press, 1976), Chapters 1–5.

Integrated Rate Laws

S. W. Benson, *The Foundations of Chemical Kinetics* (New York: McGraw–Hill, 1960), Chapters 2 and 3.
A. A. Frost and R. G. Pearson, *Kinetics and Mechanism*, 2nd ed. (New York: John Wiley, 1961), Chapters 2 and 8.

Pressure Dependence of Rate Constants

C. A. Eckert, *Ann. Rev. Phys. Chem.* **23**, 239 (1972).

5

Stationary State Mechanisms

5.1. Introduction

It is not enough to establish rate laws. The information must be interpreted to yield insights into the elementary steps by which molecules interact, react, and recombine. This is done by constructing mechanistic models consistent with the phenomenological rate laws. Such mechanisms are rarely unique and guidelines are needed to construct them. The fundamental ground rule is simplicity. With this principle it becomes possible to choose between competing models, each of which agrees with the data. It is obvious that our notion of simplicity is redefined if new intermediate species are discovered.

In many cases the rate law is a direct indication of the only important elementary step which contributes significantly to the rate of reaction. Examples occur in both gas-phase and solution kinetics. Typical of the gas-phase reactions for which this is believed to be the case are the bimolecular reactions[1]

$$NO + O_3 \rightharpoonup NO_2 + O_2$$

$$C_2H_4 + H_2 \rightharpoonup C_2H_6$$

the termolecular recombination reactions[2]

$$I + I + He \rightharpoonup I_2 + He$$

$$Br + Br + A \rightharpoonup Br_2 + A$$

[1] A tabulation of such reactions can be found in A. A. Frost and R. G. Pearson, *Kinetics and Mechanism*, 2nd ed. (New York: John Wiley, 1961), p. 104.

[2] A. A. Frost and R. G. Pearson, *Kinetics and Mechanism*, 2nd ed. (New York: John Wiley, 1961), p. 108.

and the unimolecular reactions[3]

$$N_2O_4 \rightarrow 2NO_2$$
$$(CH_3CO)_2O \rightarrow CH_3COOH + CH_2CO$$
$$cis\text{-stilbene} \rightarrow trans\text{-stilbene}$$

In solution kinetics an important distinction between solvent and reactants must be maintained. The solvent is continually interacting with the reactants; if such interactions were incorporated in the mechanistic model *all* solution mechanisms would perforce be multimolecular. By considering the solvent as a medium and *not* as a participant in the reaction (unless, of course, solvent actually takes part in a reaction step), the problem of mechanism is greatly simplified. In this sense isomerizations, rearrangements, and conformational changes, like the $\text{chair}_1 \rightleftharpoons \text{chair}_2$ interconversion in cyclohexane, are first-order reactions for which the empirical rate law is a direct indication of the only important elementary step. Most solution reactions proceed via bimolecular steps. There are countless examples for which only one such step is needed and for which the rate law reflects that process.

In this chapter we first consider a mathematically tractable model mechanism and demonstrate that, depending upon the relative magnitudes of the rate constants, there are two *chemical* approximations that may be appropriate for simplifying analysis: the preequilibrium and the steady-state assumptions. We then demonstrate how hypotheses based upon these simplifications are used to interpret rate law data and to develop chemically reasonable mechanistic descriptions for gas- and solution-phase reactions. Finally we consider the problem of catalysis, i.e., how addition of trace amounts of an intermediate permits a sluggish or kinetically forbidden reaction to become rapid if a new mechanistic pathway can be created.

5.2. Consecutive Reactions—First Order

Most reactions take place via consecutive steps. If each is first order the general mechanism is

$$A \underset{k_-}{\overset{k_+}{\rightleftharpoons}} B \underset{k_-'}{\overset{k_+'}{\rightleftharpoons}} C \underset{k_-^{(2)}}{\overset{k_+^{(2)}}{\rightleftharpoons}} \cdots \underset{k_-^{(n)}}{\overset{k_+^{(n)}}{\rightleftharpoons}} Z \qquad (5.1)$$

[3] A. A. Frost and R. G. Pearson, *Kinetics and Mechanism*, 2nd ed. (New York: John Wiley, 1961), pp. 110–111.

The corresponding rate equations are

$$\begin{aligned}
\frac{d[\mathrm{A}]}{dt} &= -k_+[\mathrm{A}] + k_-[\mathrm{B}] \\
\frac{d[\mathrm{B}]}{dt} &= +k_+[\mathrm{A}] - (k_- + k_+')[\mathrm{B}] + k_-'[\mathrm{C}] \\
&\vdots \\
\frac{d[\mathrm{Z}]}{dt} &= k_+^{(n)}[\mathrm{Y}] - k_-^{(n)}[\mathrm{Z}]
\end{aligned} \tag{5.2}$$

One is redundant; conservation of matter requires that the sum [A] + [B] + ··· + [Z] be constant. The general solution to (5.2), obtained by the matrix method outlined in Appendix A, is a superposition of n exponential relaxations

$$[\mathrm{A}] = \sum_{i=1}^{n} a_i(0) \exp(-\varkappa_i t), \quad \text{etc.} \tag{5.3}$$

where the $\varkappa_i$ are decay constants determined by the rate constants and the coefficients $a_i(0)$ are fixed by both the rate constants and the initial conditions.

Deduction of the mechanism (5.1) and of the rate equations (5.2) is greatly simplified if the various decay processes occur at vastly different rates. Analysis of the case $n = 2$ exhibits all chemically significant features of the general mechanism. To introduce the ideas of time-scale separation and of a trace intermediate assume that the reactions go to completion, i.e., $k_- \equiv k_-' \equiv 0$. Then, if initially only A is present, the concentrations of A, B, and C are found by direct integration of the rate equations

$$\begin{aligned}
[\mathrm{A}] &= A_0 \exp(-k_+ t) \\
[\mathrm{B}] &= k_+ A_0 [\exp(-k_+ t) - \exp(-k_+' t)]/(k_+' - k_+) \\
[\mathrm{C}] &= A_0 \{k_+'[1 - \exp(-k_+ t)] - k_+[1 - \exp(-k_+' t)]\}/(k_+' - k_+)
\end{aligned} \tag{5.4}$$

Further simplification occurs if there is a *separation of time scales*, i.e., if the rate constants are vastly different.

When $k_+ \gg k_+'$ the conversion of A to B is completed before B starts to decompose. Only 1% of the original amount of A is left after a time $t = 4.6/k_+$, which is much less than $1/k_+'$, the relaxation time for the further reaction of B. Then the approximate time dependences of [B] and [C] are

$$[\mathrm{B}] = A_0 \exp(-k_+' t), \quad [\mathrm{C}] = A_0 - [\mathrm{B}], \qquad k_+ \gg k_+', \quad t > 1/k_+$$

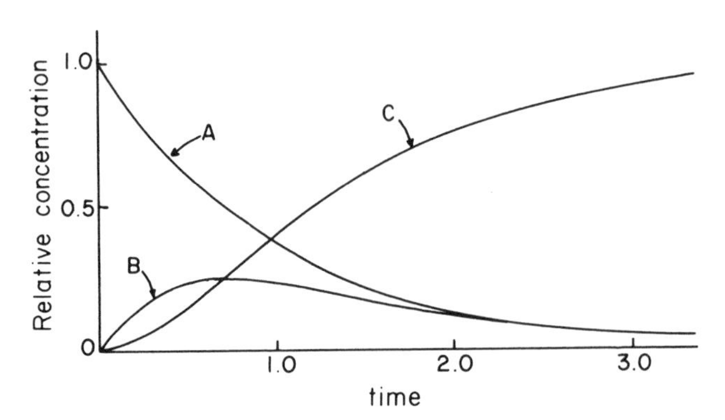

Fig. 5.1. Relative concentrations of A, B, and C for the mechanism (5.1) assuming $k_+' = 2k_+$ and $k_-' = k_- = 0$. The time scale is arbitrary.

which is readily detectable since the rapid buildup of the intermediate B should be obvious in an experiment. Furthermore the disappearance of A is much more rapid than the appearance of C.

If $k_+' \gg k_+$ then whatever B is formed reacts immediately; no significant concentration can develop. We find that

$$[\mathrm{C}] \approx A_0[1 - \exp(-k_+t)], \quad [\mathrm{B}] \approx A_0 \frac{k_+}{k_+'} \exp(-k_+t) = \frac{k_+}{k_+'} [\mathrm{A}] \quad (5.5)$$

so that [B] is always a fixed (small) fraction of [A]. Here it may be difficult to establish that the reaction proceeds in two steps since B is present in very low concentration at all times. Monitoring of neither A nor C would suggest that the reaction proceeds via an intermediate; other evidence would be needed.

If $k_+ \approx k_+'$ all three species are present simultaneously as shown graphically in Fig. 5.1. Experimentally one infers the two-step mechanism by observing C. Were only a single first-order process important, such as in the two limits considered, then $\ln[A_0 - \mathrm{C}]$ would be a linear function of t with a slope determined by the significant reaction. For the intermediate case, as shown in Fig. 5.2, this is not true until t is very large.

In general, reactions are reversible. If the rate equations (5.2) are analyzed using the matrix method of Appendix A for the special case $n = 2$, the two decay constants of (5.3) are $\varkappa_\pm$:

$$\varkappa_\pm = \tfrac{1}{2}\{(\sigma + \sigma') \pm [(\sigma - \sigma')^2 + 4k_-k_+']^{1/2}\} \quad (5.6)$$

where

$$\sigma = k_+ + k_-, \qquad \sigma' = k_+' + k_-' \quad (5.7)$$

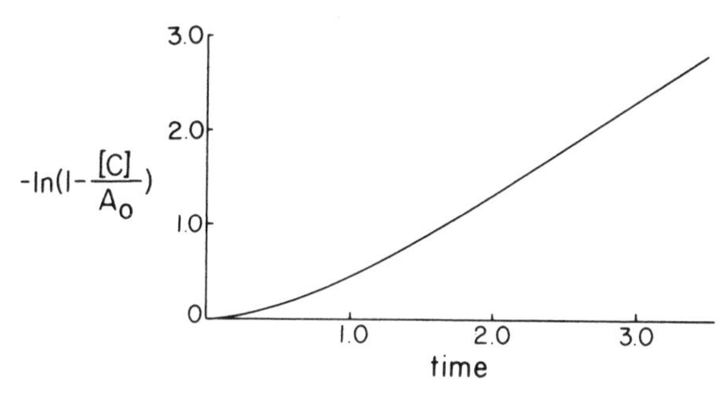

Fig. 5.2. Plot of ln $(1 - [C]/A_0)$ for the mechanism (5.1) assuming $k_+' = 2k_+$ and $k_-' = k_- = 0$. The plot is nonlinear at short times since both reactions are significant.

σ and σ' would be the decay constants for *uncoupled* $A \rightleftharpoons B$ and $B \rightleftharpoons C$ equilibria, respectively. The two decay constants $\varkappa_+$ and $\varkappa_-$ correspond to fast and slow processes, respectively. As long as the quantity under the square root in (5.6) is large, the time scales are separable, i.e., $\varkappa_+ \gg \varkappa_-$. One relaxation is complete before the other process starts, as illustrated in Fig. 5.3. There is further simplification if certain *elementary* steps of (5.1) are much faster than others.

Simplification occurs if one of the reactions equilibrates rapidly (preequilibrium) or if a trace intermediate forms (steady state). There is *preequilibration* if $\sigma \gg \sigma'$, i.e., if $A \rightleftharpoons B$ equilibration is much faster than

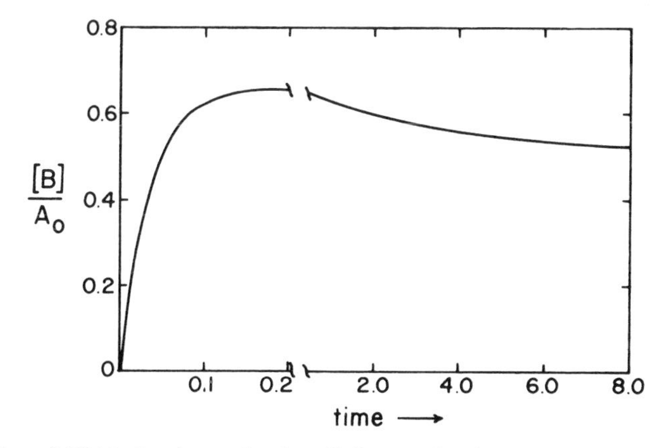

Fig. 5.3. Plot of $[B]/A_0$ for the mechanism (5.1) assuming $k_+ = 20$, $k_- = 10$, $k_+' = 0.1$, and $k_-' = 0.2$. Note the clean separation of time scales. There is rapid growth of the intermediate B governed by the fast decay ($\varkappa_+ = 30.034$) which is followed by a slow decay ($\varkappa_- = 0.266$) to equilibrium. The individual decay constants can be determined by analyzing the early growth and the late decay separately. Note that for $0.15 \lesssim t \lesssim 0.5$, [B] essentially is constant. At equilibrium $[B]/A_0 = 0.5$.

the B $\rightleftharpoons$ C equilibration. With this constraint and approximating the square root in (5.6), as done in Appendix B, the decay constants can be computed:

$$\varkappa_+ = \sigma, \qquad \varkappa_- = \sigma' - (k_- k_+'/\sigma) \tag{5.8}$$

The fast process *always* has a decay constant $k_+ + k_-$, as expected since the A $\rightleftharpoons$ B interconversion is a first-order reversible reaction. Since $k_+/k_- = K$, the equilibrium constant for the A $\rightleftharpoons$ B reaction, $\varkappa_-$ is

$$\varkappa_- = k_+' + k_-' - k_+'/(1 + K) \tag{5.9}$$

When $K \gg 1$ the equilibrium favors B and the slow process only measures the rate of B $\rightleftharpoons$ C interconversion; the decay constant is $k_+' + k_-'$. When $K \ll 1$ there is never much B present. The B $\rightleftharpoons$ C reaction is then operationally irreversible; whatever B is formed by the slow reaction of C continues to react and form A. Thus, the rate of the slow process is governed not only by the rate at which B and C interconvert but also by the *amount* of B present. In the event $\sigma' \gg \sigma$, the B $\rightleftharpoons$ C equilibration is rapid. A slow A $\rightleftharpoons$ B equilibration follows. This is the converse of the limit just considered.

When either k_+' or k_- (or both) are large compared with k_+ or k_-', B reacts as rapidly as it is formed; no significant population of this transient ever exists. The corresponding approximations to (5.6) are carried out in Appendix B with the results

$$\varkappa_+ = k_+' + k_-, \qquad \varkappa_- = \frac{k_+ k_+' + k_- k_-'}{k_+' + k_-} \tag{5.10}$$

The fast process measures the rate of disappearance of B. The smaller decay constant is a complicated function of the rate constants for the individual elementary steps.

A different derivation of (5.10) may be obtained directly from the rate law (5.1) with $n = 2$. When k_- or k_+' are large, B decomposes almost immediately upon formation. The species A and C form a reservoir which slowly convert to B. The net effect of the two opposing processes is that a *steady-state* population of B is formed after a short induction period governed by $\varkappa_+$, the fast decay. After a time period that is long compared with $\varkappa_+^{-1}$ we can make a *steady-state assumption*, that $d[B]/dt \approx 0$ in comparison with *both* $k_+[A] + k_-'[C]$ *and* $(k_- + k_+')[B]$ in (5.1). Hence

$$[B] \approx \frac{k_+[A] + k_-'[C]}{k_- + k_+'} \tag{5.11}$$

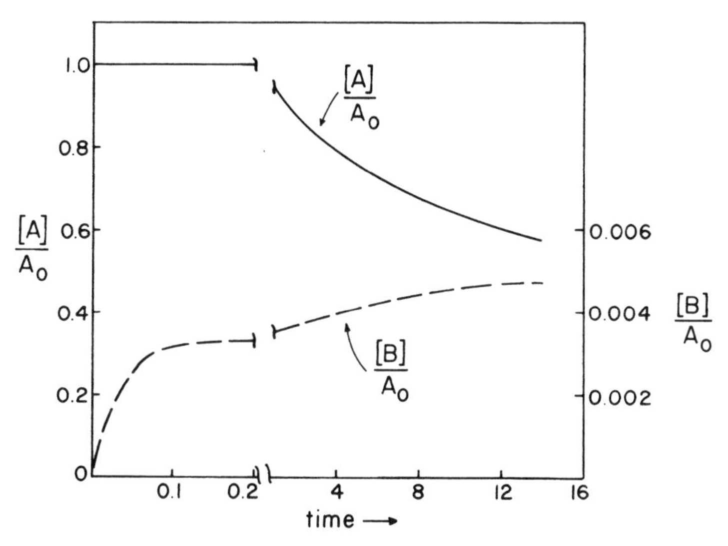

Fig. 5.4. Plot of $[A]/A_0$ and $[B]/A_0$ for the mechanism (5.1) assuming $k_+ = 0.1$, $k_- = 10$, $k_+' = 20$, and $k_-' = 0.2$. These are conditions under which the steady-state assumption is valid. There is a rapid induction period governed by the fast decay ($\varkappa_+ = 30.0$) during which a steady state is produced. This is followed by a slow approach to equilibrium ($\varkappa_- = 0.133$). During the induction period, [A] is constant. In the steady-state domain ($t \gtrsim 0.2$), [B] is *not* constant; however, there is never much B present.

It is important to understand that the steady-state assumption does *not* demand that $d[\mathrm{B}]/dt = 0$, as illustrated in Fig. 5.4. Rather, it asserts that the *net rate of change* of [B], $d[\mathrm{B}]/dt$, is much smaller than either its instantaneous rate of formation, $k_+[\mathrm{A}] + k_-'[\mathrm{C}]$, or its instantaneous rate of decomposition, $(k_- + k_+')[\mathrm{B}]$. Neither of the latter two quantities need be small, only their difference. Making use of the steady-state hypothesis and substituting (5.11) in (5.1) we find

$$\frac{d[\mathrm{A}]}{dt} = -\frac{d[\mathrm{C}]}{dt} = -\frac{k_+k_+'}{k_- + k_+'}[\mathrm{A}] + \frac{k_-k_-'}{k_- + k_+'}[\mathrm{C}] \qquad (5.12)$$

a rate law indistinguishable from one for a simple reversible first-order mechanism. The decay constant is, therefore, exactly the $\varkappa_-$ of (5.10).

The most complicated situation arises when all the rate constants are comparable in magnitude. Then $\varkappa_+$ and $\varkappa_-$ do not differ greatly. All three species are present simultaneously and experimental concentration–time plots are similar to Fig. 5.2; a single exponential decay does not account for the data. Analysis requires using the complete expressions (5.6), a complication which arises when $k_+ \approx k_-$, i.e., if A and B have similar thermodynamic stability, etc. While many chemical reactions involve species with markedly different stability there are many problems for which there is no separation of time scales.

We limited our analysis of (5.1), considering only two-step mechanisms. In a more complicated sequence there would be more relaxation times, the scale of which might be overlapping. The net result would be greater problems in interpreting the data. While time-scale separation, as we have seen, simplifies analysis because each relaxation can be observed independently, it can lead to misinterpretation if each is not observed. Reconsider the two-step process with the decay constants of (5.10). If only $\varkappa_-$ is measured one might reasonably infer a single-step mechanism. However, in some special cases the temperature dependence is sufficient to demonstrate greater complexity. Depending on the magnitudes of the four elementary rate constants, an Arrhenius plot of the observed decay constant can yield an apparent activation energy that is either negative or quite temperature dependent (see Problems 5.3 and 5.22). Such behavior is an immediate indicator that the observed decay does *not* reflect an elementary step. In general, lacking direct evidence of an intermediate, it is hard to infer its existence on kinetic grounds alone.

In general, independent of any time-scale separation, decay constants are complicated combinations of rate constants for the elementary steps. Only in the most rapid step of a reaction sequence is the decay constant a sum of rate constants, as is evident from (5.8) or (5.10). The complications increase if the mechanism involves more steps.

5.3. Consecutive Reactions—Arbitrary Order

In contrast to the first-order mechanisms just considered, the integrated rate laws for most multistep second(or higher)-order elementary processes can only be determined numerically, using computer methods. However, in many cases of interest the ideas of preequilibrium and steady state are applicable and the kinetic analysis is much simplified.

Consider the gas-phase decomposition of ozone

$$2O_3 = 3O_2$$

for which the empirical rate law is

$$\frac{d[O_3]}{dt} = -k_{\text{app}} \frac{[O_3]^2}{[O_2]} \tag{5.13}$$

A possible two-step mechanism[4] is

$$\begin{aligned} O_3 &\underset{k_{-1}}{\overset{k_1}{\rightleftharpoons}} O_2 + O \\ O + O_3 &\underset{k_{-2}}{\overset{k_2}{\rightleftharpoons}} 2O_2 \end{aligned} \tag{5.14}$$

If the first step is rapid preequilibration of atomic and molecular oxygen with ozone then [O] is determined by the values of $[O_3]$ and $[O_2]$,

$$[O] \approx K[O_3]/[O_2] \tag{5.15}$$

where $K = k_1/k_{-1}$. The rate expression for the second step is

$$\frac{d[O_3]}{dt} = -k_2[O][O_3] + k_{-2}[O_2]^2$$

which with (5.15) becomes

$$\frac{d[O_3]}{dt} = -k_2K\frac{[O_3]^2}{[O_2]} + k_{-2}[O_2]^2 \tag{5.16}$$

precisely the form observed experimentally if the back reaction in the second step does not occur. The experimental rate constant, k_{app}, is to be interpreted as the product of the equilibrium constant for the first step and the forward rate constant of the second step.

Alternatively it is reasonable to assume that the concentration of atomic oxygen is small and to apply the steady-state hypothesis to the mechanism (5.14). The result is

$$[O] = \{k_1[O_3] + k_{-2}[O_2]^2\}/\{k_{-1}[O_2] + k_2[O_3]\}$$

which when substituted in the rate expression for either step yields the approximate rate law

$$\frac{d[O_3]}{dt} = \frac{-k_1k_2[O_3]^2 + k_{-1}k_{-2}[O_2]^3}{k_{-1}[O_2] + k_2[O_3]} \tag{5.17}$$

an expression that is equivalent to (5.16) if $k_2[O_3] \ll k_{-1}[O_2]$. Stated in chemical terms, the $O + O_3$ reaction must be slow, precisely the condition

[4] We shall see in the next section that, strictly speaking, the initial decomposition of ozone must also involve a bimolecular step.

for rapid preequilibration of O_3, O_2, and O. In this example both the steady-state and preequilibrium hypotheses are consistent with the empirical rate law (5.13), and therefore adequate to demonstrate that the mechanism (5.14) is plausible. Unambiguous demonstration of (5.14) is more difficult since, in the preparation of O_3, atomic O is always present.[5]

If the proposed mechanism involves many steps or if some of the elementary reactions are third (or higher) order, exact analysis can become quite involved. The attempt at chemical simplification proceeds similarly. On chemical grounds one postulates that certain steps involve rapid equilibration and/or that some intermediates are in a steady state. The predicted rate law is then compared with the one determined experimentally. If they agree the mechanism has passed its first test.

It is not always possible to solve kinetic problems using simple approaches. Even if the mechanism only involves a few steps the data may not readily suggest the coupled rate laws. Chemical intuition is required to propose a plausible mechanism which then must be tested. The natural procedure is to integrate the mechanistic rate equations and to compare the predicted concentrations with the observed ones. If a set of kinetic parameters can be found that satisfactorily accounts for the data, the mechanism is a possible one. Such calculations always involve extensive numerical work. Thus reasonable starting points are needed, even if modern high-speed computers are used.

While closed-form solutions for irreversible two-step mechanisms involving combination of uni- and bimolecular steps have been formulated,[6] in general the rate laws must be integrated numerically. Straightforward approaches such as the Runge–Kutta methods are usually adequate to insure rapid convergence.[7] The process must be repeated for each set of rate constants to be tested. The most efficient procedure is usually to program a computer to carry out the calculations. In some instances the simple numerical integration algorithms converge slowly, or not at all. This is a particular problem for the complex mechanisms that describe systems for which there are multiple steady states, oscillations, or ex-

[5] For detailed discussion of the mechanism see S. W. Benson, *Foundations of Chemical Kinetics* (New York: McGraw–Hill, 1960), pp. 400–406.

[6] J. Y. Chien, *J. Am. Chem. Soc.* **70**, 2256 (1948); R. G. Pearson, L. C. King, and S. H. Langer, *J. Am. Chem. Soc.* **73**, 4149 (1951); A. S. Baladin and L. S. Liebenson, *Dokl. Akad. Nauk. SSSR* **39**, 22 (1943); T. Kelen, *Z. Phys. Chem. N. F.* **58**, 268 (1968); **60**, 191 (1968); *Acta Chim. Acad. Sci. Hung.* **59**, 323 (1969); **60**, 87 (1969).

[7] See D. D. McCracken and W. S. Dorn, *Numerical Methods and Fortran Programming* (New York: John Wiley, 1964), Chapter 10.

plosion.[8] Special numerical techniques, particularly suited for computer applications, have been developed to circumvent these difficulties.[9]

The study of sequential reactions is often much simplified if relaxation methods are applicable since such experiments are carried out near equilibrium where all rate expressions are first order in the progress variable. The general treatment outlined in Section 4.9 is equally applicable if there is a series of reactions. Corresponding to *each* independent reaction there is a progress variable. By analogy to (4.30) expressions for deviations from equilibrium can be written in terms of the progress variables. Then *coupled* first-order differential equations which describe the evolution of the various reactions can be found. These equations are the analogues to (4.31) and have the same mathematical form as (5.2) with progress variables replacing the concentrations. The matrix method of Appendix A can be used to obtain decay constants.[10] If there are two independent chemical reactions there are two exponential relaxations. The decay constants depend upon both the rate constants and the equilibrium concentrations. By carrying out the kinetic experiments under a variety of conditions the mechanism can often be characterized.

Even if a proposed mechanism accounts for the kinetic data, it is not necessarily correct. An important plausibility test is given by the magnitudes of the rate constants required to account for the phenomenology. In both gas- and solution-phase reactions there are natural upper bounds to the rate constants. A mechanism requiring larger values of k cannot be correct. These conditions are easily established for gas-phase reactions. The upper bound for the rate of a second-order reaction is fixed by the total rate of binary collisions, (2.21),

$$k_2[\mathrm{A}][\mathrm{B}] = \mathrm{Rate} \leq \mathrm{Z_{AB}} = N_0 \pi\, d_{\mathrm{AB}}^2 \bar{c}_{\mathrm{AB}}[\mathrm{A}][\mathrm{B}]$$

With reasonable values for molecular parameters this yields $k_2 \lesssim 10^{10}$–$10^{11}\ M^{-1}\ \mathrm{sec}^{-1}$ when $T \sim 500$ K. A similar bound can be found for k_3 if, for the reaction to occur, it is assumed that two molecules must be within a collision diameter when struck by a third. The result is $k_3 \lesssim 10^8$–10^{10} $M^{-2}\ \mathrm{sec}^{-1}$ (Problem 5.23). In solution the rate constant is limited by the rate of diffusion (Sections 5.7 and 9.13).

(8) A few of these problems are discussed in Chapter 7.

(9) C. W. Gear, *Numerical Initial Value Problems in Ordinary Differential Equations* (Englewood Cliffs, N. J.: Prentice–Hall, 1971), Chapter 11. For a general introduction see D. Edelson, *J. Chem. Ed.* **52**, 642 (1975).

(10) C. F. Bernasconi, *Relaxation Kinetics* (New York: Academic Press, 1976), Chapters 3, 4, and 7; G. W. Castellan, *Ber. Bunsen Gesell.* **67**, 898 (1963).

5.4. Unimolecular Decomposition—Gases

Numerous gas-phase decompositions follow first-order kinetics over a wide range of pressures. Typical examples are the reactions

$$N_2O_5 = 2NO_2 + O_2$$

$$\text{cyclopropane} = \text{propene}$$

Both reactions have activation energy barriers suggesting the question: How can one reconcile first-order kinetics with an activation process which must involve molecular collision? Lindemann[11] proposed the following mechanistic answer:

$$\begin{aligned} 2A &\underset{k_-}{\overset{k_+}{\rightleftharpoons}} A^* + A \\ A^* &\xrightarrow{k_d} \text{products} \end{aligned} \tag{5.18}$$

An A molecule is activated by collision with another A molecule. In the collision process the relative kinetic energy of the two colliding molecules is transformed, in part, to internal energy of one of the molecules, thus "activating" it. Such inelastic collisions (in which *kinetic* energy is *not* conserved) provide the mechanical basis for understanding the activation process.[12] The excited molecule A* is stabilized in two ways. Either it is deexcited by another inelastic collision with A or it decomposes to form products. The rate expressions deduced from (5.18) are

$$\begin{aligned} \frac{d[A^*]}{dt} &= k_+[A]^2 - k_-[A^*][A] - k_d[A^*] \\ \frac{d[\text{product}]}{dt} &= -\frac{d[A]}{dt} = k_d[A^*] \end{aligned} \tag{5.19}$$

It is reasonable to assume that the steady-state hypothesis is applicable to A* since molecular collisions are constantly occurring. The rates of production and depletion of A* greatly exceed its net rate of change. Hence

$$[A^*] = k_+[A]^2/\{k_-[A] + k_d\}$$

[11] F. A. Lindemann, *Trans. Faraday Soc.* **17**, 598 (1922).

[12] Inelastic processes are precisely those ignored by the postulates of elementary kinetic theory; see Section 1.1.

and the rate of production formation is

$$\frac{d[\text{product}]}{dt} = \frac{k_+ k_d [\text{A}]^2}{k_d + k_-[\text{A}]} \tag{5.20}$$

Depending upon the concentration of A, the predicted rate law varies between first and second order. At very low pressure $k_d \gg k_-[\text{A}]$ and the rate law is second order with an apparent rate constant k_+. However, when the pressure is high and $k_-[\text{A}] \gg k_d$ the reaction is first order with an apparent rate constant $k_\infty = k_d K$, where K is the equilibrium constant for the activation process. To test the Lindemann mechanism express the data in first-order form

$$\frac{d[\text{product}]}{dt} = k_{\text{ex}}[\text{A}]$$

and with (5.20) define a concentration dependent "first-order" rate constant

$$k_{\text{ex}} = k_\infty[\text{A}]/\{[\text{A}] + k_d/k_-\} \tag{5.21}$$

In Fig. 5.5, $\log(k_{\text{ex}}/k_\infty)$ is plotted for the decomposition of cyclopropane. While it appears to be a linear function of log[A] at low pressure, suggesting that the low-pressure limit has been reached, the slope is closer to $\frac{1}{2}$ than 1, the value predicted by (5.21). Such differences are typical; the actual falloff with pressure is more gradual than is consistent with the simple Lindemann mechanism. Improvement requires considering the specific

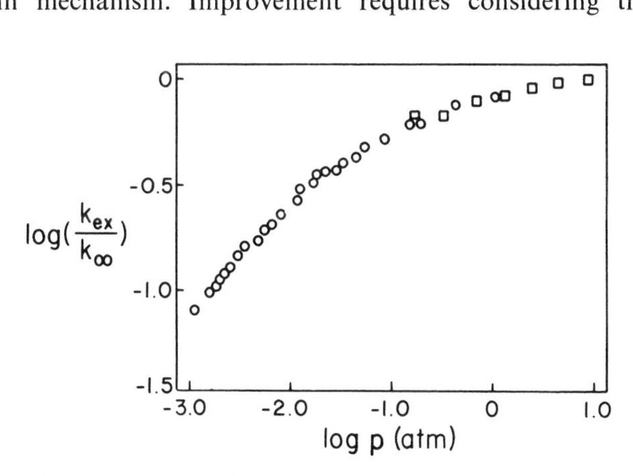

Fig. 5.5. Dependence of the isomerization rate of cyclopropane on the pressure of cyclopropane. [Adapted from H. O. Pritchard, R. G. Sowden and A. F. Trotman-Dickenson, *Proc. Roy. Soc.* **A217**, 563 (1953).]

energy dependence of the population of A* and of its decomposition rate. A theory accounting for the energy effects is discussed in Chapter 9.

Since the activation process is collisional, similar effects are found if an inert buffer gas is introduced into the system. In addition to the processes of (5.18) there is a parallel activation–deactivation reaction with the buffer gas, M,

$$\mathrm{A} + \mathrm{M} \underset{k_-'}{\overset{k_+'}{\rightleftharpoons}} \mathrm{A}^* + \mathrm{M}$$

Assuming steady-state conditions apply, the rate of product formation is

$$\frac{d[\text{product}]}{dt} = k_d \frac{k_+[\mathrm{A}]^2 + k_+'[\mathrm{A}][\mathrm{M}]}{k_-[\mathrm{A}] + k_-'[\mathrm{M}] + k_d} \tag{5.22}$$

which reduces to first-order kinetics when either the buffer or the reactant is at high pressure. At intermediate and low pressures this rate law is only qualitatively correct. Nevertheless a measure of the relative rates at which different gases activate A is given by the "first-order" rate constant derived from (5.22) at low gas pressure [the analogue to (5.21)],

$$k_{\mathrm{ex}} \approx k_+[\mathrm{A}] + k_+'[\mathrm{M}]$$

By changing both [A] and [M] the relative efficiency of activation by

Table 5.1. Relative Efficiency of Various Buffer Gases in Maintaining the Rate of Isomerization in the Cyclopropane–Propene Reaction[a]

Buffer gas	Relative efficiency
Cyclo-C_3H_6	1.0
He	0.060 ± 0.011
Ar	0.053 ± 0.007
H_2	0.24 ± 0.03
N_2	0.060 ± 0.003
CO_2	0.072 ± 0.009
CH_4	0.27 ± 0.03
H_2O	0.79 ± 0.11
Toluene	1.59 ± 0.13
Mesitylene	1.43 ± 0.26
Trifluorobenzene	1.09 ± 0.13

[a] H. O. Pritchard, R. G. Sowden, and A. F. Trotman-Dickenson, *Proc. Roy. Soc.* **A217**, 563 (1953).

different gases can be measured. An example is the cyclopropane–propene isomerization in the presence of different buffer gases. The results are given in Table 5.1. The major obvious trend is that the more complex the molecule, the more effective an activator it is. When we consider molecular collision dynamics in Chapter 10, some of the mechanical features which account for energy transfer in inelastic collisions will be clarified.

5.5. *Radical Recombination—Gases*

"Unimolecular" decomposition is a classic instance where theory demonstrates that a mechanism is more complex than the stoichiometry suggests. Another instance is in the atomic recombination reaction

$$2X = X_2$$

The experimental rate law is often second order in X but the mechanism cannot be

$$2X \xrightarrow{k_+} X_2$$

Consider a collision of two X atoms; they have a relative kinetic energy greater than zero. As the atoms interact total energy is conserved; their total energy remains positive. In Fig. 5.6 the relative potential and kinetic energy of a colliding atomic pair is plotted. At the turning point (distance of closest approach) the potential energy is necessarily positive and so the X_2 molecule is unstable as formed. The atoms fly apart and no bond is created.

Stabilization occurs if a third particle takes part in this process.[13] Some of the excess vibrational energy may be transmitted to the third particle leaving the X_2 molecule in a bound state. For the recombination of iodine atoms in the presence of a nonreactive buffer gas, M, the observed rate law is

$$\frac{1}{2}\frac{d[I]}{dt} = -k_1[I_2][I]^2 - k_2[M][I]^2 \qquad (5.23)$$

[13] Another mode of stabilization is conceivable. Before the unstable X_2 molecule dissociates it could spontaneously undergo a vibrational transition and be stabilized. This is a most unlikely process, even less common than three-body collisions in a gas. It may, however, be important in molecule formation in interstellar space; the formation of CH_2^+ from $C^+ + H_2$ appears to follow this pathway. For an introductory review see E. Herbst and W. Klemperer, *Phys. Today* **29**(6), 32 (1976).

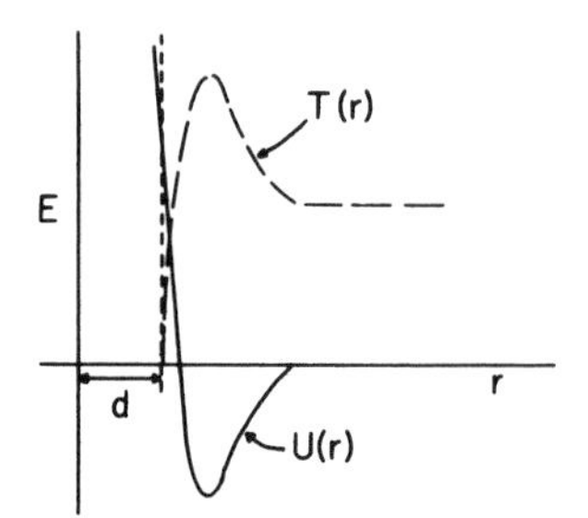

Figure 5.6. Kinetic and potential energies for atoms approaching one another along the line of centers. The total energy $T(r) + U(r)$ is conserved; d is the distance of closest approach.

suggesting the parallel mechanisms

$$2I + I_2 \xrightarrow{k_1} I_2 + I_2$$

$$2I + M \xrightarrow{k_2} I_2 + M$$

A third mechanism should also occur

$$3I \xrightarrow{k_3} I_2 + I$$

but because the concentration of I atoms is always small, it could not contribute significantly to (5.23).

The same considerations obviously apply to atomic reactions of the type

$$X + Y = XY$$

but are less important in *radical* recombination such as

$$2C_6H_5 \rightarrow \text{biphenyl}$$

The distinction is that a polyatomic molecule has many vibrational degrees of freedom. Temporary stabilization occurs if energy can be transferred from the C–C vibration to some of the other vibrational modes of the biphenyl molecule. Such a process occurs readily so that three-body collisions are not *immediately* required for the recombination of phenyl radicals. Reactions such as

$$H + OH \rightarrow H_2O$$

$$2CH_3 \rightarrow C_2H_6$$

are intermediate cases. There are fewer vibrational modes in water (3) and ethane (18) than in biphenyl (60); as a result decomposition of the energetically unstable molecule is rapid. However, in all cases the molecule is formed with excess vibrational energy; collision with another molecule is eventually required for complete stabilization.

A general mechanism for the recombination of X and Y in the presence of a buffer gas M can be proposed

$$\begin{aligned} X + Y &\underset{k_-}{\overset{k_+}{\rightleftharpoons}} XY^* \\ XY^* + M &\xrightarrow{k_2} XY + M \\ X + Y + M &\xrightarrow{k_3} XY + M \end{aligned} \tag{5.24}$$

There is a two-step pathway in which the metastable XY^* is deactivated by collision with M and the parallel one-step pathway involving direct stabilization via three-body collisions.[14] If a steady-state assumption is made on XY^*, the predicted rate law is

$$\frac{d[XY]}{dt} = \left\{k_3[M] + \frac{k_2k_+[M]}{k_- + k_2[M]}\right\}[X][Y] \tag{5.25}$$

thus the kinetics should vary from second to third order depending upon the pressure of buffer gas and the magnitude of the rate constants.[15] For small molecules the decomposition rate of XY^* must be very fast; thus $k_- \gg k_2[M]$ at reasonable pressures of M and third-order kinetics are found. For larger molecules $k_2[M] > k_-$ at moderate partial pressures of M in which case (5.25) simplifies to

$$\frac{d[XY]}{dt} = \{k_3[M] + k_+\}[X][Y]$$

Since three-body collisions (in the gas phase) are very unlikely, $k_+ > k_3[M]$ unless the partial pressure of M is quite large (Problem 5.23) and (5.25) reduces to a second-order rate law. As an indication of the effect of molecular complexity, the observed recombination rate of I atoms is always third

[14] M is sometimes termed a "chaperone," the chaperone's function being to lead partners into a stable union rather than leaving them in a highly excited state.

[15] Equation (5.24) does not account for all possible pathways. Often the following mechanism is important:

$$X + M \underset{k_-'}{\overset{k_+'}{\rightleftharpoons}} XM, \qquad XM + Y \xrightarrow{k_2'} XY + M$$

The additional term in (5.25) would be $k_+'k_2'[M][X][Y]/(k_-' + k_2'[Y])$ so that this pathway is kinetically distinguishable form those contributing to (5.24) if the shift from third- to second-order kinetics is observable.

Table 5.2. Rate Constants for the Reaction $I + I + M \rightarrow I_2 + M$[a]

Third body (M)	E_a (kJ mol^{-1})	Third body (M)	E_a (kJ mol^{-1})
Helium	−1.7	Toluene	−11
Argon	−5.4	Ethyl iodide	−10
Oxygen	−6.3	Mesitylene	−17
Carbon dioxide	−7.3	Iodine	−19
Benzene	−7.1		

[a] Data from G. Porter and J. A. Smith, *Proc. Roy. Soc.* (*London*) **A261**, 28 (1961).

order, that of phenyl radicals is always second order, and that of methyl radicals varies from third to second order as the buffer gas pressure changes from 0.0013 to 0.06 atm.[16]

At low or moderate pressure the direct three-body mechanism is not important and the apparent rate constant is, from (5.25), $k_2k_+/k_- = k_2K$, where K is the equilibrium constant for formation of vibrationally excited XY. The apparent activation energy is therefore

$$E_a = E_a^{(2)} + \Delta E^{(1)} \tag{5.26}$$

where $\Delta E^{(1)}$ is the energy of reaction for formation of XY* and $E_a^{(2)}$ the activation energy for the stabilization step. Since the first step is slightly exothermic (a weak bond is formed), $\Delta E^{(1)} < 0$; the second step is highly exothermic and requires no (or negligible) activation so $E_a^{(2)} \approx 0$. Thus we expect $E_a \approx \Delta E^{(1)} < 0$, which is the rule in recombination reactions. The same arguments apply if reaction proceeds via the weakly bonded intermediate XM (or YM). To establish which pathway is followed experiments are carried out using a range of buffer gases. Since E_a is set by the bond energy of the intermediate it is independent of buffer if XY* is the transient and variable if XM (or YM) is the transient. The effect of buffer gas in the iodine recombination is illustrated in Table 5.2; clearly the intermediates are IM. Comparable data for the recombination of H atoms or of Cl atoms or for the addition of H to NO show no effect of buffer on E_a suggesting that XY* is the likely intermediate.[17]

The mechanism (5.24) is, in its deactivation step, the converse of the Lindemann mechanism (5.18). Whereas unimolecular decomposition poses

[16] G. B. Kistiakowsky and E. K. Roberts, *J. Chem. Phys.* **21**, 1637 (1953).
[17] V. N. Kondratiev, *Rate Constants of Gas Phase Reactions*, translation by L. J. Holtschlag (Washington, D.C.: National Bureau of Standards, 1972).

the problem of adding energy to a molecule, bimolecular recombination is a problem in energy removal. The sequence (5.24) does not account for quantitative aspects of recombination kinetics. Unlike unimolecular reactions, which are well understood in terms of more elaborate theories, no truly adequate treatment of the termolecular processes has yet been formulated.[18]

5.6. Complex Mechanisms—Gaseous Chain Reactions

The classic example of a chain mechanism is the $H_2 + Br_2$ reaction first studied by Bodenstein.[19] The experimental rate law is

$$\frac{d[\mathrm{HBr}]}{dt} = \frac{k'[\mathrm{H_2}]^{1.0}[\mathrm{Br_2}]^{0.5}}{1 + k''[\mathrm{HBr}]^{1.0}/[\mathrm{Br_2}]^{1.0}} \tag{5.27}$$

where $k'' \sim 0.1$ and $k' \propto e^{-20{,}100/T}$ for temperatures from 500 to 600 K. The result defied interpretation for a number of years. Independently Christiansen, Herzfeld, and Polanyi[20] considered the following reaction sequence:

$$\begin{aligned}
&(1)\quad \mathrm{M + Br_2} \xrightarrow{k_1} \mathrm{2Br + M}\\
&(2)\quad \mathrm{Br + H_2} \xrightarrow{k_2} \mathrm{HBr + H}\\
&(3)\quad \mathrm{H + Br_2} \xrightarrow{k_3} \mathrm{HBr + Br}\\
&(4)\quad \mathrm{H + HBr} \xrightarrow{k_4} \mathrm{H_2 + Br}\\
&(5)\quad \mathrm{M + 2Br} \xrightarrow{k_5} \mathrm{Br_2 + M}
\end{aligned} \tag{5.28}$$

which is a typical chain mechanism.[21] Reactive intermediates, in this

[18] For an abbreviated discussion of these theories see W. C. Gardiner, *Rates and Mechanisms of Chemical Reactions* (New York: Benjamin, 1969), pp. 141–148. A more rigorous treatment is given by D. L. Bunker, *Theory of Elementary Gas Reaction Rates* (Oxford: Pergamon Press, 1966), Chapter 4. A recent review article is H. O. Pritchard, *Acc. Chem. Res.* **9**, 99 (1976).

[19] M. Bodenstein and S. C. Lind, *Z. Phys. Chem.* **57**, 168 (1907).

[20] J. A. Christiansen, *Kgl. Danske Vid. Sels. Mat.-Fys. Medd.* **1**, 14 (1919); K. F. Herzfeld, *Ann. Physik* **59**, 635 (1919); M. Polanyi, *Z. Electrochem.* **26**, 50 (1920).

[21] Since the mechanism (5.28) was postulated *before* Lindemann's rationalization of unimolecular decomposition, buffer gas was not included in steps (1) or (5) of the original proposal. Since there is preequilibration of Br_2 and Br, the rate law does not reflect the participation of buffer gas.

case H and Br, at low concentrations are postulated in order to continue the cycle. Steps (2), (3), and (4) maintain the concentration of the intermediates although only (2) and (3) lead to HBr formation; step (4) involves inhibition by HBr which is necessary in view of the experimental rate law (5.27). Step (1) is required to initiate the chain. Step (5) is a chain termination reaction. The buffer gas M in steps (1) and (5), required for activation or deactivation, may be either H_2 or Br_2 or an inert additive.

The rate of production of HBr is

$$\frac{d[\mathrm{HBr}]}{dt} = k_2[\mathrm{Br}][\mathrm{H}_2] + k_3[\mathrm{H}][\mathrm{Br}_2] - k_4[\mathrm{H}][\mathrm{HBr}] \tag{5.29}$$

To determine the concentration of Br and H atoms we apply the steady-state hypothesis. Since

$$\frac{d[\mathrm{Br}]}{dt} = -k_2[\mathrm{Br}][\mathrm{H}_2] + k_3[\mathrm{H}][\mathrm{Br}_2] + k_4[\mathrm{H}][\mathrm{HBr}] + 2k_1[\mathrm{Br}_2][\mathrm{M}] - 2k_5[\mathrm{Br}]^2[\mathrm{M}] \approx 0 \tag{5.30a}$$

$$\frac{d[\mathrm{H}]}{dt} = k_2[\mathrm{Br}][\mathrm{H}_2] - k_3[\mathrm{H}][\mathrm{Br}_2] - k_4[\mathrm{H}][\mathrm{HBr}] \approx 0 \tag{5.30b}$$

we find from (5.30b)

$$[\mathrm{H}] = k_2[\mathrm{H}_2][\mathrm{Br}]/\{k_3[\mathrm{Br}_2] + k_4[\mathrm{HBr}]\} \tag{5.31}$$

and, by adding (5.30b) to (5.30a),

$$[\mathrm{Br}] = (k_1/k_5)^{1/2}[\mathrm{Br}_2]^{1/2} \tag{5.32}$$

Combining (5.31) and (5.32) yields

$$[\mathrm{H}] = k_2(k_1/k_5)^{1/2}[\mathrm{H}_2][\mathrm{Br}_2]^{1/2}/\{k_3[\mathrm{Br}_2] + k_4[\mathrm{HBr}]\} \tag{5.33}$$

Substituting (5.32) and (5.33) into (5.29) then yields, after slight rearrangement,

$$\frac{d[\mathrm{HBr}]}{dt} = \frac{2k_2(k_1/k_5)^{1/2}[\mathrm{H}_2][\mathrm{Br}_2]^{1/2}}{1 + (k_4/k_3)[\mathrm{HBr}]/[\mathrm{Br}_2]} \tag{5.34}$$

in exact accord with the empirical expression (5.27) if $k' = 2k_2(k_1/k_5)^{1/2}$ and $k'' = k_4/k_3$.

It is possible to evaluate k_2 since $k_1/k_5 = K_{eq}$ for the dissociation of Br_2. Then, considering the temperature dependence of k' and K_{eq}, the activation energy for step (2) in the mechanism (5.28) is found to be 73.6 kJ mol^{-1}, which is certainly reasonable. Since this step is endothermic by 68.2 kJ mol^{-1}, a comparable activation energy is required.

The mechanism (5.28) is consistent with the empirical rate law (5.27). The activation energy for step (2) is quite reasonable. However, it is also important to show that other possible steps such as

$$
\begin{array}{lc}
(6) & M + H_2 \rightharpoonup 2H + M \\
(7) & M + HBr \rightharpoonup H + Br + M \\
(8) & Br + HBr \rightharpoonup H + Br_2 \\
(9) & M + 2H \rightharpoonup H_2 + M \\
(10) & M + H + Br \rightharpoonup HBr + M
\end{array}
$$

are not plausible. If these were significant reaction pathways the rate law would not be (5.34). There are also other reasons for eliminating them. The bond energy of both H_2 and HBr is much larger than that of Br_2. Thus the activation energy of either (6) or (8) is much greater than that of step (1) of (5.28); the dissociation rate of either H_2 or HBr is very slow in comparison with the dissociation rate of Br_2. Bond energy arguments indicate that the Br + HBr reaction (8) has an activation barrier of at least 173.2 kJ mol^{-1}; thus it is not competitive with steps (2) or (4) of (5.28), each of which has a much lower activation energy. The chain termination steps involving atomic H, (9) or (10), can be eliminated because the H-atom concentration is too small to be effective (Problem 5.7).[22]

The $H_2 + Br_2$ reaction is an example of a linear chain reaction. In each propagation step one atom of reactive intermediate is produced for each atom that is used up. From (5.34) we see that, if the temperature is constant, the *rate* of HBr production is greatest near the beginning of the reaction and drops off as H_2 and Br_2 are consumed. If the system is not thermostatted, the temperature will increase because the reaction is exothermic, and the reaction is thus accelerated. If heating occurs with sufficient rapidity, depletion of reactants will not be sufficiently fast to keep the reaction in check and the conditions for a *thermal explosion* are present. While not occurring in the $H_2 + Br_2$ reaction, because heating is relatively slow, an explosion can occur in the photoinitiated $H_2 + Cl_2$ reaction if the light intensity is high enough.

[22] A. A. Frost and R. G. Pearson, *Kinetics and Mechanism*, 2nd ed. (New York: John Wiley, 1961), p. 239.

A completely different cause of explosion is found in reactions which proceed by a branched-chain mechanism. Here more than one atom of intermediate is produced in the chain propagation scheme for each atom that reacts. The result, if chain carrier is produced faster in the initiation and propagation steps than it is consumed in termination steps, is a rapid increase in the concentration of chain carriers. As a consequence the reaction velocity accelerates rapidly and explosions can occur. The best-known mechanism of this type is the fission reaction in an atomic bomb. Chemical examples abound and include the reactions of oxygen with hydrogen, phosphorous, carbon monoxide, etc. Heating is an after effect of explosion, not the cause.[23]

The simplest branched-chain mechanism requires initiation, branching, and termination steps. If we assume a first-order chain termination step, such as reaction of the chain carrier R at the walls of the container, and let ... indicate any reactant or product other than chain carrier, the mechanism is

$$\begin{array}{llll} \textit{Initiation:} & \cdots \rightharpoonup n\mathrm{R} + \cdots & r_i & \\ \textit{Branching:} & \mathrm{R} + \cdots \rightharpoonup \alpha\mathrm{R} + \cdots & r_b[\mathrm{R}], \quad \alpha > 1 & \\ \textit{Termination:} & \mathrm{R} + \cdots \rightharpoonup \cdots & r_t[\mathrm{R}] & \end{array} \qquad (5.35)$$

The various rates r may depend upon the concentrations of any reactant species *other than the chain carrier*. As long as the rate of formation of R is small, steady-state kinetics are applicable and

$$\frac{d[\mathrm{R}]}{dt} = nr_i + (\alpha - 1)r_b[\mathrm{R}] - r_t[\mathrm{R}] \approx 0$$

so that

$$[\mathrm{R}] = nr_i/[r_t - (\alpha - 1)r_b] \qquad (5.36)$$

Since $\alpha > 1$ the denominator in (5.36) is zero if $r_b = r_t/(\alpha - 1)$. Then $[\mathrm{R}] \rightharpoonup \infty$ and the steady-state hypothesis cannot apply. However, the important qualitative feature is that the concentration of chain carriers increases very rapidly under such conditions. As a consequence there is an enormous increase in reaction velocity with the possibility of explosion.

Explosion is a phenomenon which cannot be analyzed using the stationary state hypothesis. Due to chemical heating or unchecked population growth an instability is created. Similar instabilities can lead to concentration

[23] The two mechanisms for explosion are discussed at greater length in Chapter 7.

oscillations (rather than relaxations) during the course of chemical reaction. It is even possible for spontaneous, periodic spatial inhomogeneities to form during chemical reaction. Further discussion of these topics is reserved for Chapter 7.

5.7. Simple Mechanisms—Solution

The qualitative features of reaction mechanisms in solutions are substantially different from those in gases. Unimolecular processes still occur via collisional activation. However, solvent molecules which can affect activation are always in high concentration. In terms of a rate law such as (5.22) the experiments are being carried out under conditions where $k_-'[M] \gg k_d$ or $k_-[A]$ and $k_+'[M] \gg k_+[A]$; here [M] represents the concentration of solvent, always in large excess. There is significant short-range order in liquids as solvent molecules are loosely bound to one another and form transient structures which reduce the mobility of the products of a decomposition. Thus the rate at which products separate by diffusion must limit the rate of unimolecular reaction in solution.[24] Unimolecular decomposition may still be considered a two-step process:

$$A \underset{k_-}{\overset{k_+}{\rightleftharpoons}} BC^* \xrightarrow{k_d} B + C$$

The first step involves rapid equilibration of A with a BC* complex; in the second step B and C diffuse apart. The rate law, assuming a steady state of BC*, is always first order in A with an apparent rate constant $k_+k_d/(k_- + k_d)$.

Isomerization, rearrangement, and conformational changes do not require diffusional separation of the products. Instead, a reasonable mechanism for the conversion of A to A′ is

$$A \underset{k_-}{\overset{k_+}{\rightleftharpoons}} A^* \underset{k_-'}{\overset{k_+'}{\rightleftharpoons}} A'$$

[24] The solvent structure is often described as a "cage," a somewhat misleading term since the "solvent cage" has no permanence. Different solvent molecules diffuse in and out of the vicinity of the products so that the structure is being continuously broken and reformed. However, the important feature which limits mobility is that the decomposition products cannot separate unless solvent molecules are displaced. The idea of a solvent cage was introduced by E. Rabinowitch and W. C. Wood, *Trans. Faraday Soc.* **32**, 1381 (1936).

The corresponding rate law, assuming a steady state of the excited intermediate A*, is

$$\frac{d[A]}{dt} = -\frac{k_+k_+'}{k_- + k_+'}[A] + \frac{k_-k_-'}{k_- + k_+'}[A'] \tag{5.37}$$

typical of a reversible first-order reaction.

Most solution reactions take place via bimolecular steps. Collisional deactivation of products occurs easily because of interaction with solvent. Chemical reaction may be viewed as a multistep process

$$\begin{aligned} A + B &\rightleftharpoons (A \cdots B) \quad \text{(diffusion)} \\ (A \cdots B) &\rightleftharpoons (C \cdots D) \quad \text{(reaction)} \\ (C \cdots D) &\rightleftharpoons C + D \quad \text{(diffusion)} \end{aligned} \tag{5.38}$$

in which reaction is initiated by the formation of the encounter complex (A ··· B). The rate of encounter (or separation) is governed by the rate of diffusion of reactants (or products). After encounter an (A ··· B) complex is constrained by the surrounding solvent and the reactants undergo many collisions before reacting or separating. The contrast with gas-phase reaction is striking; there encounter and collision are identical. In solution, reaction may occur rapidly after the formation of the encounter complex; in that case the kinetics of the overall process are limited not by the chemical step but by the diffusive ones.

It is possible to make quite accurate estimates of the rate constant for the diffusion process

$$A + B \xrightarrow{k_{\text{diff}}} (A \cdots B)$$

by solving the diffusion equation (2.23) under appropriate boundary conditions. This was done by Debye[25]; k_{diff} depends upon the diffusion coefficents of the reactants and thus, by Walden's rule, upon the solvent viscosity. If A and B are ions, k_{diff} is a very sensitive function of the solvent dielectric constant and the product of the ionic charges.[26] In aqueous solution most solvated species have similar size and comparable diffusion coefficients. It is found that k_{diff} is most sensitive to changes in the *charge* of the reacting species; it drops precipitously for reactions involving similarly charged ions.

[25] P. Debye, *Trans. Electrochem. Soc.* **82**, 265 (1942).

[26] We shall consider the Debye theory in Chapter 9.

Since k_{diff} provides an *upper bound* to the overall rate of reaction, any kinetic scheme from which one deduces a rate constant greater than k_{diff} must be excluded from further consideration. Thus such estimates are extremely valuable. A listing of rate constants for diffusion-limited reactions involving very different species (free radicals, solvated electrons, ions) is given in Table 5.3. To be exactly comparable the data should all be taken at the same ionic strength and temperature. At higher temperatures and higher ionic strengths the importance of ionic charge is somewhat reduced. Nonetheless the charge product effect remains striking. The $H^+ + SO_4^{2-}$ reaction (charge product, -2) is 200 times as fast as the $OH^- + HP_2O_7^{3-}$ reaction (charge product, $+3$). The effect of diffusion coefficient, for reactions involving species with the same charge product, is illustrated by comparing the rate constant for reactions involving H^+ ($D \sim 10^{-8}$ m^2 sec^{-1}), OH^- ($D \sim 5 \times 10^{-9}$ m^2 sec^{-1}), and all other solvated species ($D \sim 2 \times 10^{-9}$ m^2 sec^{-1}). The comparison is most striking for the rates of the $H^+ + OH^-$ reaction and the H + OH reaction. The former is 20 times faster, which is *not* an ionic effect (from the data presented, the difference in charge product might account for a factor of 2).

An example where the criterion of the diffusion limit was used to eliminate a possible mechanism was a study of carbonyl addition reactions similar to that discussed as an application of temperature jump in Section 4.9. Reactions of the type

$$\mathrm{RNH_2 + R'CHO \rightleftharpoons R{-}\underset{\substack{|\\ H}}{N}{-}\overset{\substack{H\\ |}}{\underset{\substack{|\\ OH}}{C}}{-}R'}$$

were shown *not* to involve the protonated form of the aldehyde as an intermediate.[27] From available experimental data the rate constant for the hypothetical step

$$\mathrm{RNH_2 + R'CHOH^+ \rightarrow R{-}\overset{\substack{H\\ |}}{\underset{\substack{|\\ H}}{N^{\pm}}}{-}\overset{\substack{H\\ |}}{\underset{\substack{|\\ OH}}{C}}{-}R'}$$

$$\mathrm{R = NH_2NHCO{-}}, \qquad \mathrm{R' = NO_2{-}C_6H_4{-}}$$

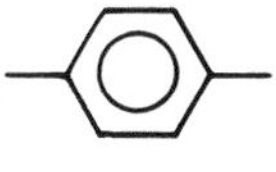

[27] E. H. Cordes and W. P. Jencks, *J. Am. Chem. Soc.* **84**, 4319 (1962).

Table 5.3. Rate Constants for Some Diffusion-Limited Reactions in Aqueous Solution

Reaction	Temperature (K)	Ionic strength (M)	Rate constant (M^{-1} sec^{-1})	Reference
$H^+ + SO_4^{2-} \rightarrow HSO_4^-$	293	0.1	$\sim 1.0 \times 10^{11}$	a
$H^+ + HCO_3^- \rightarrow H_2CO_3$	298	0	4.7×10^{10}	a
$OH^- + NH_4^+ \rightarrow NH_4OH$	293	0	3.4×10^{10}	a
$H^+ + OH^- \rightarrow H_2O$	298	0	1.4×10^{11}	a
$H + OH \rightarrow H_2O$	298	—	7×10^{9}	b
$H^+ + NH_3 \rightarrow NH_4^+$	298	—	4.3×10^{10}	a
$2\ CH_3\dot{C}HOH \rightarrow$ 2,3-Butanediol	298	—	1.4×10^{9}	c
$OH^- + C_6H_5OH \rightarrow C_6H_5O^- + H_2O$	298	—	1.4×10^{10}	a
$e_{aq}^- + ClCH_2CO_2H \rightarrow Cl^- + \cdot CH_2CO_2H$	298	—	6.5×10^{9}	d
$e_{aq}^- + ClCH_2CO_2^- \rightarrow Cl^- + \cdot CH_2CO_2^-$	298	0.1	1.2×10^{9}	e
$H^+ + UO_2(OH)^+ \rightarrow UO_2^{2+} + H_2O$	298	0	$\sim 1.6 \times 10^{10}$	f
$OH^- + HCO_3^- \rightarrow CO_3^{2-} + H_2O$	293	1	$\sim 6 \times 10^{9}$	a
$H^+ + [Co(NH_3)_5OH]^{2+} \rightarrow [Co(NH_3)_5OH_2]^{3+}$	283	0.1	5.0×10^{9}	a
$e_{aq}^- + Fe(CN)_6^{2-} \rightarrow Fe(CN)_6^{3-}$	298	0.1	3×10^{9}	g
$OH^- + HP_2O_7^{3-} \rightarrow P_2O_7^{4-} + H_2O$	285	0.1	4.7×10^{8}	a

[a] M. Eigen, W. Kruse, G. Maass, and L. de Maeyer, *Prog. React. Kinetics* **2**, 285 (1964), where original references are given.
[b] J. K. Thomas, *Trans. Faraday Soc.* **61**, 702 (1965).
[c] M. S. Matheson and L. M. Dorfman, *Trans. Faraday Soc.* **61**, 156 (1965).
[d] E. Hayon and H. A. Allen, *J. Phys. Chem.* **65**, 2181 (1961).
[e] M. Anbar and E. J. Hart, *J. Phys. Chem.* **69**, 271 (1965).
[f] D. L. Cole, E. M. Eyring, D. T. Rampton, A. Silzars, and R. P. Jensen, *J. Phys. Chem.* **71**, 2771 (1967).
[g] M. S. Matheson and L. M. Dorfman, *Pulse Radiolysis* (Cambridge: MIT Press, 1969), p. 104.

would have to have been $2.7 \times 10^{12}\ M^{-1}\ sec^{-1}$ to account for the observed kinetics. This is 50 times as large as the rate constant for *any* reaction in Table 5.3 with the same charge product. It is 400 times as large as the rate constant for reactions with zero charge product which involve neither H^+ nor OH^-. Since solution reactions cannot have $k \gg k_{diff}$, this pathway was excluded. The temperature-jump study of similar reactions carried out 9 years later provided direct confirmation of this deduction.

The solvent affects reactivity in other ways than just reducing mobility. A case in point is the photodissociation of iodine which occurs when iodine is irradiated with sufficiently high frequency light,

$$I_2 + h\nu \rightarrow 2I$$

In the gas phase the quantum yield (the number of iodine molecules dissociated per quantum of light absorbed) is 1; each photon dissociates a molecule.[28] In solution the quantum yields drop dramatically, a result that can be interpreted in terms of the effect of solvent structure.[29] Upon absorbing light the molecule dissociates. However, separation is not immediate. The iodine atoms are trapped within a transient cage of solvent molecules to which they may transfer any excess kinetic energy. They undergo many collisions before separating. The observed quantum yield is determined by competition between the rate of diffusion and recombination. It depends upon both the mass of the solvent molecule (which affects the efficiency of energy transfer from iodine atoms to solvent) and the viscosity of the solvent (which governs the rate of separation of the atoms).

5.8. Complex Mechanisms—Solution

Many rate laws for solution reactions are consistent with diffusion control. In general, knowledge of the chemistry of the reactants is the key to unraveling the mechanism. We consider two examples.

(28) J. Franck and E. Rabinowitch, *Trans. Faraday Soc.* **30**, 120 (1934).

(29) E. Rabinowitch and W. C. Wood, *Trans. Faraday Soc.* **32**, 1381 (1936); R. M. Noyes, *J. Chem. Phys.* **18**, 658 (1950); F. W. Lampe and R. M. Noyes, *J. Am. Chem. Soc.* **76**, 2140 (1954); H. Rosman and R. M. Noyes, *J. Am. Chem. Soc.* **80**, 2410 (1958); D. Booth and R. M. Noyes, *J. Am. Chem. Soc.* **82**, 1868 (1960); L. F. Meadows and R. M. Noyes, *J. Am. Chem. Soc.* **82**, 1872 (1960).

5.8.1. The Fe(III) + Eu(II) System—The Effect of pH

In Section 4.10 the rate law for the electron-transfer reaction

$$\text{Eu(II)} + \text{Fe(III)} = \text{Eu(III)} + \text{Fe(II)}$$

was determined for values of the pH between 1 and 3. More elaborate data analysis indicates that the rate law is

$$-\frac{d[\text{Eu(II)}]}{dt} = k[\text{Fe(III)}][\text{Eu(II)}] \tag{5.39a}$$

with the pH-dependent rate constant well represented by[30]

$$k = \left(k_0 + \frac{k_{-1}}{[\text{H}^+]}\right)\left(\frac{[\text{H}^+]}{[\text{H}^+] + K_a}\right) \tag{5.39b}$$

To interpret this requires considering the equilibrium chemistry of the ferric ion. Even in acid solution (see Problem 4.10) Fe(III) exists in the two rapidly equilibrated forms Fe^{3+} and $FeOH^{2+}$:

$$\text{Fe}^{3+}_{\text{aq}} + \text{H}_2\text{O} = \text{FeOH}^{2+}_{\text{aq}} + \text{H}^+ \qquad \text{(equilibrium, } K\text{)}$$

Using the equilibrium condition and the stoichiometry we find

$$[\text{Fe}^{3+}_{\text{aq}}] = \frac{[\text{H}^+][\text{Fe(III)}]}{[\text{H}^+] + K}, \qquad [\text{FeOH}^{2+}_{\text{aq}}] = \frac{K[\text{Fe(III)}]}{[\text{H}^+] + K} \tag{5.40}$$

Assuming parallel binary reactions between Eu(II) and each of the ferric ion species, the postulated rate law is

$$-\frac{d[\text{Eu(II)}]}{dt} = k'[\text{Eu(II)}][\text{Fe}^{3+}_{\text{aq}}] + k''[\text{Eu(II)}][\text{FeOH}^{2+}_{\text{aq}}] \tag{5.41}$$

Substituting from (5.40) yields the experimental form (5.39) if we identify

$$k' = k_0, \qquad k''K = k_{-1}, \qquad K = K_a$$

At 1.4°C and unit ionic strength the parameters are $k_0 = 3.38 \times 10^3$ $M^{-1}\,\text{sec}^{-1}$, $k_{-1} = 2.24 \times 10^3\,\text{sec}^{-1}$, and $K_a \sim 7 \times 10^{-4}\,M$ so that $k'' \sim$ $3.2 \times 10^6\,M^{-1}\,\text{sec}^{-1}$. Under the experimental conditions (pH 1–3), $[\text{Fe}^{3+}_{\text{aq}}] >$

[30] D. W. Carlyle and J. H. Espenson, *J. Am. Chem. Soc.* **90**, 2272 (1968).

$[FeOH^{2+}_{aq}]$ and only at the highest pH is the $FeOH^{2+}_{aq}$ concentration significant. Nonetheless, since the hydroxy complex is so much more reactive, the pathway involving that species predominates at all but the lowest pH. Note that neither k' nor k'' are comparable to their diffusion limits (about $2 \times 10^7\ M^{-1}\ \text{sec}^{-1}$ and $2 \times 10^8\ M^{-1}\ \text{sec}^{-1}$, respectively). The *chemical* process of electron-transfer controls the rate of reaction of either Fe^{3+}_{aq} or $FeOH^{2+}_{aq}$.

5.8.2. The Fe(III) + V(III) Reaction—Inhibition

In studies of the electron-transfer process

$$\text{V(III)} + \text{Fe(III)} = \text{V(IV)} + \text{Fe(II)}$$

initial kinetic experiments were consistent with the rate law

$$-\frac{d[\text{Fe(III)}]}{dt} = k_0[\text{Fe(III)}][\text{V(III)}] \tag{5.42}$$

More detailed study indicated greater complexity. The reaction rate depends upon the *product* concentration; it is accelerated by V(IV) and inhibited by Fe(II). The complete rate law is[31]

$$-\frac{d[\text{Fe(III)}]}{dt} = \left\{k' + k''\frac{[\text{V(IV)}]}{[\text{Fe(II)}]}\right\}[\text{Fe(III)}][\text{V(III)}] \tag{5.43}$$

which cannot be observed if, initially, no products are present. Under such conditions their concentrations are *always equal* and (5.43) reduces to (5.42) with $k_0 = k' + k''$.

As is the case in the Fe(III) + Eu(II) system the rate constants k' and k'' are pH dependent, which in all probability reflects the different reaction rates of the various hydroxy forms of Fe(III). Even ignoring this complication the rate law (5.43) requires a mechanism that postulates *parallel* reactions. The first term in the rate law is consistent with the one-step electron-transfer reaction

$$\text{Fe(III)} + \text{V(III)} \xrightarrow{k'} \text{Fe(II)} + \text{V(IV)} \tag{5.44}$$

Since the second term in the rate law is minus first order in Fe(II), it must

[31] W. C. E. Higginson, D. R. Rosseinsky, J. B. Stead, and A. G. Sykes, *Disc. Faraday Soc.* **29**, 49 (1960).

involve more than a single step. In some step V(IV) must appear as a *reactant*. Furthermore Fe(II) must appear as a product in a *reversible* step in order for it to act as an inhibitor. Recognizing that vanadium can exist in all oxidation states from +2 to +5, a simple pathway that incorporates these constraints is

$$\begin{aligned} \mathrm{V(IV)} + \mathrm{Fe(III)} &\underset{k_{-2}}{\overset{k_2}{\rightleftharpoons}} \mathrm{V(V)} + \mathrm{Fe(II)} \\ \mathrm{V(V)} + \mathrm{V(III)} &\xrightarrow{k_3} 2\mathrm{V(IV)} \end{aligned} \tag{5.45}$$

Assuming the steady-state approximation can be applied to [V(V)] the rate law derived from (5.45) is

$$-\frac{d[\mathrm{Fe(III)}]}{dt} = \frac{k_3 k_2 [\mathrm{Fe(III)}][\mathrm{V(IV)}][\mathrm{V(III)}]}{k_{-2}[\mathrm{Fe(II)}] + k_3[\mathrm{V(III)}]}$$

which, if $k_{-2}[\mathrm{Fe(II)}] \gg k_3[\mathrm{V(III)}]$, reduces to the second term of (5.45) with $k'' = k_3 k_2 / k_{-2}$. In (5.45) the V(IV) formed by direct reaction of Fe(III) and V(III) is oxidized still further. The +5 and +3 forms of vanadium then react to reform the V(IV). Such pathways, where the oxidation (or reduction) of one reactant temporarily proceeds beyond the thermodynamically stable state during the course of reaction, are common for species which exist in many oxidation states. There is always a further step which reduces (or oxidizes) the intermediate to the stable state.

5.9. Homogeneous Catalysis

The rate of a chemical reaction may be enhanced in many ways. These methods include:

(1) raising the temperature;
(2) irradiating with light;
(3) changing the solvent;
(4) adding trace amounts of a reagent.

The first two methods are *not* catalytic since they are based on adding energy to the system. The third involves gross alteration of the reaction medium. Only the fourth, in which *trace* quantities of reagent markedly affect the reaction rate, is catalysis.

Conventionally a catalyst is defined as a substance that alters the speed of a chemical reaction without undergoing any chemical changes itself. Thus it should not appear in the overall stoichiometry of the reaction nor should it combine other than transiently with either products or reactants. In practice this is a somewhat restricted use of the term; catalysts often combine with reaction products or are slowly used up in side reactions which occur along with the process that gives them their catalytic activity.[32]

A case where a catalyst accelerates reaction is the commercial process for sulfuric acid. The first step is the oxidation of SO_2 to SO_3; uncatalyzed this proceeds via the slow termolecular reaction

$$2SO_2 + O_2 \rightharpoonup 2SO_3 \tag{5.46}$$

In the presence of NO a two-step pathway may be followed (the termolecular NO–O_2 reaction has zero activation energy):

$$\begin{aligned} 2NO + O_2 &\rightharpoonup 2NO_2 \\ NO_2 + SO_2 &\rightharpoonup NO + SO_3 \end{aligned} \tag{5.47}$$

Each of these reactions is reasonably fast; the net rate of production of SO_3 is greatly enhanced by addition of NO. The overall chemical result of the sequence (5.47) is exactly the same as the sequence (5.46). The catalyst provides a *parallel* pathway to reaction, one that is much faster than the uncatalyzed one.

An example which illustrates a different aspect of catalysis is the Wacker process for oxidizing ethylene to acetaldehyde.[33] The earlier technique involved two steps and the isolation of an intermediate. However, consideration of the aqueous reactions[34]

$$\begin{aligned} C_2H_4 + PdCl_2 + H_2O &= CH_3CHO + 2HCl + Pd(s) \\ 2CuCl + Pd(s) &= 2CuCl + PdCl_2 \end{aligned}$$

suggests that Cu(II) could be used to oxidize ethylene as long as both

[32] An example of a catalyst that is slowly destroyed in the course of its catalytic action is the catalytic converter in new automobiles. The Pt catalyst accelerates the conversion of CO to CO_2. At the same time it also aids in the formation of SO_3 and NO_2, products which, at elevated temperatures, will combine with the Pt. In addition it is poisoned by accumulation of carbon on the surface.

[33] J. Smidt, W. Hafner, R. Jira, J. Sedlmeier, R. Sieber, R. Rüttinger, and H. Kojer, *Angew. Chem.* **71**, 176 (1959).

[34] The specific species which participate are various complexes of Pd(II), Cu(II), and Cu(I). They are not indicated.

Pd(II) and Pd(0) were present

$$C_2H_4 + 2CuCl_2 + H_2O \overset{Pd}{=\!=} CH_3CHO + 2CuCl + 2HCl$$

Then, since CuCl is easily oxidized in HCl solution

$$2CuCl + 2HCl + \tfrac{1}{2}O_2 = 2CuCl_2 + H_2O$$

one could imagine that by properly adjusting the conditions trace quantities of Cu and Pd could catalyze a one-step oxidation in aqueous HCl

$$C_2H_4 + \tfrac{1}{2}O_2 \underset{HCl}{\overset{Cu,Pd}{=\!=}} CH_3CHO \tag{5.48}$$

Thus judicious use of catalysts may provide a mechanism that is both faster and requires isolating fewer intermediates. The elementary steps in the Wacker process are quite involved; a succession of organopalladium intermediates is formed.[35] The chemistry of the reaction (5.48) is complex but the individual steps are rapid.

Catalysis is, in many cases, irreversible; the mechanism may be represented as

$$\begin{aligned} A + B &\xrightarrow{k_1} \text{products} && \text{slow (uncatalyzed)} \\ A + \text{catalyst} &\xrightarrow{k_2} X && \text{fast (catalyzed)} \\ X + B &\xrightarrow{k_2'} \text{products} + \text{catalyst} && \text{fast} \end{aligned} \tag{5.49}$$

Assuming a steady state in the intermediate X, the reaction rate is

$$\frac{d[\text{product}]}{dt} = k_1[A][B] + k_2[A][\text{catalyst}]$$

so that, if $k_2 \gg k_1$, a small amount of catalyst greatly accelerates the reaction. In addition (5.49) predicts that the rate law is first order in catalyst, a common (though not universal) experimental situation. A counterexample is the N_2O_5 catalyzed decomposition of ozone which is $\frac{2}{3}$ order in both N_2O_5 and O_3.[36]

[35] For their characterization see A. Aquilo, *Adv. Organomet. Chem.* **5**, 321 (1967).

[36] H. J. Schumacher and G. Sprenger, *Z. Physik. Chem.* **A140**, 281 (1929); **B2**, 267 (1929).

5.10. Enzyme Catalysis

In biochemical systems a small amount of enzyme (E) catalyzes conversion of a substrate (S) to product (P).[37] The rate law observed, in many cases, is

$$\frac{d[\mathrm{P}]}{dt} = \frac{kE_0[\mathrm{S}]}{K_m + [\mathrm{S}]} \tag{5.50}$$

where E_0 is the *total* enzyme concentration. The simplest model consistent with (5.50) is the Michaelis–Menton mechanism[38]

$$\mathrm{E} + \mathrm{S} \underset{k_{-1}}{\overset{k_1}{\rightleftharpoons}} \mathrm{ES} \xrightarrow{k_2} \mathrm{E} + \mathrm{P} \tag{5.51}$$

Assuming that ES is in a steady state we find

$$[\mathrm{ES}] = k_1[\mathrm{E}][\mathrm{S}]/(k_{-1} + k_2)$$

and the rate of product formation is

$$\frac{d[\mathrm{P}]}{dt} = \frac{k_2[\mathrm{E}][\mathrm{S}]}{K_m} \tag{5.52}$$

where $K_m = (k_2 + k_{-1})/k_1$ is known as the *Michaelis constant*. Since the uncomplexed enzyme concentration is generally not experimentally accessible, (5.52) is more useful when expressed in terms of the total enzyme concentration E_0. Since $E_0 = [\mathrm{E}] + [\mathrm{ES}]$ we find

$$[\mathrm{E}] = K_m E_0/(K_m + [\mathrm{S}])$$

and hence

$$\frac{d[\mathrm{P}]}{dt} = \frac{k_2 E_0[\mathrm{S}]}{K_m + [\mathrm{S}]} \tag{5.53}$$

precisely the form observed experimentally (5.50). By varying the substrate concentration both k_2, known as the *turnover number*, and the Michaelis constant can be determined. At high substrate concentration there is no free enzyme (the enzyme is *saturated*) and (5.53) is exceptionally simple;

[37] For an introduction to the biochemistry of enzyme mechanisms see A. L. Lehninger, *Biochemistry*, 2nd ed. (New York: Worth, 1975), Chapters 8 and 9; J. T. Wong, *Kinetics of Enzyme Mechanisms* (London: Academic Press, 1975).

[38] L. Michaelis and M. L. Menton, *Biochem. Z.* **49**, 333 (1913) used a preequilibrium formulation valid when $k_2 \ll k_{-1}$. The first steady-state treatment was by G. E. Briggs and J. B. S. Haldane, *Biochem. J.* **19**, 338 (1925).

the rate of product formation is a constant, $V = k_2E_0$. The Michaelis constant can then be determined from the integrated form of (5.53)

$$-K_m \ln\{[\mathrm{S}]/S_0\} = [\mathrm{S}] - S_0 + Vt$$

by plotting ln[S] vs. [S] + Vt; the slope is $-K_m$. Alternatively the reaction velocity $v = d[P]/dt$ can be plotted as a function of substrate concentration in a variety of ways. The most common are

$$v^{-1} = 1/V + (K_m/V)[\mathrm{S}]^{-1} \quad \text{(Lineweaver–Burk)}$$
$$v = V - K_m(v/[S]) \quad \text{(Eadie)}$$
$$[\mathrm{S}]/v = K_m/V + [\mathrm{S}]/V \quad \text{(Hanes)}$$

Of these the Hanes transformation is most reliable for extracting values of kinetic parameters.[39]

The Michaelis–Menton mechanism (5.51) is the simplest member of a set of series–parallel reactions,

$$\begin{array}{ccccc} \mathrm{E} + \mathrm{S} & \rightleftharpoons & \mathrm{X}_1 & \rightleftharpoons & \mathrm{E} + \mathrm{P} \\ \mathrm{H}^+ \upharpoonleft\downharpoonright & & \upharpoonleft\downharpoonright & & \upharpoonleft\downharpoonright \mathrm{H}^+ \\ \mathrm{EH}^+ + \mathrm{S} & \rightleftharpoons & \mathrm{X}_2 & \rightleftharpoons & \mathrm{EH}^+ + \mathrm{P} \\ \upharpoonleft\downharpoonright & & \upharpoonleft\downharpoonright & & \upharpoonleft\downharpoonright \\ \vdots & & \vdots & & \vdots \end{array} \tag{5.54}$$

This particular mechanism takes into account the fact that enzymes can be protonated which alters the specific rate constants for the various elementary steps. Similar, but more complicated, schemes arise if either substrate or product (or both) can be protonated. In addition substances other than hydrogen ion may complex with enzyme, substrate, or product. The reaction network would be analogous to (5.54). While a direct steady-state treatment of (5.54) is complicated, a systematic method for determining the rate law in terms of the specific rate constants has been worked out.[40]

[39] J. T. Wong, *Kinetics of Enzyme Mechanisms* (London: Academic Press, 1975), pp. 239–240.

[40] E. L. King and C. Altman, *J. Phys. Chem.* **60**, 1375 (1956). A clear exposition of the method is given by K. M. Plowman, *Enzyme Kinetics* (New York: McGraw-Hill, 1977), pp. 30–38, 156–164. For a formulation that lends itself to computer treatment see A. R. Schulz and D. D. Fisher, *Can. J. Biochem.* **48**, 922 (1970).

In addition to varying pH, enzyme kinetics can be studied by adding other substances which react with the enzyme in a variety of ways. The use of *inhibitors* is often an aid to understanding the nature of the enzymatic activity. In competitive inhibition the inhibitor complexes the enzyme and renders it inactive. Assuming the dissociation equilibrium

$$\mathrm{EI} \rightleftharpoons \mathrm{E} + \mathrm{I}, \qquad K_{\mathrm{I}} = \frac{[\mathrm{E}][\mathrm{I}]}{[\mathrm{EI}]}$$

competes with the first step in (5.51), the analog to (5.53) is

$$\frac{d[\mathrm{P}]}{dt} = \frac{k_2 E_0 [\mathrm{S}]}{[\mathrm{S}] + K_m\{1 + [\mathrm{I}]/K_{\mathrm{I}}\}} \tag{5.55}$$

Comparison with (5.53) indicates that the dependence of reaction rate on substrate concentration is unaltered; however, the apparent Michaelis constant is now a function of inhibitor concentration.

In uncompetitive inhibition the inhibitor is presumed to complex the reaction intermediate ES,

$$\mathrm{ESI} \rightleftharpoons \mathrm{ES} + \mathrm{I}, \qquad K_{\mathrm{I}}' = [\mathrm{ES}][\mathrm{I}]/[\mathrm{ESI}]$$

and thus compete with the product formation step in (5.51). The reaction rate is now

$$\frac{d[\mathrm{P}]}{dt} = \frac{k_2 E_0 [\mathrm{S}]}{[\mathrm{S}]\{1 + [\mathrm{I}]/K_{\mathrm{I}}'\} + K_m} \tag{5.56}$$

Again the dependence on [S] is unaffected; however, the limiting velocity at high substrate concentration is now a function of inhibitor concentration. The rate laws (5.55) and (5.56) are kinetically distinguishable and permit differentiation between these possible modes of inhibition. Another important, distinguishable process is noncompetitive inhibition (Problems 5.13 and 5.14).

5.11. Heterogeneous Catalysis

When metal catalysts are used to accelerate gaseous reactions the mechanism involves five steps:

(1) diffusion or convection of reactant to the surface;
(2) adsorption of reactant;

(3) elementary reactions of adsorbate, either with gas molecules or with a nearby adsorbed molecule;
(4) desorption of product;
(5) diffusion or convection of product from the surface.

Any of these may be rate limiting. Measurements of diffusion rates allow (1) and (5) to be characterized separately. Adsorption and desorption may often be studied separately as well, allowing the surface reaction (3) to be studied separately.

Except at very low pressure diffusion does not limit the rate of reaction; however, adsorption is very important. If the surface has a total of S_0 adsorption *sites* a general mechanism is

$$\begin{aligned} \mathrm{A} + \mathrm{S} &\rightleftharpoons \mathrm{A} \cdot \mathrm{S} \qquad \text{(equilibrium, } K\text{)} \\ \mathrm{A} \cdot \mathrm{S} + \cdots &\xrightarrow{k_2} \text{products} \end{aligned} \tag{5.57}$$

Whether A · S itself reacts or whether it combines with a gas-phase molecule to yield product, (5.57) is mechanistically equivalent to the Michaelis–Menton mechanism (5.51); it differs from the mechanism (5.49) in that the adsorption process is *reversible*. The surface sites replace enzyme molecules as catalysts; if [A] is too large the surface is saturated. Adapting the arguments that led to (5.53) yields the Langmuir adsorption isotherm[41] which is the fraction of occupied sites $\theta \equiv [\mathrm{A} \cdot \mathrm{S}]/S_0$,

$$\theta = K[\mathrm{A}]/\{1 + K[\mathrm{A}]\} = bp/(1 + bp) \tag{5.58a}$$

where, assuming the ideal gas law applies,

$$b = K/kT \tag{5.58b}$$

Pursuing the enzymatic analogy, inhibition arises if another constituent of the gas is adsorbed also,

$$\mathrm{B} + \mathrm{S} \rightleftharpoons \mathrm{B} \cdot \mathrm{S} \qquad \text{(equilibrium, } K'\text{)}$$

The same analysis which led to (5.58) now indicates that

$$\theta_{\mathrm{A}} = K[\mathrm{A}]/\{1 + K[\mathrm{A}] + K'[\mathrm{B}]\} \tag{5.59}$$

where θ_{A} is the fraction of sites occupied by A.

[41] I. Langmuir, *J. Am. Chem. Soc.* **38**, 2221 (1916); **40**, 1361 (1918).

A variety of mechanisms for the reaction of adsorbate can be imagined:

$$A \cdot S \xrightarrow{k_2} \text{product} \tag{5.60a}$$

$$A \cdot S + B \xrightarrow{k_2} \text{product} \tag{5.60b}$$

$$A \cdot S + B \cdot S \xrightarrow{k_2} \text{product} \tag{5.60c}$$

In (5.60a) adsorbate undergoes a unimolecular transformation; an example is the formation of atomic H at the walls of a high-temperature reaction vessel containing H_2. In (5.60b) adsorbate reacts with a gas molecule as is found in some surface-catalyzed combinations of atoms and free radicals.[42] In (5.60c) adsorbed molecules on neighboring sites interact, as seems to be the case in the exchange reaction between D_2 and NH_3.[43] Corresponding to the mechanisms are the rate equations

$$R \equiv -\frac{d[A]}{dt} = \frac{k_2 K S_0 [A]}{1 + K[A] + K'[B]} \tag{5.61a}$$

$$= \frac{k_2 K S_0 [A][B]}{1 + K[A] + K'[B]} \tag{5.61b}$$

$$= \frac{k_2 K K' S_0 [A][B]}{(1 + K[A] + K'[B])^2} \tag{5.61c}$$

In every case a steady-state concentration of adsorbate is assumed. The pressure dependence of the reaction rates is extreme. For the case involving reaction of adsorbed species the rate varies from second order to zeroth order as surface coverage increases. If species A is preferentially adsorbed the reaction rate, at high [A], is

$$R = \frac{k_2 K'}{K} S_0 \frac{[B]}{[A]} \tag{5.62}$$

If a nonreactive inhibitor is added the situation is more complex (see Problem 5.19). For the cases considered desorption of product is presumed to be very fast. If it is not, there is product-induced inhibition which must be overcome for practical catalytic application (see Problem 5.18).

Reactions at surfaces are dependent on the phenomenon of adsorption. The treatment in the preceding paragraphs has assumed all surface sites

[42] One example is the surface-catalyzed recombination of H and OH, W. V. Smith, *J. Chem. Phys.* **11**, 110 (1943).

[43] J. Weber and K. J. Laidler, *J. Chem. Phys.* **19**, 1089 (1951).

are similar which is not true. Elegant experiments, in which various gases are used to etch polished metal spheres, have shown that different crystallographic surfaces are etched at very different rates.[44] In an extreme example, where copper was exposed to oxygen, the {100} face reacted 17 times as fast as the {311} face.[45] With the development of ultrahigh vacuum techniques ($p \lesssim 10^{-12}$ atm) it has become possible to study the details of the adsorption process. At such ultralow pressures, adsorption is slow enough that reproducible surface crystallographic studies of an adsorbed monolayer are feasible using low-energy electron diffraction (LEED).[46]

Depending upon the binding energy, adsorption is described as physical adsorption or chemisorption. In physical adsorption binding is weak (10–30 kJ mol^{-1}) and the adsorbate–substrate separation is large, typical of van der Waals interaction. In chemisorption, binding is stronger (>60 kJ mol^{-1}) and the separations are short, typical of chemical bonds. The surface structures formed by physically adsorbed and chemisorbed species differ markedly.

Physical adsorption of inert gases on metals is studied at temperatures between 10 and 78 K. At too high a temperature the adsorbed layer boils off the surface. At too low a temperature an adsorbed gas molecule does not migrate on the surface after striking it; the surface structure is random and does not anneal to reflect the energetics of adsorbate–substrate interaction. In the temperature range for which surface equilibration occurs, the structure of the surface layer is independent of the inert gas adsorbed and of the metal surface exposed. Whether Xe is adsorbed on graphite,[47] Pd,[48] Ir,[49] or Cu[50] the surface layer develops a structure of hexagonal symmetry, an arrangement which corresponds to the closest-packed plane of crystalline Xe. Similar results are found in the binding of Ne or Ar to the {100} face of Nb.[51] In each instance the structure of the surface layer is determined by adsorbate–adsorbate interaction; the adsorbate–substrate binding is too weak to have a significant effect.[52]

[44] A. T. Gwathney and R. E. Cunningham, *Adv. Catal.* **10**, 57 (1958).
[45] R. W. Young, Jr., J. V. Cathcart, and A. T. Gwathney, *Acta Met.* **4**, 145 (1946).
[46] For an introductory review see J. C. Buchholz and G. A. Somorjai, *Acc. Chem. Res.* **9**, 333 (1976).
[47] J. J. Lander and J. Morrison, *Surf. Sci.* **6**, 1 (1967).
[48] P. W. Palmberg, *Surf. Sci.* **25**, 598 (1971).
[49] A. Ignatiev, A. V. Jones, and T. N. Rhodin, *Surf. Sci.* **30**, 573 (1972).
[50] M. A. Chesters and A. Pritchard, *Surf. Sci.* **28**, 460 (1971).
[51] J. M. Dickey, H. H. Farrell, and M. Strongin, *Surf. Sci.* **23**, 448 (1970).
[52] The surface layer is oriented in a limited number of ways with respect to the underlying metal, indicating some specificity in the adsorbate–adsorbent interaction.

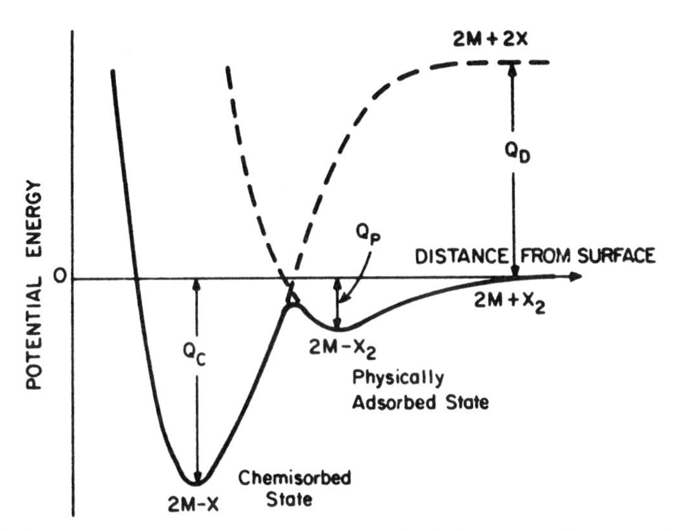

Figure 5.7. Schematic diagram for the variation of potential energy as a function of distance from a metal surface for dissociative adsorption of a molecule X_2. Q_P is the heat of physical adsorption for the molecule, Q_C is the heat of chemisorption, and Q_D is the heat of dissociation of the free molecule. The binding energy per chemisorbed atom is $(Q_D + Q_C)/2$. Note that for $|Q_C|$ sufficiently small, chemisorption no longer occurs. [T. N. Rhodin and D. L. Adams, *Treatise on Solid State Chemistry*, N. B. Hannay, ed. (New York: Plenum Press, 1976), Vol. 6A, p. 346, reprinted by permission.]

Catalysis, which typically requires bond rupture of the adsorbate (e.g., $H_2 \rightharpoonup 2H$), is a problem in chemisorption. Molecular dissociation is described as a two-step process

$$\begin{aligned} &M + X_2 \rightharpoonup M--X_2 \quad \text{(physical adsorption)} \\ &M--X_2 \rightharpoonup 2M - X \quad \text{(chemisorption)} \end{aligned} \tag{5.63}$$

In the first step X_2 is weakly bound at large separation; if M—X interaction is sufficiently large, dissociation occurs and a chemisorbed species is formed. The energetics of the model, with its two distinct steps, is illustrated in Fig. 5.7.[53] The binding energies of the chemisorbed species are large; as a result the equilibrium surface structure reflects the periodicity of the metal surface. Binding is very site specific and chemisorbed atoms, to maximize their interaction with substrate atoms, occupy positions of high coordination number.[54] Bonding appears to be basically covalent. A comparison of

[53] The model was proposed by J. E. Lennard-Jones, *Trans. Faraday Soc.*, **28**, 28 (1932).

[54] S. Anderson and J. B. Pendry, *J. Phys. C.* **5**, L41 (1972); J. E. Demath, D. W. Jepsen, and P. M. Marcus, *Phys. Rev. Lett.* **31**, 540 (1973); **32**, 1182 (1974).

adsorbate–metal atom distances with those estimated by summing covalent radii indicate few significant differences.[55]

Nondissociative molecular chemisorption has been less exhaustively studied. In the one case where a full LEED analysis is available, it appears that geometric constraints are very significant. Acetylene binds to the Pt{111} face at a triangular site[56]; here the π-orbitals interact effectively with exposed Pt atoms without molecular deformation being required.[57]

Chemisorption is too effective if adsorbate–substrate binding is stronger than metal atom binding at the surface. In such cases the adsorbate atom is actually incorporated in a reconstructed metal surface so that adsorbate and metal atoms are intermixed in the surface layer. Such processes destroy catalytic activity. They may be the first step in a solid-state reaction leading to oxide, carbide, or nitride formation. Surface reconstruction has been observed when C or O are chemisorbed on W,[58] Ni,[59] or Fe[60] at high temperatures.

Modern surface chemistry remains *terra incognita.* Recent work, designed to identify surface features that promote catalytic activity, suggests that there are significant differences between smooth and stepped surfaces of transition metals.[61] On smooth Pt surfaces, like the {100} and {111} faces, neither H_2 nor O_2 chemisorb readily; however, on staircaselike surfaces both appear to chemisorb (and dissociate) at relatively low temperatures.[62] Similar effects may control the rate at which C—C and C—H bonds are broken using Pt catalyst.[63] The interior and exposed edges of stepped surfaces may bind molecules preferentially.[64] Other discontinuities which affect the electron densities of the surface atoms could also effect their binding properties. As a result, dislocations might alter catalytic afficiency.[65]

(55) J. C. Buchholz and G. A. Somorjai, *Acc. Chem. Res.* **9**, 333 (1976).
(56) L. L. Kesmodel, P. C. Stair, R. C. Baetzold, and G. A. Somorjai, *Phys. Rev. Lett.* **36**, 1316 (1976).
(57) Such structures need not be simple. Recent work indicates that benzene, when bound on the Pt {111} face, does not bond parallel to the surface but is tilted [P. C. Stair and G. A. Somorjai, *J. Chem. Phys.* **67**, 4361 (1977)].
(58) M. Boudart and D. F. Ollis, *The Structure and Chemistry of Solid Surfaces*, G. A. Somorjai, ed. (New York: John Wiley, 1969).
(59) A. U. MacRae, *Science* **139**, 379 (1963).
(60) J. E. Boggio and H. E. Farnsworth, *Surf. Sci.* **3**, 62 (1964).
(61) B. Lang, R. W. Joyner, and G. A. Somorjai, *J. Catal.* **27**, 405 (1972).
(62) J. C. Buchholz and G. A. Somorjai, *Acc. Chem. Res.* **9**, 333 (1976).
(63) J. C. Buchholz and G. A. Somorjai, *Acc. Chem. Res.* **9**, 333 (1976).
(64) B. Lang, R. W. Joyner, and G. A. Somorjai, *Surf. Sci.* **30**, 454 (1972); M. Salmeron, R. J. Gale, and G. A. Somorjai, *J. Chem. Phys.* **67**, 5324 (1977).
(65) K. Besocke and H. Wagner, *Surf. Sci.* **53**, 351 (1975).

Appendix A. Matrix Solution to the Coupled Rate Equation (5.2)

The set of coupled equations (5.2) can be readily solved using matrix methods.[66] The advantage of this approach is that it can easily be generalized to treat any number of coupled first-order (or pseudo-first-order) rate equations. For simplicity consider a two-step process and define a column vector constructed from the concentrations

$$\mathbf{X} = \begin{pmatrix} \mathrm{A} \\ \mathrm{B} \\ \mathrm{C} \end{pmatrix} \tag{5A.1}$$

Then (5.2) can be written as

$$\frac{d\mathbf{X}}{dt} = \mathbf{K} \cdot \mathbf{X} \tag{5A.2}$$

where the corresponding rate constant matrix is

$$\mathbf{K} = \begin{pmatrix} -k_+ & k_- & 0 \\ k_+ & -k_- - k_+' & k_-' \\ 0 & k_+' & -k_-' \end{pmatrix} \tag{5A.3}$$

From the results of matrix algebra it is known that a transformation matrix $\mathbf{T}$ can always be constructed to generate a new vector

$$\mathbf{Y} = \mathbf{T} \cdot \mathbf{X} \tag{5A.4}$$

which has the property that

$$\frac{d\mathbf{Y}}{dt} = -\mathbf{\Lambda} \cdot \mathbf{Y} \tag{5A.5}$$

where $\mathbf{\Lambda}$ is *diagonal*, i.e.,

$$\mathbf{\Lambda} = \begin{pmatrix} \lambda_1 & 0 & 0 \\ 0 & \lambda_2 & 0 \\ 0 & 0 & \lambda_3 \end{pmatrix} \tag{5A.6}$$

The transformation $\mathbf{T}$ has decoupled the rate equations and the components

[66] See, e.g., H. Margenau and G. M. Murphy, *The Mathematics of Physics and Chemistry* (Princeton: Van Nostrand, 1956), Chapter 10.

of (5A.5) are the simple first-order equations

$$\frac{dy_i}{dt} = -\lambda_i y_i$$

which have the solution

$$y_i(t) = y_i(0) \exp(-\lambda_i t) \tag{5A.7}$$

so that the λ_i are the decay constants.

The rules for determining $\mathbf{\Lambda}$ and $\mathbf{T}$ are well known; $\mathbf{\Lambda}$ is found by solving the algebraic equation which is formed by expanding the determinant

$$|\mathbf{K} + \lambda \mathbf{I}| = 0 \tag{5A.8}$$

where $\mathbf{I}$ is the unit matrix,

$$\mathbf{I} = \begin{pmatrix} 1 & 0 & 0 \\ 0 & 1 & 0 \\ 0 & 0 & 1 \end{pmatrix}$$

The roots of (5A.8) are just the λ_i. To determine $\mathbf{T}$ multiply (5A.2) by $\mathbf{T}$ which yields, with (5A.4),

$$\frac{d\mathbf{Y}}{dt} = \mathbf{T} \cdot \mathbf{K} \cdot \mathbf{X}$$

Introducing (5A.4) into (5A.5) we find another relation between $d\mathbf{Y}/dt$ and $\mathbf{X}$

$$\frac{d\mathbf{Y}}{dt} = -\mathbf{\Lambda} \cdot \mathbf{T} \cdot \mathbf{X}$$

Comparing these two equations yields a matrix equation for $\mathbf{T}$

$$\mathbf{T} \cdot \mathbf{K} = -\mathbf{\Lambda} \cdot \mathbf{T} \tag{5A.9}$$

which is really a set of simultaneous linear equations in the unknown elements of the $\mathbf{T}$ matrix, t_{ij}. Introducing the expression for $\mathbf{\Lambda}$, (5A.6), this becomes

$$\sum_m t_{im} k_{mj} = -\lambda_i t_{ij} \tag{5A.10}$$

Thus once the λ_i are determined from (5A.8) the t_{ij} can be found from (5A.10).

In most instances only the decay constants are of interest. Combining (5A.3) and (5A.8) and expanding the determinant, the equation which determines the decay constants for the rate equations (5.2) is

$$\lambda^3 - (k_+ + k_- + k_+{}' + k_-{}')\lambda^2 + (k_+k_+{}' + k_-k_-{}' + k_+k_-{}')\lambda = 0$$

This has three roots, one obviously $\lambda = 0$, and the others are found by applying the quadratic formula. Defining the quantities

$$\sigma = k_+ + k_-, \qquad \sigma' = k_+{}' + k_-{}'$$

the roots are

$$\begin{aligned} \lambda_1 &= 0 \\ \lambda_2 &= \{\sigma + \sigma' - [(\sigma - \sigma')^2 + 4k_-k_+{}']^{1/2}\}/2 \\ \lambda_3 &= \{\sigma + \sigma' + [(\sigma - \sigma')^2 + 4k_-k_+{}']^{1/2}\}/2 \end{aligned} \tag{5A.11}$$

The root $\lambda = 0$ is a restatement of the condition of conservation of mass. From (5A.7) a certain linear combination of concentrations, $y_i(t)$, is constant. The other roots determine the temporal behavior of the system; in terms of the notation of Section 5.2,

$$\lambda_2 = \varkappa_-, \qquad \lambda_3 = \varkappa_+$$

and $\varkappa_-$, being the smaller decay constant, governs the long-term behavior of the system.

If, for some reason, the concentrations $x_i(t)$ are also needed, they may readily be computed. From (5A.4) we obtain

$$\mathbf{X} = \mathbf{T}^{-1} \cdot \mathbf{Y} \equiv \mathbf{S} \cdot \mathbf{Y} \tag{5A.12}$$

so that with (5A.7)

$$x_i(t) = \sum_j s_{ij} y_j(0) \exp(-\lambda_j t) \tag{5A.13}$$

To find $y_j(0)$ apply (5A.4) at $t = 0$

$$y_j(0) = \sum_k t_{jk} x_k(0)$$

and using (5A.13) we have

$$x_i(t) = \sum_{j,k} s_{ij} \exp(-\lambda_j t) t_{jk} x_k(0) \tag{5A.14}$$

or, in matrix notation,

$$\mathbf{X}(t) = \mathbf{T}^{-1} \cdot \mathbf{F}(t) \cdot \mathbf{T} \cdot \mathbf{X}(0) \tag{5A.15}$$

where

$$[\mathbf{F}(t)]_{ij} = \begin{cases} \exp(-\lambda_j t), & i = j \\ 0, & i \neq j \end{cases} \tag{5A.16}$$

Appendix B. Approximations to $\varkappa_+$ and $\varkappa_-$

If $\sigma \gg \sigma'$, (5.6) or (5A.11) can be greatly simplified. We can write

$$\varkappa_\pm = \frac{\sigma}{2}\left\{\left(1 + \frac{\sigma'}{\sigma}\right) \pm \left[1 - \frac{2\sigma'}{\sigma} + \left(\frac{\sigma'}{\sigma}\right)^2 + \frac{4k_-k_+'}{\sigma^2}\right]^{1/2}\right\} \tag{5B.1}$$

Since the terms under the square root are much less than 1 we can use the approximation

$$(1 + x)^{1/2} = 1 + \frac{x}{2} - \frac{x^2}{8} + \cdots \tag{5B.2}$$

and find that quantity to be

$$1 + \frac{1}{2}\left[-\frac{2\sigma'}{\sigma} + \left(\frac{\sigma'}{\sigma}\right)^2 + \frac{4k_-k_+'}{\sigma^2}\right] - \frac{1}{8}\left(-\frac{2\sigma'}{\sigma}\right)^2 \tag{5B.3}$$

The second term in this expansion is needed because there are quadratic terms in the quantity being approximated. Consolidating (5B.3) yields

$$1 - \frac{\sigma'}{\sigma} + \frac{2k_-k_+'}{\sigma^2}$$

which may be combined with (5B.1)

$$\begin{aligned} \varkappa_+ &\approx \frac{\sigma}{2}\left(1 + \frac{\sigma'}{\sigma} + 1 - \frac{\sigma'}{\sigma} + \frac{2k_-k_+'}{\sigma^2}\right) = \sigma \\ \varkappa_- &\approx \frac{\sigma}{2}\left(1 + \frac{\sigma'}{\sigma} - 1 + \frac{\sigma'}{\sigma} - \frac{2k_-k_+'}{\sigma^2}\right) = \sigma' - \frac{k_-k_+'}{\sigma} \end{aligned} \tag{5B.4}$$

Since $\sigma' \ll \sigma$ by hypothesis, the term k_-k_+'/σ satisfies the following inequalities

$$\frac{k_-k_+'}{\sigma} < \frac{k_+'\sigma}{\sigma} = k_+' < \sigma'$$

and thus has been neglected in comparison to σ in $\varkappa_+$.

The approximation to (5.6) when k_+' or k_- (or both) are large is found in a similar way. Define

$$q = k_+' + k_-, \qquad r = k_+ + k_-'$$

so that with (5.7) the expressions for the decay constants are

$$\varkappa_\pm = \frac{q}{2}\left\{1 + \frac{r}{q} \pm \left[1 - \frac{2r}{q} + \frac{r^2}{q^2} + \frac{4(k_+k_- + k_+'k_-' - k_+k_-')}{q^2}\right]^{1/2}\right\} \tag{5B.5}$$

Approximating the square root using (5B.2) yields

$$1 - \frac{r}{q} + \frac{2(k_+k_- + k_+'k_-' - k_+k_-')}{q^2}$$

Combining this with (5B.5) leads to

$$\varkappa_+ = q$$

$$\begin{aligned} \varkappa_- &= \frac{q}{2}\left[1 + \frac{r}{q} - 1 + \frac{r}{q} - \frac{2(k_+k_- + k_+'k_-' - k_+k_-')}{q^2}\right] \\ &= \frac{rq - k_+k_- - k_+'k_-' + k_+k_-'}{q} \approx \frac{k_+k_+' + k_-k_-'}{k_+' + k_-} \end{aligned} \tag{5B.6}$$

The term $k_+k_-'/(k_+' + k_-)$ has been dropped in the expression for $\varkappa_-$ because of the inequality

$$k_+k_+' + k_-k_-' \gg k_+k_-'$$

which must be satisfied since, by hypothesis,

$$\text{either } k_+' \text{ or } k_- \gg \text{both } k_+ \text{ and } k_-'$$

Problems

5.1. Assume a two-step mechanism of the form (5.1) for which $k_+ = 10^6\ \text{sec}^{-1}$, $k_- = 10^3\ \text{sec}^{-1}$, $k_+' = 10^3\ \text{sec}^{-1}$, and $k_-' = 1\ \text{sec}^{-1}$. Assume initially only A is present.

(a) Calculate $\varkappa_+$ and $\varkappa_-$ using (5.6).

(b) Determine A(t), B(t), and C(t) by substituting expressions of the form $\text{A}(t) = a_+ \exp(-\varkappa_+ t) + a_- \exp(-\varkappa_- t) + a_{\text{eq}}$ into the rate equations.

5.2. The reversible decomposition of phosgene $COCl_2(g)$ into $CO(g)$ and $Cl_2(g)$ has been interpreted according to two distinct mechanisms[67]

$$Cl_2 = 2Cl \qquad \text{(equilibrium, } K_1\text{)}$$

$$\text{(i)} \quad Cl + CO = COCl \qquad \text{(equilibrium, } K_2\text{)}$$

$$COCl + Cl \underset{k_{-3}}{\overset{k_3}{\rightleftharpoons}} COCl_2$$

$$\text{(ii)} \quad Cl + Cl_2 = Cl_3 \qquad \text{(equilibrium, } K_4\text{)}$$

$$Cl_3 + CO \underset{k_{-5}}{\overset{k_5}{\rightleftharpoons}} COCl_2 + Cl$$

Derive the expression for the rate of formation of $COCl_2$ according to both mechanisms. Are they kinetically distinguishable?

5.3. The activation energy for the reaction

$$2NO + O_2 = 2NO_2$$

is -41 kJ. Suggest a mechanism to account for this.

5.4. Bimolecular association reactions involving simple species generally have Arrhenius A-factors much less than the collision frequency. Interpret this in terms of the discussion in Section 5.5. [For typical A-factors see S. W. Benson, *The Foundations of Chemical Kinetics* (New York: McGraw–Hill, 1960), p. 302.]

5.5. The pyrolysis of ethane may be accounted for by the Rice–Herzfeld mechanism [F. O. Rice and K. Herzfeld, *J. Amer. Chem. Soc.* **56**, 284 (1934)]:

$$\text{(i)} \quad C_2H_6 \rightarrow 2CH_3 \qquad \text{(initiation)}$$

$$\left.\begin{array}{ll} \text{(ii)} & CH_3 + C_2H_6 \rightarrow CH_4 + C_2H_5 \\ \text{(iii)} & C_2H_5 \rightarrow C_2H_4 + H \\ \text{(iv)} & H + C_2H_6 \rightarrow C_2H_5 + H_2 \end{array}\right\} \text{(propagation)}$$

$$\text{(v)} \quad H + C_2H_5 \rightarrow C_2H_6 \qquad \text{(termination)}$$

(a) Show that $d[C_2H_4]/dt \propto d[H_2]/dt \propto -d[C_2H_6]/dt \propto [C_2H_6]$.
(b) How can this mechanism be *experimentally* distinguished from unimolecular decomposition?
(c) Use the bond energies for C—C, C=C, C—H, and H—H (342, 614, 415, 435 kJ mol^{-1}, respectively) to estimate lower bounds for the activation energy in each step.
(d) Use these estimates to show that

$$-d[C_2H_6]/dt = d[H_2]/dt = d[C_2H_4]/dt = (k_1k_3k_4/k_5)^{1/2}[C_2H_6]$$

(e) Estimate the apparent activation energy in the overall rate law.

[67] After I. Amdur and G. G. Hammes, *Chemical Kinetics* (New York: McGraw–Hill, 1966), p. 84.

5.6. Another pyrolysis whose kinetics is described by the Rice–Herzfeld mechanism is the decomposition of acetaldehyde:

$$\begin{array}{lll} \text{(i)} & CH_3CHO \rightarrow CH_3 + CHO & \text{(initiation)} \\ \text{(ii)} & CH_3 + CH_3CHO \rightarrow CH_4 + CH_2CHO & \multirow{2}{*}{(propagation)} \\ \text{(iii)} & CH_2CHO \rightarrow CO + CH_3 & \\ \text{(iv)} & 2CH_3 \rightarrow C_2H_6 & \text{(termination)} \end{array}$$

(a) Determine the rate laws for the formation of the *stable* products and the decomposition of acetaldehyde.
(b) Use the bond energy data of Problem 5.5 and estimate the activation energy for each of the apparent rate constants determined in (a).

5.7. In the formation of HBr, termination steps involving atomic hydrogen were ignored because [H] is small. Make whatever approximations you feel are reasonable and demonstrate the truth of this assertion.

5.8. In the Christiansen–Herzfeld–Polanyi mechanism only *chemical* processes were included. In addition to recombination of Br atoms, the chain may be terminated by adsorption at the walls of the reaction vessel. Consider the additional steps in (5.28)

$$Br \xrightarrow{k_6} \text{adsorbed atom}$$

$$H \xrightarrow{k_7} \text{adsorbed atom}$$

and derive the analogue to (5.34). What reaction conditions would emphasize adsorption?

5.9. The observed second-order rate constant for the reaction

$$\text{semicarbazide} + RCHO \rightarrow \text{semicarbazone}$$

is pH dependent. At low pH, $k_{obs} = k_0 + 10^{4.33}[H^+]\ M^{-1}\ sec^{-1}$ while at high pH, $k_{obs} = k_0' + 10^{6.13}[H^+]\ M^{-1}\ sec^{-1}$ for reaction with *p*-chlorobenzaldehyde [E. H. Cordes and W. P. Jencks, *J. Amer. Chem. Soc.* **84**, 4319 (1962)]. The equilibrium constant for the protonation reaction

$$RCHO + H^+ = RCHOH^+$$

is $K = 10^{-7.26}$.

(a) Assume the slow step in the reaction is

$$\text{semicarbazide} + RCHOH^+ \xrightarrow{k_2} \text{protonated semicarbazone}$$

and show that the pH-dependent term in k_{obs} would be $k_2K[H^+]$.
(b) Evaluate the k_2 required if the pathway in (a) is important at low pH, high pH. Compare your values with those for diffusion limited reactions. Is the pathway in (a) tenable?

5.10. A proposed mechanism for the $I^- + ClO^-$ exchange reaction is

$$ClO^- + H_2O = HClO + OH^- \quad \text{(equilibrium, } K_1\text{)}$$

$$HClO + I^- \underset{k_{-2}}{\overset{k_2}{\rightleftharpoons}} HIO + Cl^-$$

$$HIO + OH^- = IO^- + H_2O \quad \text{(equilibrium, } K_3\text{)}$$

(a) Show that

$$d[Cl^-]/dt = \{k_2K_1[ClO^-][I^-] - k_{-2}K_3^{-1}[IO^-][Cl^-]\}/[OH^-]$$

(b) At 298 K, $K_1 = 3.1 \times 10^{-7}$ and $K_3^{-1} = 4.3 \times 10^{-4}$. The observed rate law in basic media is

$$d[Cl^-]/dt = \varkappa[ClO^-][I^-]/[OH^-], \qquad \varkappa = 60 \text{ sec}^{-1}$$

What can be deduced about k_2 and k_{-2}?

5.11. Suggest a *plausible* mechanism for each of the following reactions based upon the rate law. All are in aqueous solution. In some instances there may be more than one reasonable mechanism.[68]

(a) [J. H. Baxendale and P. George, *Trans. Faraday Soc.* **46**, 736 (1950).]

$$Fe(H_2O)_6^{3+} + 3\text{bipy(ridine)} = Fe(\text{bipy})_3^{3+} + 6H_2O$$
$$d[Fe(\text{bipy})_3^{3+}]/dt = k[Fe(H_2O)_6^{3+}][\text{bipy}]^3$$

(b) [A. B. Hoffman and H. Taube, *Inorg. Chem.* **7**, 1971 (1968).]

$$(NH_3)_5CoO_2Co(NH_3)_5^{4+} + 4V(II) + 14H^+ = 2Co(II) + 10NH_4^+ + 4V(III) + 2H_2O$$
$$-d[\text{Co complex}]/dt = k[\text{Co complex}]$$

(c) [O. J. Parker and J. H. Espenson, *Inorg. Chem.* **8**, 1523 (1969).]

$$Fe(III) + Cu(I) = Fe(II) + Cu(II)$$
$$-d[Fe(III)]/dt = k[Fe(III)][Cu(I)]/[H^+]$$

(d) [W. C. E. Higginson, D. R. Rosseinsky, J. B. Stead, and A. G. Sykes, *Disc. Faraday Soc.* **29**, 49 (1960).]

$$2Hg(II) + 2V(III) = Hg_2(I) + 2V(IV)$$
$$-d[V(III)]/dt = [Hg(II)][V(III)]^2\{(a[V(IV)] + b[V(III)])^{-1} + (c[V(IV)] + d[Hg(II)])^{-1}\}$$

[68] After R. G. Wilkins, *The Study of Kinetics and Mechanism of Reactions of Transition Metal Compounds* (Boston: Allyn and Bacon, 1974), pp. 117–118.

5.12. The N_2O_5-catalyzed decomposition of O_3 follows the unusual rate law

$$-d[O_3]/dt = k_a[O_3]^{2/3}[N_2O_5]^{2/3}$$

[H. J. Schumacher and G. Sprenger, *Z. Physik. Chem.* **A140**, 281 (1929); **B2**, 267 (1929)]. Recognizing that N_2O_5 is in equilibrium with NO_2 and NO_3 suggest a mechanism to account for the observed rate law. (The simplest one involves *two* binary kinetic steps in addition to the N_2O_5 equilibrium.)

5.13. When an inhibitor acts simultaneously competitively and uncompetitively, this is known as *noncompetitive* inhibition. Derive the analog to (5.55) and (5.56). Is this kinetically distinguishable?

5.14. The invertase-catalyzed hydrolysis of sucrose was studied by measuring the initial rate of reaction as a function of [sucrose] [A. M. Chase, H. C. V. Meier, and V. J. Menna, *J. Cell Comp. Physiol.* **59**, 1 (1962)]. The following data were obtained[69]:

[Sucrose] (M):	0.0292	0.0584	0.0876	0.117	0.146	0.175	0.234
Initial rate (M sec^{-1}):	0.182	0.265	0.311	0.330	0.349	0.372	0.371
Initial rate (M sec^{-1}) in $2M$ urea:	0.083	0.111	0.154	0.182	0.186	0.192	0.188

(a) Determine the apparent Michaelis constants and the saturation rate for the two sets of experiments {plot (initial rate)$^{-1}$ vs. [sucrose]$^{-1}$}.
(b) What is the nature of the urea inhibition? Estimate K_I and/or K_I'.

5.15. In general the reverse reaction

$$E + P \xrightarrow{k_{-2}} X$$

cannot be ignored in the Michaelis–Menton scheme. Show that if this is included in the mechanism (5.51) the reaction rate is

$$\frac{-d[S]}{dt} = \frac{d[P]}{dt} = \frac{(k_2[S]/K_m - k_{-1}[P]/K_m')E_0}{1 + [S]/K_m + [P]/K_m'}$$

5.16. The fumarase-catalyzed reaction

$$\text{fumarate} + \text{water} = l\text{-malate}$$

follows the general Michaelis–Menton mechanism of Problem 5.15. At 25°C, pH 7, and ionic strength of 0.01 the following parameters were found:

$$V_+/E_0 = 1.2\times10^3 \text{ sec}^{-1}, \qquad K_M = 4.7\times10^{-6}\ M$$
$$V_-/E_0 = 0.93\times10^3 \text{ sec}^{-1}, \qquad K_M' = 15.9\times10^{-6}\ M$$

[69] After W. J. Moore, *Physical Chemistry*, 4th ed. (Englewood Cliffs, N. J.: Prentice-Hall, 1972), pp. 418–419.

[R. A. Alberty and W. H. Pierce, *J. Am. Chem. Soc.* **79**, 1526 (1957)], where V_+ and V_- are the saturation velocities for the forward and reverse reactions, respectively. Compute ΔG_{obs} and as many of the four rate constants as the data permit.[70]

5.17. Demonstrate (5.58a) and (5.59).

5.18. Consider the possibility that product desorption is significant in catalysis. The simplest possible mechanism is

$$A + S \underset{k_{-1}}{\overset{k_1}{\rightleftharpoons}} A \cdot S$$

$$A \cdot S \xrightarrow{k} B \cdot S$$

$$B \cdot S \underset{k'_{-1}}{\overset{k_1'}{\rightleftharpoons}} B + S$$

Derive the rate law for the disappearance of A and compare with the result of Problem 5.15. Why do the results differ?

5.19. Consider the mechanism (5.60c) with the additional complication that there is an inhibitor present which adsorbs on the surface according to the reaction

$$In_2 + S \rightleftharpoons 2In \cdot S \qquad K_I$$

Derive the analogue to (5.61c). What is the apparent rate law when $[In_2]$ is large?

5.20. In Problem 4.20, three reverse rate laws were shown to be consistent with the stoichiometry and the forward rate law for the reaction

$$2Hg(II) + 2Fe(II) = Hg_2^{+2} + 2Fe(III)$$

Deduce possible mechanisms to account for each of the reverse rate laws.

5.21. Laser T-jump has been used to study the reaction

$$I_2 + I^- \underset{k_-}{\overset{k_+}{\rightleftharpoons}} I_3^-$$

[D. H. Turner, G. W. Flynn, N. Sutin, and J. V. Beitz, *J. Am. Chem. Soc.* **94**, 1554 (1972)]; $k_+ = 6\times10^9\ M^{-1}\ \text{sec}^{-1}$. Would you expect the rate constant for the reaction

$$I_2 + OH^- \rightarrow I_2OH^-$$

to be similar to, greater than, or less than k_+? Explain.

[70] After F. Daniels and R. A. Alberty, *Physical Chemistry*, 4th ed. (New York: John Wiley, 1975), p. 355.

5.22. Show that for the mechanism (5.1) under conditions where the steady state applies, the decay constant $\varkappa_-$ is significantly temperature dependent if
(a) at some T, $k_+' \approx k_-$ but $E_+' \neq E_-$.
(b) $A_+' \approx A_-$ and $E_+' \approx E_-$.

5.23. (a) Show that the ratio of ternary to binary collisions in a gas is $Z_3/Z_2 \approx d/\lambda$, where λ is the mean free path.
(b) Assuming all three-body collisions are reactive obtain a numerical upper bound for k_3 in a gas.
(c) At what pressure does termolecular reaction compete with bimolecular reaction, assuming in both cases that reaction occurs with each collision? To what density does this correspond?

General References

Gas-Phase Reactions

I. Amdur and G. G. Hammes, *Chemical Kinetics* (New York: McGraw–Hill, 1966), Chapter 3.
S. W. Benson, *The Foundations of Chemical Kinetics* (New York: McGraw–Hill, 1960), pp. 252–264, 290–305, 308–312, 319–424.
J. Nicholas, *Chemical Kinetics* (New York: Halstead, 1976), Chapters 6 and 7.
A. A. Westenberg, *Ann. Rev. Phys. Chem.* **24**, 77 (1973).

Solution Reactions

S. W. Benson, *The Foundations of Chemical Kinetics* (New York: McGraw–Hill, 1960), Chapter 14 and 15.
K. J. Laidler, *Chemical Kinetics*, 2nd ed. (New York: McGraw–Hill, 1965), Chapter 10.
R. G. Wilkins, *Kinetics and Mechanism of Reactions of Transition Metal Complexes* (Boston: Allyn and Bacon, 1974), Chapter 2.

Catalysis

I. Amdur and G. G. Hammes, *Chemical Kinetics* (New York: McGraw–Hill, 1966), Chapter 7.
K. J. Laidler, *Chemical Kinetics*, 2nd ed. (New York: McGraw–Hill, 1965), pp. 256–286, Chapter 9.
P. B. Weisz, *Ann. Rev. Phys. Chem.* **21**, 175 (1970).
J. T. Wong, *Kinetics of Enzyme Mechanisms* (London: Academic Press, 1975).

Other Topics

Crystallization Kinetics:

M. Kahlweit, *Ann. Rev. Phys. Chem.* **27**, 59 (1976).

Kinetics of Heterogeneous Reactions:

M. E. Wadsworth, *Ann. Rev. Phys. Chem.* **23**, 355 (1972).

6

Photochemistry

6.1. Introduction

In ordinary kinetic studies the required activation energy is introduced via intermolecular collisions. When the activation barrier is large, reaction must be run at high temperature or progress is imperceptible. Moreover, thermal activation is quite undiscriminating. Energy is partitioned among the various internal degrees of freedom of the reactants according to a Boltzmann distribution. There is no procedure to concentrate the energy of the reagents in the vibrational degree of freedom which most favors reaction. Furthermore, thermal processes always proceed by low-energy pathways. It is thus not possible, by heating, to select another mode of reaction requiring vastly greater activation energy.

Photochemical activation is a powerful kinetic tool. A large amount of energy is pumped into a molecule when it undergoes an electronic transition. Photoexcited molecules are far from equilibrium and their excess energy is lost as the system tries to equilibrate. Such molecules are exceptionally reactive. Irradiation may accelerate reaction; it may permit reaction by one (or many) new pathways.

A problem with ordinary photochemical activation is that too much energy is absorbed. As a consequence the possibilities for reaction are usually greatly augmented but there are often so many photochemical pathways that this approach can be quite unselective. A new method, based on *vibrational* excitation of specific reagent normal modes, gives promise of allowing the excitation energy to be introduced in such a way that enhances selectivity. By tuning a CO_2 laser to a specific absorption frequency of the molecule being studied the vibrational energy might be introduced in a way that favors a particular reactive pathway.

6.2. Measurement of Intensity

The fundamental photochemical variables are the wavelength λ and the intensity $I_0(\lambda)$ of the incident light. According to the Beer–Lambert law, light is attenuated as it passes through a sample. The intensity a distance x into the sample cell is

$$I(\lambda; x) = I_0(\lambda)e^{-x\gamma(\lambda)} \tag{6.1}$$

where

$$\gamma(\lambda) = \sum \alpha_i(\lambda)[Y_i] \tag{6.2}$$

$[Y_i]$ is the concentration of species i and $\alpha_i(\lambda)$ is the molar absorption coefficient of that species. The decrease in intensity for a cell of path length L is $I_a(\lambda) = I_0(\lambda) - I(\lambda; L)$ and the attenuation attributable to a particular species j is

$$I_a^{(j)}(\lambda) = \{\alpha_j(\lambda)[Y_j]/\gamma(\lambda)\}[I_0(\lambda) - I(\lambda; x)] \tag{6.3}$$

An ordinary spectrophotometer measures the relative attenuation of the beam $I(\lambda; L)/I_0(\lambda)$, *not* the absolute intensities of incident and transmitted light. However, in photochemical applications absolute intensity is important since the fundamental quantity of interest is the quantum yield φ_λ,

$$\varphi_\lambda = \frac{\text{rate of primary product formation}}{\left(\begin{array}{c}\text{rate of absorption of light quanta of wavelength } \lambda \\ \text{by the species of interest}\end{array}\right)} \tag{6.4}$$

Since the quantum flux J_λ is proportional to the intensity $J_\lambda = I(\lambda)/(hc_0/\lambda)$, where c_0 is the speed of light, the *absorption rate per unit volume*, $R_a(\lambda)$, for a beam of area A passing through a cell of volume V is

$$R_a(\lambda) = \frac{A}{V}\frac{\lambda}{hc_0}[I_0(\lambda) - I(\lambda; L)] = \frac{A\lambda I_a(\lambda)}{Vhc_0} \tag{6.5}$$

Accurate measurement of the *output* of a photochemical source may be particularly difficult since $I(\lambda; L)$ is often much less than I_0. Reliable measurements are made using *chemical actinometers*, systems whose photochemical conversion rates are accurately known as a function of wavelength. In the range 200 nm $< \lambda <$ 550 nm the ferrioxalate actinometer is an accurate quantum counter.[(1)] Irradiation of a solution of potassium

[(1)] For a discussion of actinometry see J. N. Demas, in *Creation and Detection of the Excited State*, W. R. Ware, ed. (New York: Dekker, 1976), Vol. 4, pp. 6–31.

ferrioxalate leads to the reduction

$$\text{Fe(III)} \xrightarrow{h\nu} \text{Fe(II)}$$

and the evolution of CO_2. The frequency-dependent quantum yield of Fe(II) is well known so that analysis for Fe(II), which can be done with great accuracy even at low [Fe(II)], determines the rate of absorption of light quanta from (6.4). Then, from (6.5), $I_a(\lambda)$ may be calculated and, since $I(\lambda; L)/I_0(\lambda)$ is known, one can determine the light intensity incident on the actinometer.

6.3. Spectroscopic Review

In conventional photokinetic studies, radiation effects an *electronic* transition from the ground state of the molecule to some electronically excited state. Coupled with the electronic transition there are changes in the vibrational and rotational state of the molecule. For a transition to occur the Bohr frequency condition, $\Delta\epsilon = h\nu = hc_0/\lambda$, must be obeyed. Absorption at a specific wavelength is determined by the molar absorption coefficient which, for photochemically interesting problems, is proportional to the square of the *electric dipole* transition moment

$$\alpha(\lambda) \propto \left| \int \psi_f^* \hat{\mathbf{M}} \psi_i \, d\tau \right|^2 \tag{6.6}$$

ψ_i and ψ_f are the wave functions for the initial and final states, respectively, and $\hat{\mathbf{M}}$ is the electric dipole moment operator.[2] The magnitude of the transition moment is thus a function of the initial and final states.

There are selection rules which determine whether the transition moment is nonzero.[3] In most photochemical applications three features of the wave functions are of major importance. The first is spin multiplicity. As long as the molecule to be photoexcited is composed of lighter atoms (atomic number $\lesssim 36$) spin conservation in a transition is a good approximation. Only singlet $\leftrightarrow$ singlet or triplet $\leftrightarrow$ triplet transitions are intense enough to be accessible photochemically; excitation is between states in

[2] For a precise determination of the relationship between $\alpha(\lambda)$ and the transition moment see W. Kauzmann, *Quantum Chemistry* (New York: Academic Press, 1957), pp. 577–583, 645–646.

[3] For an introductory treatment see I. N. Levine, *Molecular Spectroscopy* (New York: John Wiley, 1975), Chapter 7.

the singlet (or triplet) *manifold.*[4] In molecules containing heavier atoms spin conservation is no longer rigorous; it is less reliable as atomic number increases, breaking down totally for $Z \gtrsim 54$.

In addition to spin, the transition moment is strongly influenced by the geometric symmetry of the initial and final states. In molecules of high symmetry, e.g., benzene, quinone, oxygen, relatively few electronic transitions are allowed, even within a spin manifold. A particularly important restriction affects molecules with a center of symmetry. In these molecules the electronic wave function either changes sign (is odd) or remains unaltered (is even) when the positions of the nuclei are inverted through the center of symmetry. Only odd $\leftrightarrows$ even transitions are permitted.

While the occurrence of a photochemical excitation is fundamentally governed by electronic considerations, the shape of the absorption band and the nature of the primary photoproduct is determined by the vibrational part of the wave functions.[5] According to the Franck–Condon principle electronic transitions take place *without significant changes in nuclear structure.* A semiclassical justification follows from the fact that electrons are very light compared with nuclei. The electron distribution may thus be altered in an interval that is short compared to the time required for significant vibrational motion.[6]

In photochemical studies photoproducts are often characterized by their emission spectra. The same qualitative considerations govern emission and absorption; however, since emission does not require an energy input, it may occur spontaneously. The larger the population of excited-state molecules the greater the emission intensity. There are usually many ways to deexcite a species A*; each radiative process may be described in chemical terms as

$$A^* \xrightarrow{k(\lambda)} A + h\nu$$

where k is a wavelength-dependent rate constant. The analog to (6.5) for emission is the rate law

$$R_e(\lambda) = k(\lambda)[A^*] = \tau^{-1}(\lambda)[A^*] \tag{6.7}$$

[4] If, as in the case of NO, the ground state is a doublet, excitation is within the doublet manifold.

[5] Sometimes electronically forbidden transitions are vibronically allowed, i.e., the coupling of electronic and vibrational motion relaxes a symmetry constraint. Such transitions are never intense. In no case does vibronic coupling affect spin considerations.

[6] For an in-depth discussion of the Franck–Condon principle see G. Herzberg, *Spectra of Diatomic Molecules*, 2nd ed. (Princeton, N. J.: Van Nostrand, 1950), pp. 194–203, 420–432.

$\tau(\lambda)$ is known as the *intrinsic lifetime* for emission. When the transition dipole moment is large emission is rapid and τ is short; for strongly allowed transitions τ is in the range 1 μsec to 1 nsec. As the transition moment decreases τ becomes correspondingly larger until, for spin-forbidden transitions, intrinsic lifetimes of the order of seconds are not uncommon. Formulas have been developed that permit τ to be calculated if the transition is also observed in absorption.[7] If a transition is sharp, its intrinsic lifetime is determined by the extinction coefficient and the line width. If transition is from a broad band, as is common in photochemistry, it is no longer $k(\lambda)$ that is significant but a rate constant found by integrating over the whole absorption band. The extinction coefficient and band shape determine the corresponding τ. The Strickler–Berg result, valid for broad bands, is

$$k = \tau^{-1} = 1.07 \times 10^{-40} n^2 \langle \nu^{-3} \rangle^{-1} (g_l/g_u) \int \epsilon(\nu)\, d \ln \nu$$

where

$$\langle \nu^{-3} \rangle = \int I(\nu) \frac{1}{\nu^3}\, d\nu$$

Here n is the index of refraction of the medium, ν the frequency, g_l and g_u the degeneracies of the lower and upper electronic states, $\epsilon(\nu)$ the extinction coefficient, and $I(\nu)$ the *normalized* fluorescence intensity.

6.4. Primary Processes

Photoexcitation adds a substantial amount of energy to a molecule; in its decay to stable photoproducts all the excess energy must be accounted for. The distinction between primary and secondary relaxation processes is a bit arbitrary. Primary processes involve deexcitation or dissociation of the excited molecule. Secondary processes are the *chemical* reactions of the excited molecule or of its dissociation fragments. A simplified energy-level scheme (Jablonski diagram) illustrating excitation and some of the primary pathways for energy release is given in Fig. 6.1.

6.4.1. Vibrational Relaxation

Because of Franck–Condon requirements photoexcited molecules are usually produced in a distribution of vibrational states. Collisions induce

[7] S. J. Strickler and R. A. Berg, *J. Chem. Phys.* **37**, 814 (1962).

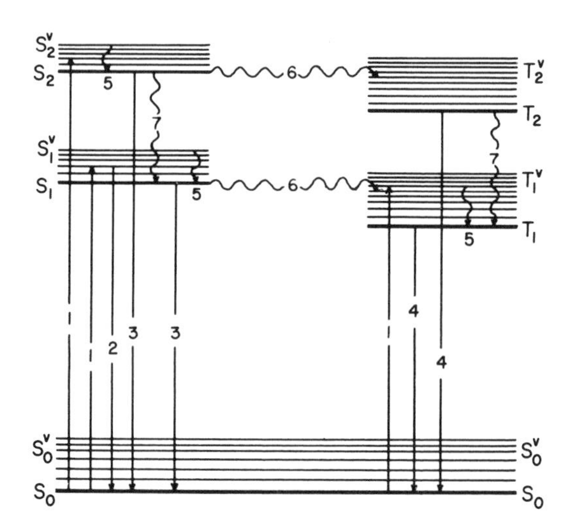

Fig. 6.1. A simplified Jablonski diagram illustrating some pathways for energy release. The ground state is a singlet, S_0. Excited states are either singlet, S_i, or triplet, T_i ($i \geq 1$). Vibrational excitation is indicated by a superscript v, i.e., $S_0{}^v$. Radiative processes are denoted by straight lines, radiationless ones by wavy lines. Those illustrated are: (1) absorption; (2) resonance radiation; (3) fluorescence; (4) phosphorescence; (5) vibrational relaxation; (6) intersystem crossing; (7) internal conversion followed by vibrational relaxation.

vibrational–translational (V–T) and vibrational–vibrational (V–V) energy transfer, leading, if other processes do not occur, to a Maxwellian distribution of energies among the vibrational levels of the excited electronic state. The V–T process is most inefficient, particularly if the excited molecule is small, because of the large changes in translational state that are required. The probability per collision for the process

$$CO(v{=}1) + Ar \rightarrow CO(v{=}0) + Ar$$

is $< 1.8 \times 10^{-8}$ at room temperature.[8] However, V–V energy transfer occurs much more easily, as few as 10–1000 collisions being required in near-resonant processes;[9] the more complex the molecules the more readily thermalization occurs by either pathway.

As a result, in solution where the collision frequency is $\sim 10^{13}\ sec^{-1}$,

[8] W. H. Green and J. K. Hancock, *J. Chem. Phys.* **59**, 4326 (1973).

[9] C. B. Moore, *Adv. Chem. Phys.* **23**, 41 (1973). The closer the vibrational changes are to being isoenergetic, the less the translational states must change. At resonance V–T transfer is not required.

excited molecules are vibrationally thermalized within their various electronic states before any other process occurs. In the gas phase large molecules provide their own heat bath and a redistribution of vibrational energy occurs after vibronic excitation. However, for small molecules at low pressures collisions are infrequent; the initial vibrational energy distribution does not change unless the intrinsic lifetime is long.

6.4.2. *Fluorescence and Phosphorescence*

The simplest way to release the excess energy is for the molecule to re-emit light. In *resonance radiation* the emitted and absorbed quanta have the same frequency; the molecule returns directly to its initial state. Except in dilute gases this is uncommon. More often there has been vibrational relaxation.[10] If both ground and excited state are within the same spin manifold the radiative lifetime is short, 1 nsec $\lesssim \tau \lesssim$ 1 μsec; the emission process is termed *fluorescence*. When the transition required for deexcitation does not conserve spin τ is much larger, 1 msec $\lesssim \tau \lesssim$ 10 sec; the process is *phosphorescence*. All three radiative processes are illustrated in Fig. 6.1.

In many photochemical problems the reactions of a photoexcited species are studied. As we shall see, in order to characterize the kinetics the measured lifetime *in the absence of reaction*, τ_0, must be known. To establish τ_0 one measures the quantum yield for fluorescence, $\varphi_f(\lambda)$, defined as

$$\varphi_f(\lambda) = R_f / R_a(\lambda) \tag{6.8}$$

where R_f is the total rate of fluorescence emission throughout the spectrum and $R_a(\lambda)$ is the absorption rate at the irradiation wavelength, (6.5). Even under the most favorable circumstances $\varphi_f(\lambda)$ is rarely unity. From Fig. 6.1 it is clear that excitation may be to any vibronic state. However, due to radiationless relaxation processes to be discussed later fluorescence (phosphorescence) from a state other than $S_1(T_1)$ is rarely seen. Assuming the kinetic pathways for deexcitation of an excited molecule A* can be represented as

$$\begin{aligned} &\mathrm{A}^*(S_1) \xrightarrow{k_f} \mathrm{A}(S_0) + h\nu_f \quad &&\text{(fluorescence)} \\ &\mathrm{A}^*(S_1) \xrightarrow{k'} \mathrm{A}(S_0) &&\text{(all other paths)} \end{aligned} \tag{6.9a}$$

[10] This process is often called *internal conversion*. However, we shall limit use of the latter term to describe radiationless singlet–singlet electronic deexcitation.

Table 6.1. Fluorescence Quantum Efficiencies of Selected Molecules

Molecule	Excitation wavelength (nm)	φ_f	Process reducing φ_f
Benzene ($p < 1.3 \times 10^{-3}$ atm)	~350	0.39[a]	Intersystem crossing
NO_2 (gas phase)	<398	0[b]	Predissociation
Anthracene (benzene solution)	~350	0.26[c]	Intersystem crossing

[a] E. M. Anderson and G. B. Kistiakowsky, *J. Chem. Phys.* **48**, 4787 (1968); A. E. Douglas and C. W. Mathews, *J. Chem. Phys.* **48**, 4788 (1968).
[b] R. G. W. Norrish, *J. Chem. Soc.* **1929**, 1611.
[c] Quoted in W. C. Gardiner, *Rates and Mechanisms of Chemical Reaction* (New York: Benjamin, 1969), p. 262.

while the excitation kinetics are

$$A(S_0) + h\nu_a \xrightarrow{k_a} A^*(S_1{}^v) \xrightarrow[\text{relaxation}]{\text{rapid vib.}} A^*(S_1) \tag{6.9b}$$

the steady-state population of $A^*(S_1)$ is

$$[A^*(S_1)] = [k_a/(k_f + k')][A] \tag{6.10}$$

If there is little attenuation of the exciting radiation, the absorption rate $R_a = k_a[A]$ so that with (6.7) and (6.8) the fluorescence quantum efficiency is

$$\varphi_f = k_f[A^*(S_1)]/k_a[A] = k_f/(k_f + k') \tag{6.11}$$

Since the transition probabilities for spontaneous emission and stimulated absorption are proportional[(11)] and since $k_a(\lambda)$ is related to the absorption coefficient $\gamma(\lambda)$,[(12)] the absorption spectrum can be used to establish k_f. The fluorescence quantum yield then determines k'. The measured lifetime in the particular experiment is, using the definition (6.7),

$$\tau_0 = (k_f + k')^{-1} \tag{6.12}$$

Some quantum efficiencies are given in Table 6.1. Note that even at low pressures molecules may be deexcited without fluorescing.

[(11)] W. Kauzmann, *Quantum Chemistry* (New York: Academic Press, 1957), pp. 647–650.
[(12)] J. P. Simons, *Photochemistry and Spectroscopy* (London: Wiley-Interscience, 1971), p. 174.

6.4.3. Radiationless Conversion

From Fig. 6.1 it is apparent that molecules in different vibronic states may have the same energy. For example, the ground vibrational state of S_1 is nearly isoenergetic with a high vibrational state (possibly in the continuum) of T_1. Similarly $S_2(v=0)$ is isoenergetic with S_1^v (v large); again, the vibronic state of S_1 may be in the continuum. Since energy is conserved, it is possible for excited molecules to undergo nonradiative transitions such as

$$\begin{aligned} S_1(v=0) &\rightharpoonup T_1^v(v \text{ large}) \\ S_2(v=0) &\rightharpoonup S_1^v(v \text{ large}) \end{aligned} \tag{6.13}$$

The first, involving a spin change, is called *intersystem crossing*; the second, conserving spin, is *internal conversion*. Since vibrational relaxation from highly excited vibrational states is relatively efficient (the separation of adjacent levels being small), collisional deactivation rapidly makes the reverse processes nonresonant. Both processes of (6.13) compete with fluorescence and reduce φ_f.

The considerations governing the rate of internal conversion and that of intersystem crossing are similar.[13] These rates are proportional to the density of vibronic levels in the final state $\varrho_f(E)$ and the square of the interaction matrix element that couples the two vibronic states, V,

$$k = 2\pi V^2 \varrho_f(E)/h \tag{6.14}$$

Here V is computed as

$$V = \int d\tau \psi_i^* \hat{H} \psi_f$$

where $\hat{H}$ is the complete Hamiltonian of the system and ψ_i and ψ_f are the wave functions of the initial and final vibronic states.[14] In polyatomic molecules $\varrho_f(E)$ may be extremely large since there are many possible

[13] M. Bixon and J. Jortner, *J. Chem. Phys.* **48**, 715 (1968).

[14] Radiationless transitions occur even in dilute gases indicating that such processes are possible in *isolated* molecules. If a molecule were excited to a *stationary* state of the Hamiltonian the matrix element V would, of necessity, be zero. However, excitation at a precise energy is not possible unless exceptionally monochromatic radiation is used (as could be the case with laser excitation). Hence, the energetic distinction between nearly degenerate vibronic states is lost if the line width of the exciting radiation is sufficiently large. As a result transition may be to a nonstationary state. See A. H. Zewail, T. E. Orlowski, and K. E. Jones, *Proc. Nat. Acad. Sci. U.S.A.* **74**, 1310 (1977).

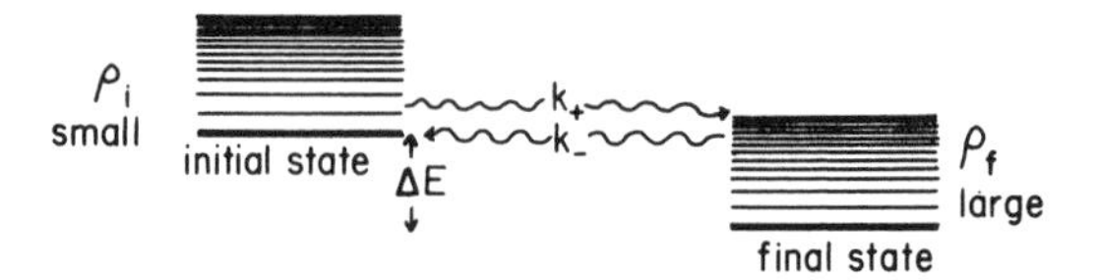

Fig. 6.2. Radiationless conversion in a molecule with a high density of vibronic states. As $\varrho_f \gg \varrho_i$, $k_+ \gg k_-$. See text for estimates of ϱ_f as a function of the number of atoms in a molecule.

vibrational modes. A prototype calculation assuming conversion to a vibronic state with 96 kJ mol^{-1} vibrational energy where the mean vibrational separation is 12 kJ mol^{-1} was carried out for molecules with different numbers of atoms. Varying N from 3 to 4 to 5 to 6 to 10 increases $\varrho(E)$ by the factors of 1 to 70 to 800 to 7000 to 7×10^6.[15]

An energy level diagram for radiationless conversion is illustrated in Fig. 6.2. Since $\varrho_f(E)$ increases and V decreases with increasing E, the dependence of the rate constant k on the separation energy ΔE is not immediately apparent. It is usually found that transition is more likely when ΔE is small. A case in point is intersystem crossing in anthracene for which the energy level diagram is illustrated in Fig. 6.3. The transition $S_1 \rightharpoonup T_2{}^v$ is favored over the transition $S_1 \rightharpoonup T_1{}^v$ even though $\varrho(E)$ for the latter case is much larger.[16]

6.4.4. Dissociation

Figure 6.4 illustrates for a diatomic molecule, the possible excited-state potential energy curves that permit dissociation. As mentioned previously (the Franck–Condon principle) electronic transitions occur so rapidly that the nuclei are, in effect, clamped during transition.[17] Quantum mechanics requires that the transition occur between states which have large probability amplitudes at the same internuclear separation. The vibrational wave function of the ground state is maximum at the center of the well; those of excited states have major maxima near the classical

[15] M. Bixon and J. Jortner, *J. Chem. Phys.* **48**, 715 (1968).

[16] As quoted in J. P. Simons, *Photochemistry and Spectroscopy* (London: Wiley-Interscience, 1971), p. 185.

[17] If there are nearly degenerate vibronic states, the concept of a well-defined nuclear configuration is not precise. This, as already mentioned, is important in understanding radiationless pathways in isolated molecules. For an analysis of the applicability of the idea of "molecular structure," see R. G. Woolley, *J. Am. Chem. Soc.* **100**, 1073 (1978).

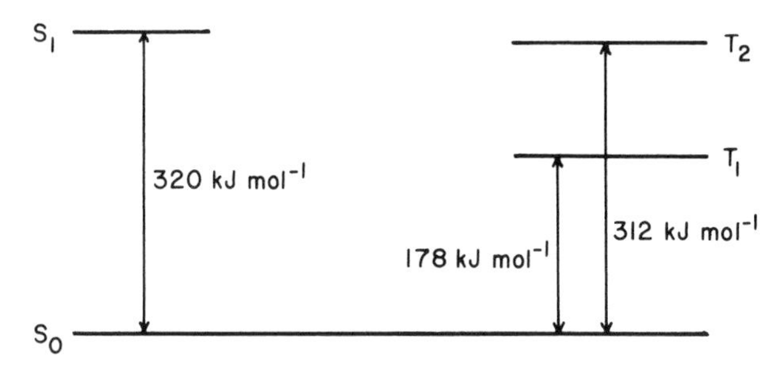

Fig. 6.3. Electronic energy levels in anthracene. The radiationless conversion rate $S_1 \rightarrow T_2$ is much larger than that for $S_1 \rightarrow T_1$ reflecting the effect of energy differences.

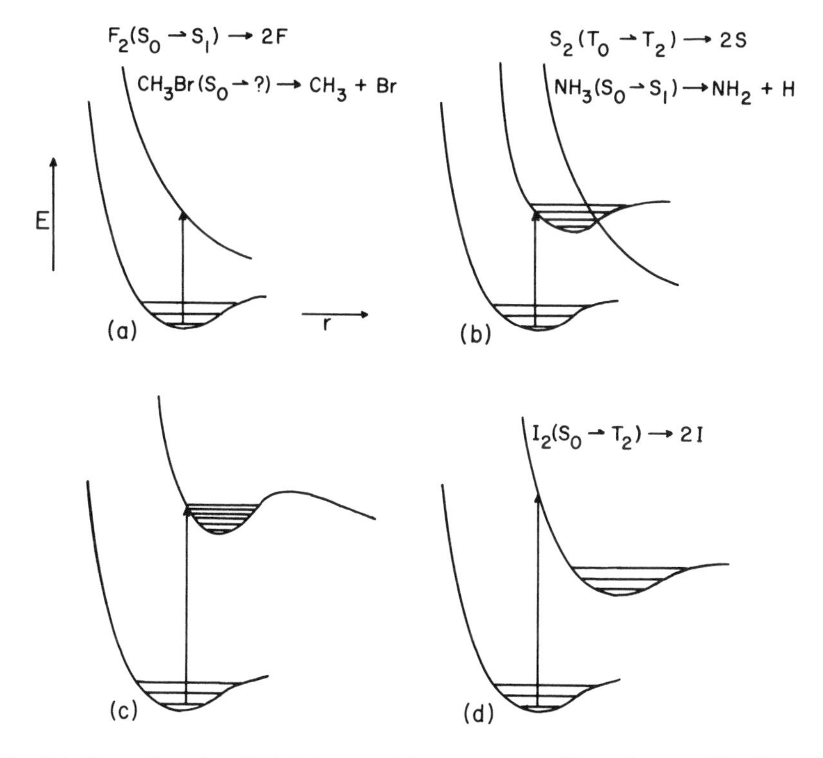

Fig. 6.4. Ground- and excited-state potential energy curves illustrating possible dissociation pathways. For a diatomic molecule r is the internuclear distance; in a polyatomic molecule r is a normal mode displacement. Vibrational separations are greatly exaggerated. The vertical arrows correspond to Franck–Condon allowed transitions. Where possible, molecular examples are indicated. (a) Repulsive excited state—immediate dissociation. (b) Crossing of bound and repulsive excited states—intersystem crossing leads to predissociation. (c) Metastable excited state—tunneling leads to dissociation. (d) Bound state—excitation energy greater than dissociation limit.

Table 6.2. Some Photodissociation Reactions

NO_2	$\rightharpoonup$	$NO + O$
$CHBr_3$	$\rightharpoonup$	$CHBr_2 + Br$
H_2O	$\rightharpoonup$	$HO + H$
Cyclobutanone	$\rightharpoonup$	CO + Cyclopropane
H_2O_2	$\rightharpoonup$	$OH + OH$
CH_4	$\nearrow$	$CH_2 + H_2$ (preferred)
	$\searrow$	$CH_3 + H$
Octatetraene	$\rightharpoonup$	Benzene + C_2H_2

turning points.[18] To accommodate the various constraints the excitation arrows in Fig. 6.4 are drawn at the equilibrium internuclear separation of the ground state; they extend only to the edge of an upper potential curve, the point which corresponds to the classical turning point.

If excitation is to a repulsive state (Fig. 6.4a), dissociation is immediate, occurring within one vibrational period, $t \lesssim 10^{-13}$ sec. Another rapid process involves predissociation (Fig. 6.4b). Here the molecule is excited to a bound state which, because there is a resonant continuum level of the repulsive electronic state, undergoes radiationless conversion followed by immediate dissociation. Excitation could also be to a metastable vibronic state (Fig. 6.4c) which dissociates via tunneling. The dissociation probability would be very sensitive to the height and the width of the barrier to be overcome, and the decomposition rate greatly enhanced by vibrational excitation in the excited state. Finally, as in Fig. 6.4d, excitation may be to a bonding state but into a continuum level above the dissociation limit; decomposition is again immediate.[19]

Figure 6.4 may be transcribed for polyatomic molecules. A similar diagram exists for *each* vibrational mode where a normal mode displacement corresponds to the interparticle distance. The situation is complicated further because the vibrational energy is rapidly partitioned among many

[18] For a discussion of the properties of the nuclear wave functions see G. Herzberg, *Spectra of Diatomic Molecules*, 2nd ed. (Princeton, N. J.: Van Nostrand, 1950), pp. 76–79, 93–94.

[19] Dissociation and predissociation in diatomic molecules are discussed in detail in G. Herzberg, *Spectra of Diatomic Molecules*, 2nd ed. (Princeton, N.J.: Van Nostrand, 1950), pp. 387–434.

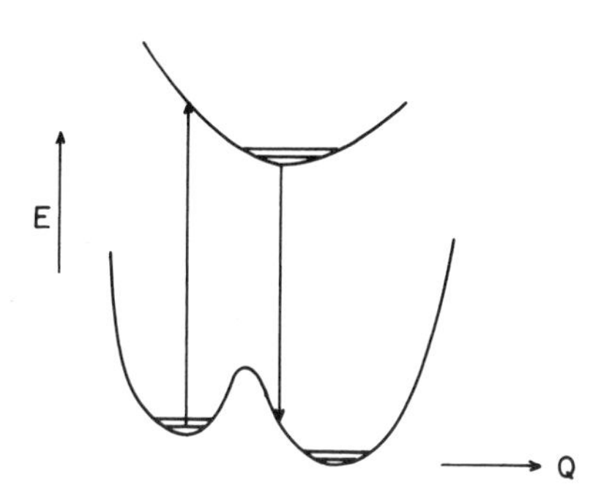

Fig. 6.5. Schematic diagram indicating a method whereby photoexcitation may lead to isomerization. Transitions are assumed to obey the Franck–Condon principle. Q is a normal mode which distorts molecule from one isomer to the others, e.g., rotation about C=C in an olefin. Vibrational separations are much exaggerated.

vibrational modes. Polyatomic illustrations of Figs. 6.4c and d have not yet been convincingly established.[20]

Photodissociation is a common primary process. Some examples are given in Table 6.2. The photoproducts may be characterized in many ways. Stable species are observed by chemical analysis. Excited ones are identified spectroscopically and reactive ones by their participation in secondary reactions.

6.4.5. Isomerization

The relaxation mechanisms that lead to photoinduced isomerizations such as

cis-olefin $+ h\nu \rightarrow \rightarrow$ *trans*-olefin (numerous steps)

$+ h\nu \rightarrow \rightarrow$ (numerous steps)

involve vibrational relaxation of the excited state sometimes preceded by radiationless conversion. In the ground electronic state the isomers are separate vibrational minima where the vibrational barrier to interconversion is large. Excitation often leads, either directly or indirectly, to a state with a different equilibrium geometry, as schematically shown in Fig. 6.5. If the molecule fluoresces isomerization is assured because of Franck–Condon requirements. If it is stabilized by radiationless conversion, the consequent vibrational relaxation greatly favors the formation of the more stable isotope.

[20] For a discussion of dissociation and predissociation in polyatomic molecules, see G. Herzberg, *Electronic Spectra of Polyatomic Molecules* (Princeton, N.J.: Van Nostrand, 1966), pp. 445–484.

6.5. Secondary Processes

6.5.1. Quenching Measurements

When a photoexcited molecule undergoes chemical reaction the fluorescence quantum efficiency is reduced, i.e., it is *quenched.* Fluorescence quenching provides an excellent example of how competition between chemical and physical processes may be used to establish both rate law and mechanism. The reaction of photoexcited acridine (A*) with amines in aqueous solution has been thoroughly studied.[21] Addition of amines reduces the fluorescence efficiency. Data for an analogous system are presented as a Stern–Volmer plot in Fig. 6.6. The relative fluorescence efficiency, $\varphi_f(c=0)/\varphi_f(c)$, is equal to the ratio of fluorescence intensities without and with quenchers $I_f(c=0)/I_f(c)$. If this latter quantity is graphed against amine concentration, c, it appears that

$$I_f(0)/I_f(c) = \varphi_f(0)/\varphi_f(c) = 1 + \alpha c \tag{6.15}$$

a result that is readily interpreted if we assume that A* reacts with amine

$$\text{A}^* + \text{amine} \xrightarrow{k_2} \text{products} \tag{6.16}$$

Modifying the arguments which led to (6.11) to account for the chemical process yields

$$\varphi_f(c) = k_f/(k_f + k' + k_2 c) \tag{6.17}$$

or with (6.12)

$$\varphi_f(0)/\varphi_f(c) = (k_f + k' + k_2 c)/(k_f + k') = 1 + k_2 \tau_0 c \tag{6.18}$$

precisely the Stern–Volmer form (6.15). It is now obvious why determination of τ_0 is important. Only if it is known can rate constants for reaction be established; otherwise only relative rate constants are accessible. For aqueous acridine τ_0 is 13.5 nsec and values for k_2 may be calculated.[22]

There are two significant limitations in using competition experiments, of which fluorescence quenching is representative, to establish rate laws and mechanisms. The methods monitor the depletion of a single species, in this case photoexcited acridine. As such they provide information about the first step in a reaction and, by themselves, do not characterize the

[21] A. Weller, *Prog. React. Kinetics* **1**, 187 (1961).

[22] A. Weller, *Prog. React. Kinetics* **1**, 187 (1961). Absolute lifetimes were determined using the method due to A. Schmillen, *Z. Phys.* **135**, 294 (1953).

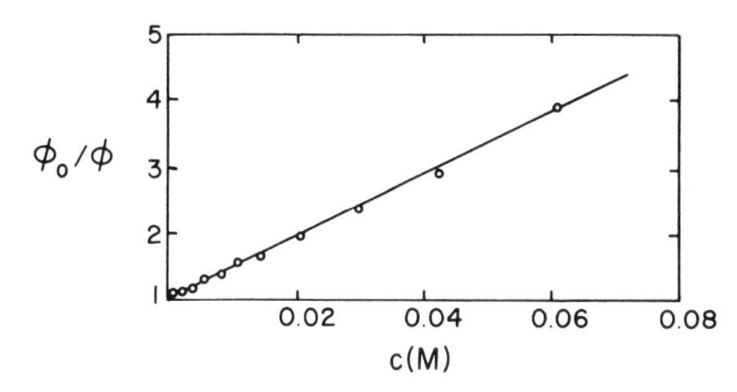

Fig. 6.6. Stern–Volmer plot of quantum efficiency in quenching of anthracene fluorescence by N–N diethylaniline in toluene solvent [H. Knibbe, Ph. D. Thesis (Free University of Amsterdam: 1969), p. 44].

products or whether further chemical transformation takes place.[23] In addition the rate of the competition reaction must be comparable to the rate of the physical process. For example, if $k_2 < 7\times10^5\ M^{-1}\ \text{sec}^{-1}$, the quenching ratio (6.18) would not drop to 0.95 unless $c \gtrsim 5\ M$. Thus, using excited acridine as a probe, it is impossible to study processes with $k_2 \lesssim 10^6\ M^{-1}\ \text{sec}^{-1}$ since enormous quencher concentrations are required. The converse may also prove to be a problem, i.e., if a reaction is too rapid quenching may be essentially total at negligible quencher concentrations. However, this is *not* a limitation in the acridine quenching (see Problem 6.3) experiments.

6.5.2. Sensitization and Excitation Transfer

Many molecules that quench the fluorescence of a primary photoproduct undergo a transition to an excited electronic state. The general mechanism is, using D for donor and A for acceptor,

$$\text{D}^* + \text{A} \rightharpoonup \text{D} + \text{A}^* \tag{6.19}$$

where the D* is the molecule which absorbed the radiation. This process is extremely useful; photosensitization permits the production of A* molecules in cases where direct photoexcitation of A is either impossible (a forbidden transition) or inconvenient. The requirements for radiationless energy transfer are:

(1) the process is nearly resonant (small increments of internal energy can be accommodated by changes in translational energy);

[23] The reaction of amines with photoexcited acridine actually follows a multistep pathway, in part diffusion controlled. See A. Weller, *Prog. React. Kinetics* **1**, 187 (1961).

(2) spin angular momentum is conserved (not necessary if the photosensitizer is a heavy atom).

In sensitization involving atoms and/or diatomic molecules the resonance condition severely limits the processes which may occur. However, if both D and A are polyatomic molecules the energy restriction reduces to the constraint that the electronic energy of D* be greater than that needed to form A*; any difference is made up by vibrational and rotational excitation of A*.

The rate of excitation transfer is very dependent on whether donor or acceptor change spin states. If both molecules remain in their original spin manifolds, coupling of the transition dipoles leads to very efficient energy transfer. The long-range dipole–dipole interaction is significant even at distances as great as 6–7 nm. Molecules need not be in immediate contact for the reaction

$$^1D^* + {}^1A \xrightarrow{k} {}^1D + {}^1A^*$$

to occur.[24] Diffusion no longer limits the rate constants; k is not viscosity dependent and, in the most efficient aqueous processes, is $\sim 3 \times 10^{11}$ M^{-1} sec^{-1}, many times the ordinary diffusion limit of $\sim 10^{10}$ M^{-1} sec^{-1} (Section 5.7). A quantitative theory for k yields

$$k \propto \frac{1}{R^6} \int I(\nu)\epsilon(\nu) \frac{1}{\nu^4} \, d\nu$$

where $\epsilon(\nu)$ is the extinction coefficient of the acceptor, $I(\nu)$ is the *normalized* fluorescence of the donor, and R is the distance between donor and acceptor.[25] The R^{-6} dependence has been directly verified by constructing systems in which donor and acceptor are substituent groups within the same molecule, but widely separated by a rigid, nonconjugated system.[26] Values of R determined from k are in good agreement with those deduced from measurements on molecular models.

Reactions such as

$$^3D^* + {}^1A \rightarrow {}^1D + {}^3A^*$$

$$^3D^* + {}^3A \rightarrow {}^1D + {}^1A^*$$

are spin allowed. However, the molecules' electron distributions must

[24] T. Förster, *Disc. Faraday Soc.* **27**, 7 (1959); *Naturwiss.* **33**, 166 (1946).
[25] T. Förster, *Disc. Faraday Soc.* **27**, 7 (1959).
[26] S. A. Latt, H. T. Cheung, and E. R. Blout, *J. Am. Chem. Soc.* **87**, 995 (1965).

interact strongly before energy transfer is probable.[27] The sensitization rate is now diffusion limited and thus viscosity dependent; an example where the limit is reached occurs in glycerol solvent with naphthalene as donor and 1-iodonaphthalene as acceptor.[28]

The photosensitized acceptor may relax by any primary process as well as by further reaction. Some instances where A* is formed in a weakly bound, dissociative, or predissociative state are[29]

$$Hg(T_2) + H_2(S_0) \rightarrow\rightarrow Hg(S_0) + 2H\cdot$$
$$Hg(T_2) + H_2O(S_0) \rightarrow\rightarrow Hg(S_0) + H\cdot + \cdot OH$$
$$Hg(T_2) + CH_3CHO \rightarrow\rightarrow Hg(S_0) + CH_3\cdot + \cdot CHO$$

In addition to excitation transfer, the donor and acceptor may associate and form a stable excited complex, an *excimer*, which is characterized by a fluorescence spectrum different from any observable in D* or A*. The phenomenon was discovered while studying the fluorescence of P(yrene).[30] At low concentrations the fluorescence is due to the $S_1 \rightarrow S_0$ transition. At high concentrations a new, lower-frequency fluorescence develops which is associated with the reactions

$$P(S_1) + P(S_0) \rightarrow P_2^*(\text{singlet}) \rightarrow 2P(S_0) + h\nu'$$

6.5.3. Some Reactions

The photochemical literature is replete with photolysis studies. The rates are measured via quenching of either fluorescence or phosphorescence. The photoproducts are characterized either chemically or spectroscopically. A few examples suffice to suggest the scope of such investigations.

6.5.3.1. Photolysis of Ketene

$$CH_2CO + h\nu(\lambda < 473.5\text{ nm}) \rightarrow CH_2CO(S_1{}^v)$$

$$CH_2CO(S_1{}^v) \nearrow CH_2(S_1 \text{ or } T_0) + CO \quad \text{(predissociation)}$$
$$CH_2CO(S_1{}^v) \searrow CH_2CO \quad \text{(vibrational relaxation)}$$

$$CH_2 + CH_2CO \rightarrow C_2H_4 + CO$$

[27] D. L. Dexter, *J. Chem. Phys.* **21**, 836 (1953).
[28] A. D. Osborne and G. A. Porter, *Proc. Roy. Soc.* **A284**, 9 (1965).
[29] See J. P. Simons, *Photochemistry and Spectroscopy* (London: Wiley-Interscience, 1971), pp. 231–247 for details.
[30] T. Förster and K. Kaspar, *Z. Elektrochem.* **59**, 976 (1955).

In this example no resolvable structure is found in the absorption spectrum of CH_2CO as long as $\lambda < \lambda_0 = 473.5$ nm, suggesting rapid predissociation. As the frequency of excitation increases the quantum yield for predissociation, φ_{diss}, increases; it is unity regardless of pressure for $\lambda < \lambda_1 = 270$ nm. When $\lambda_1 < \lambda < \lambda_0$, φ_{diss} is pressure dependent, dropping as p increases. As $p \to 0$ one would expect $\varphi_{\text{diss}} = 1$ since the unstructured spectrum indicates predissociation. However, this is not observed: at $\lambda = 366$ nm, $\varphi_{\text{diss}} \to 0.04$ as $p \to 0$. No totally convincing explanation of the pressure dependence of φ_{diss} exists. The net decomposition efficiency is twice φ_{diss} since CH_2 reacts further with ketene.[31]

6.5.3.2. Photolysis of NO_2

$$NO_2 + h\nu(320 < \lambda < 397.9 \text{ nm}) \to NO(D_0) + O(T_0)$$

$$O(T_0) + NO_2 \to NO_3^* \begin{cases} \nearrow NO + O_2(v \lesssim 12) \\ \searrow NO_3 \end{cases}$$

Irradiation produces excited NO_2 which predissociates into *doublet* NO and atomic oxygen. The O atoms react further with NO_2 either producing more NO or transient NO_3. When NO_2 produced in automobile exhaust emissions is photolyzed the further course of atmospheric reaction is very complex. Some of the reactions are[32]:

$$O + O_2 \to O_3, \qquad O_3 + NO \to NO_2 + O_2$$

$$O_3 + NO_2 \to NO_3 + O_2, \qquad NO_3 + NO_2 \rightleftharpoons N_2O_5$$

$$N_2O_5 + H_2O \to 2HNO_3$$

The unimolecular reactions and the recombination reactions of course

[31] Recent studies indicate that the mechanism of ketene photodissociation is wavelength dependent; as the energy of the exciting radiation increases, new primary processes become energetically allowed [W. L. Hase and P. M. Kelley, *J. Chem. Phys.* **66**, 5093 (1977)].

[32] The kinetics of smog production is complicated, involving products not only of nitrogen oxide photolysis but also of hydrocarbon oxidation. With a catalog of the elementary reactions that may be involved in the overall process it is possible, using the numerical methods mentioned in Section 5.3, to solve the coupled rate equations. Comparison of calculation and experiment then provides a test of whether the proposed reaction scheme is plausible. See T. A. Hecht and H. J. Seinfield, *Envir. Sci. Tech.* **6**, 47 (1972). It is worth reiterating that finding a multistep mechanism which accounts for the data does *not* demonstrate uniqueness; it only indicates consistency.

require participation of an inert molecule to satisfy energy-transfer constraints.

6.5.3.3. *Isomerization of Mesitylene*

Irradiating mesitylene with near-ultraviolet light leads to a shift of methyl groups. Isotopic labeling experiments have shown that the apparent methyl migration is really the ring rearrangement[33]

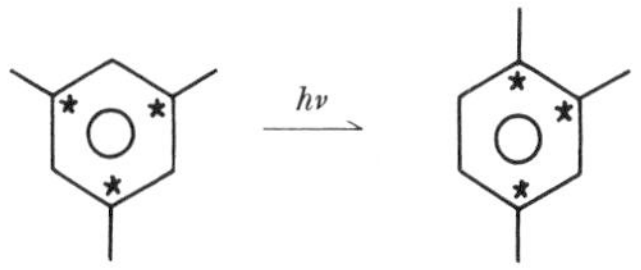

thus suggesting as intermediate a form of Dewar benzene

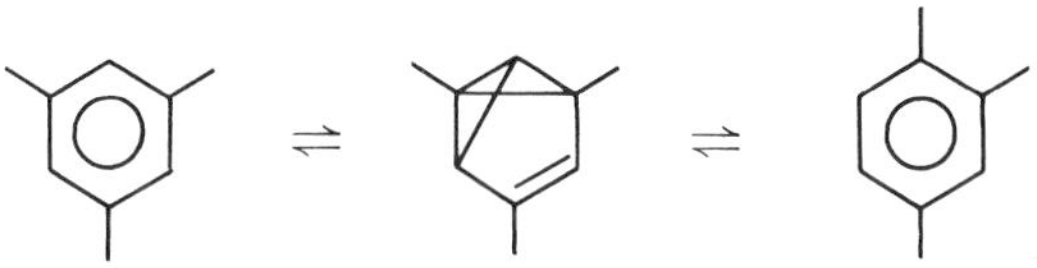

Similar experiments on benzene-1,3,5 d_3 showed the same migration pattern.

6.5.3.4. *Photolysis of Trapped Species*

Gas- or solution-phase photolysis is always complicated by the fact that the primary photoproducts diffuse and undergo secondary reactions. To avoid such difficulties the technique of matrix isolation was developed. Here a dilute mixture of, e.g., CH_4 in N_2 is condensed from the vapor phase onto a transparent window maintained at temperatures of 4–20 K. Then the CH_4 can be photolyzed and the resultant primary photoproduct, $\cdot CH_3$, studied at leisure.[34] A variant of this technique is to use a reactive matrix. When HBr or HI, condensed in a CO matrix, is photolyzed the primary process is dissociation of the excited HX. The H atom then reacts and HCO is observed.[35]

[33] L. Kaplan, K. E. Wilzbach, W. G. Brown, and S. S. Yang, *J. Am. Chem. Soc.* **87**, 675 (1965); K. E. Wilzbach, A. L. Harkness, and L. Kaplan, *J. Am. Chem. Soc.* **90**, 1116 (1968); and L. Kaplan and K. E. Wilzbach, *J. Am. Chem. Soc.* **90**, 3291 (1968).

[34] D. E. Milligan and M. E. Jacox, *J. Chem. Phys.* **47**, 5146 (1967).

[35] G. E. Ewing, W. E. Thompson, and G. C. Pimentel, *J. Chem. Phys.* **32**, 927 (1960).

6.6. Chemiluminescence

Chemiluminescence is, in a general sense, the opposite of photodissociation. Here a molecule is formed in an excited state by means of chemical reaction. The new species then fluoresces to deexcite. It is the process which accounts for the characteristic colors of flames. An example is found in the combustion of CO where atomic oxygen is produced, most likely as the first step in the oxidation of CO. Then the following sequence of reactions occurs[36]:

$$O(T_0) + CO(S_0) + M \rightarrow CO_2(T_1) + M$$

$$CO_2(T_1) \rightarrow CO_2(S_1) \rightarrow CO_2(S_0^v) + h\nu$$

The final step in the relaxation is emission of chemiluminescent radiation of wavelength ~310–380 nm.

The chemiluminescence spectrum contains considerable information. When highly exothermic reactions of the type

$$A + BC \rightarrow AB^* + C$$

are studied the product AB* is always formed in a vibrationally excited state which, in order to be stabilized, must either emit radiation or be collisionally deactivated. At normal pressures the latter process dominates since the radiative lifetime for *vibrational* fluorescence is $\sim 10^{-3}$ sec. At low pressure (10^{-3}–10^{-5} atm) the inherent inefficiency of V–T transfer allows observation of the *infrared* fluorescence of AB* with an internal energy distribution produced during the bond-forming process. The chemiluminescence spectrum may be analyzed since the vibrational spectrum of AB is known; then the vibrational and rotational energy distribution in AB* formed in the collisional process may be deduced. If the exothermicity of the reaction is known, the fraction which appears as translational, rotational, or vibrational energy of the products can be estimated. Some data are given in Table 6.3. The differences are considerable. As we shall see in Chapter 10 product energy distributions may be used to determine gross features of the ABC interaction potential.

[36] A. G. Gaydon, *The Spectroscopy of Flames* (London: Chapman and Hall, 1957), Chapter 6; R. N. Dixon, *Proc. Roy. Soc.* (*London*) **A275**, 431 (1963).

Table 6.3. The Fraction of Reaction Exothermicity Appearing in Various Degrees of Freedom[a]

Reaction	Exothermicity (kJ mol^{-1})	Mean fraction of exothermicity appearing in t(ranslation), r(otation), and v(ibration)		
		f_t	f_r	f_v
$H + Cl_2 \rightarrow HCl + Cl$	203	0.54	0.07	0.39
$D + Cl_2 \rightarrow DCl + Cl$	208	0.50	0.10	0.40
$H + Br_2 \rightarrow HBr + Br$	183	0.39	0.05	0.56
$F + H_2 \rightarrow HF + H$	145	0.26	0.07	0.67
$F + D_2 \rightarrow DF + D$	144	0.25	0.06	0.69
$Cl + HI \rightarrow HCl + I$	142	0.16	0.13	0.71
$Cl + DI \rightarrow DCl + I$	142	0.15	0.14	0.71

[a] T. Carrington and J. C. Polanyi, in *MTP International Review of Science, Physical Chemistry*, Series 1, Vol. 9, J. C. Polanyi, ed. (London: Butterworths, 1972), pp. 149–150.

6.7. Orbital Symmetry Correlations: Woodward–Hoffman Rules

Both thermal and photochemical isomerization of disubstituted *trans*-butadienes to disubstituted cyclobutenes are stereospecific; however, the products are different:

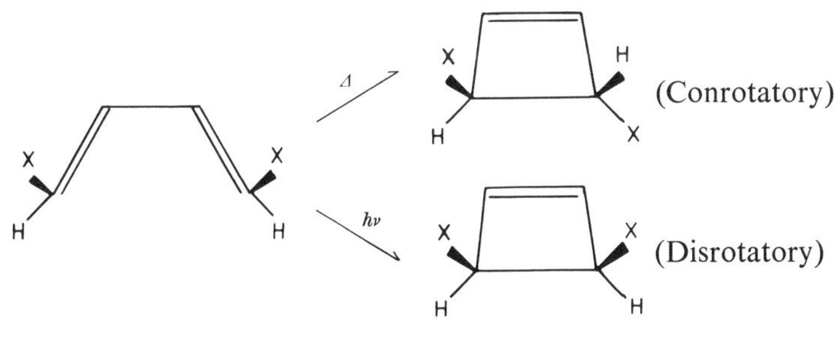

Similar stereochemical control is found in countless reactions, both unimolecular and bimolecular. The explanation is based upon three assump-

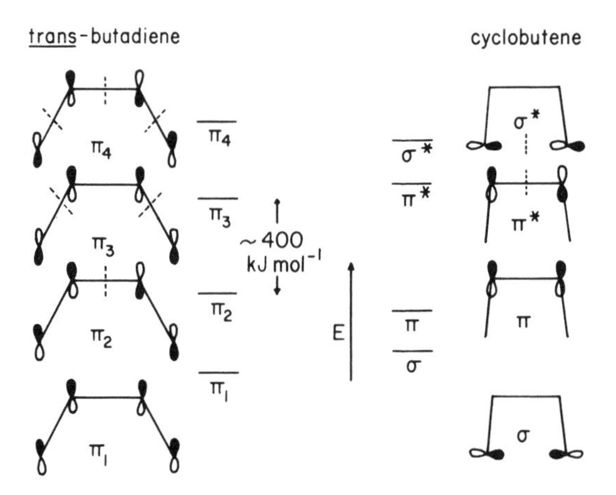

Fig. 6.7. Energy levels and nodal structure of one-electron molecular orbitals (MO's) which control the *trans*-butadiene ⇀ cyclobutene isomerization. The electron density in the various π-MO's is perpendicular to the carbon skeleton. The electron density in the σ-MO's is coplanar with the carbon skeleton. The spacing of energy levels is approximate. The dashed lines indicate the position of the nodes in the various orbitals.

tions. (1) Reaction proceeds in a concerted, single-step pathway: the σ-bond and the π-bond in cyclobutene are formed simultaneously. (2) An adiabatic principle is obeyed: as the nuclei move the electron distribution adjusts instantaneously. (3) The activation energies associated with the different concerted nuclear motions are enormously different: the higher-energy pathways are absolutely forbidden.[37]

To illustrate the molecular basis for this selectivity consider the molecular orbitals of butadiene and cyclobutene which are involved in the electronic rearrangement.[38] These are illustrated in Fig. 6.7; their relative energies are $\pi_1 < \pi_2 < \pi_3 < \pi_4$ and $\sigma < \pi < \pi^* < \sigma^*$. Now consider disrotatory motion around the C—C bonds (rotation in the opposite direction) as shown in Fig. 6.8. The two orbitals π_1 and π_3 must correlate with two orbitals of cyclobutene. One of these is obviously π since both π_1 and π_3 are in phase and therefore bonding between C_2 and C_3; the other is σ since in both π_1 and π_3 disrotatory motion creates a *bonding* σ-orbital between C_1 and C_4. Similar arguments can be used to establish the pair-

[37] R. B. Woodward and R. Hoffman, *J. Am. Chem. Soc.* **87**, 395, 2046, 2511, 4388, 4389 (1965); H. C. Longuet-Higgins and E. W. Abrahamson, *J. Am. Chem. Soc.* **87**, 2045 (1965).

[38] For details on molecular orbital calculations see F. L. Pilar, *Elementary Quantum Chemistry* (New York: McGraw–Hill, 1968), Chapter 18.

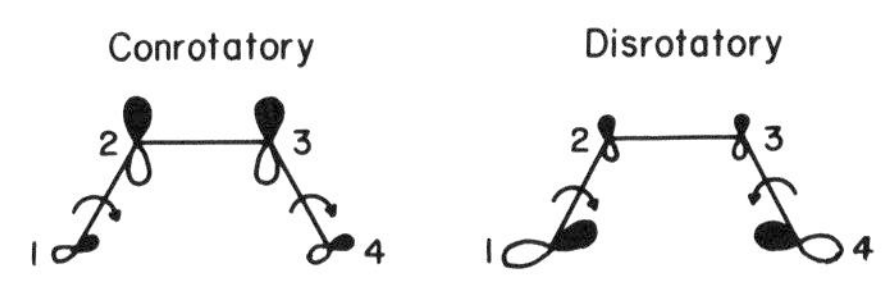

Fig. 6.8. Effect of different concerted rotations about the C_1—C_2 and C_3—C_4 bonds on the π_1-MO of *trans*-butadiene. Conrotation distorts $\pi_1 \rightharpoonup \pi$; disrotation distorts $\pi_1 \rightharpoonup \sigma$. Orbital correlation follows a *minimum* energy pathway.

wise *orbital correlations*

Conrotatory: $(\pi_1, \pi_3) \rightharpoonup (\pi, \sigma^*)$, $(\pi_2, \pi_4) \rightharpoonup (\sigma, \pi^*)$

Disrotatory: $(\pi_1, \pi_3) \rightharpoonup (\sigma, \pi)$, $(\pi_2, \pi_4) \rightharpoonup (\pi^*, \sigma^*)$

The orbital structure changes continuously as rotation occurs and the energy levels of a correlated pair cannot cross since the orbitals have the same symmetry properties.[39] Thus the individual orbital correlations are

Conrotatory: $(\pi_1 \rightharpoonup \pi, \pi_2 \rightharpoonup \sigma, \pi_3 \rightharpoonup \sigma^*, \pi_4 \rightharpoonup \pi^*)$

Disrotatory: $(\pi_1 \rightharpoonup \sigma, \pi_2 \rightharpoonup \pi^*, \pi_3 \rightharpoonup \pi, \pi_4 \rightharpoonup \sigma^*)$

Now consider the two isomerizations. In the thermal process *trans*-butadiene is in its ground electronic configuration $\pi_1^2\pi_2^2$. Conrotatory motion yields cyclobutene in the ground electronic configuration $\sigma^2\pi^2$ while disrotatory motion leads to the highly excited configuration $\sigma^2(\pi^*)^2$; the activation barrier for disrotatory motion forbids this path. The lowest photoexcited state of *trans*-butadiene is $\pi_1^2\pi_2\pi_3$ which, assuming conrotatory displacements, correlates with the highly excited $\sigma\pi^2\sigma^*$ state of cyclobutene. In a disrotatory displacement cyclobutene is formed in its first excited state $\sigma^2\pi\pi^*$. The latter requires little or no activation; the former requires a great deal and is forbidden. The stereochemical discrimination found in (6.20) is thereby accounted for.[40]

[39] This is a direct consequence of quantum mechanics. Any analysis which suggests that the ordering of the energy levels of one-electron orbitals of the same symmetry may be inverted as a molecule distorts is approximate. Important interactions must have been ignored if such a result is predicted. If these interactions are accounted for the levels no longer cross. See W. Kauzmann, *Quantum Chemistry* (New York: Academic Press, 1957), pp. 536–539 for a discussion of a similar problem: the change from ionic to covalent binding in NaCl as the Na—Cl separation is varied.

[40] For an extended discussion see R. Hoffman and R. B. Woodward, *Science* **167**, 825 (1970).

Similar arguments have been developed to explain stereochemical control in other systems. In each case the fundamentals are the same. Reaction proceeds via a single-step pathway. The orbital structure adjusts instantaneously as the nuclei move. The energy levels of correlated pairs of orbitals do not cross.

6.8. Laser Activation

Ordinary photochemical excitation, initiated by absorption in the visible or ultraviolet, introduces an enormous amount of internal energy into a molecule. Radiation at 400 nm corresponds to an energy of 300 kJ mol^{-1}, enough to activate most reactions. As we have seen, such activation is rarely discriminating. A range of products is formed depending upon the relative efficiency of the various primary and secondary processes. The ideal approach to selective bond breaking would be to pump energy into the particular vibrational mode associated with the chemical process of interest. With the development of high-power infrared lasers, it is possible to imagine accomplishing this task.

The unique features of a laser, its high intensity and monochromaticity, have led to its application in countless problems of scientific interest. We shall only consider one. Infrared laser photochemistry is a very new field (it began to develop in the early 1970s); as such there are numerous unresolved problems as well as ambiguities and contradictions. The following discussion is designed to suggest the scope of the field, not to be comprehensive.[41]

The most versatile infrared source is the CO_2 laser; it can be tuned to emit light at about 100 discrete wavelengths between 9.2 and 11.0 μm ($\sim$12 kJ mol^{-1}).[42] If used to provide a continuous source of photons the power output is $\sim$100 W cm^{-2}. However, if properly focused, a CO_2 laser can emit pulses of nanosecond duration with a power flux of up to 100 MW cm^{-2}; at these wavelengths the pulse contains $\sim 10^{19}$ photons cm^{-2}, which corresponds to a flux of $\sim 10^{28}$ cm^{-2} sec^{-1}! While the pulse is on, the photon density in the sample is $\sim 10^{17}$ cm^{-3}, which is comparable to molecular densities when $p \sim 0.01$ atm and $T \sim 300$ K. In addition to

[41] It is based on two state-of-the-art reviews: A. L. Robinson, *Science* **193**, 1230 (1976); **194**, 45 (1976).

[42] For a discussion of laser sources see S. R. Leone and C. B. Moore, in *Chemical and Biochemical Applications of Lasers*, C. B. Moore, ed. (New York: Academic Press, 1974), Vol. 1, Chapter 1.

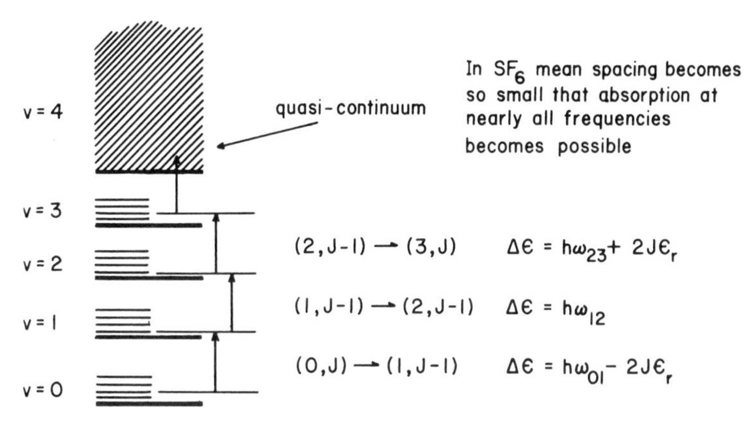

Fig. 6.9. Vibrational–rotational energy levels for a polyatomic molecule. Resonance is established by change in rotational level to compensate for vibrational anharmonicity. Rotational separation is exaggerated.

providing enormous photon fluxes the laser, because of its very narrow line widths, provides highly selective excitation. A properly tuned CO_2 laser has been used to selectively excite either $^{10}BCl_3$ or $^{11}BCl_3$.[43] At present only two other regions of the infrared spectrum are accessible with high-power lasers—the CO and HF lasers which emit in the vicinity of 5 μm (~24 kJ mol^{-1}) and 2.7 μm (~45 kJ mol^{-1}), respectively.

In laser-induced gas-phase reactions the first step is absorption of a vibrational quantum which, at moderate pressures, would only heat the gas since V–T energy transfer occurs. If this process were dominant the laser would be no more than a fancy (and expensive) Bunsen burner. At low pressures conditions are more favorable for multiple-photon absorption. The photon flux is large enough so that more than one quantum can be absorbed before significant V–T transfer occurs. The only trick is to maintain resonance between the laser frequency and the various transitions to be excited. If only pure vibrational transitions were involved anharmonicity would ensure that a laser tuned to the fundamental frequency ($\nu_{0\to 1}$) would be nonresonant for other transitions. However, since there are also changes in rotational quantum number resonance may be reestablished. The resonance conditions are illustrated in Fig. 6.9.

An example of multiple-photon absorption is found in the laser-induced decomposition of D_3BPF_3 irradiated with a continuous source

[43] R. V. Ambartzumian, N. V. Chekalin, V. S. Doljikov, V. S. Letokhov, and E. A. Ryabov, *Chem. Phys. Lett.* **25**, 515 (1974).

producing $\sim$150 W cm^{-2} at a photon energy of 11.7 kJ mol^{-1}. At pressures between 2×10^{-6} and 7×10^{-6} atm the room-temperature decomposition

$$2D_3BPF_3 = B_2D_6 + 2PF_3, \qquad E_a = 123 \text{ kJ mol}^{-1}$$

appears to proceed via sequential absorption of three photons.[44] Instead of requiring the collisional activation energy 123 kJ mol^{-1}, the apparent collisional activation energy in the irradiated sample was reduced to 14.5 kJ mol^{-1}. While the precise meaning of activation energy for nonequilibrium energy distributions is far from obvious, it does appear that laser activation leads to far more efficient energizing of a molecule.

One important conclusion emerges. A small amount of *vibrational* excitation may be exceptionally effective in enhancing reactivity. Under thermal conditions $\sim$120 kJ mol^{-1} are needed if collision is to induce reaction. This process, which might be termed T–V transfer, is neither efficient nor selective. Even if molecules have a high collision energy there is no reason, as we shall see in Chapter 10, for it to be readily converted into internal energy. Furthermore, the internal energy absorbed need not go into vibrational modes favoring B–P bond weakening. In the laser-induced reaction the energy apparently excites P–F vibration and then is redistributed among modes with similar frequency weakening the P–B bond; a small amount of collisional energy transfer then completes the job. Thus relatively little energy, introduced selectively, is much more effective than a thermal sledgehammer.[45]

At higher pressures and in liquids collisions are frequent. Energy deposited in a single vibrational mode does not remain localized. Collisional processes first redistribute energy within the molecule and then convert it into relative translational energy. By using ultrahigh-power lasers the absorbing molecules can be multiply excited *before* collision is probable. In addition to lowering the overall activation barrier for reaction the multiple-photon process may be used to excite molecules to selected vibrational states which then react to produce specific products. An example is the laser-driven reaction

$$H_2 + BCl_3 = HBCl_2 + HCl$$

[44] K-R. Chien and S. H. Bauer, *J. Phys. Chem.* **80**, 1405 (1976); E. R. Lory, S. H. Bauer, and T. Manuccia, *J. Phys. Chem.* **79**, 545 (1975).

[45] For more recent work, emphasizing the effectiveness of laser activation in the reaction $NO + O_3 \rightarrow$ products, see K.-W. Hui and T. A. Cool, *J. Chem. Phys.* **68**, 1022 (1977) and references cited therein.

The thermal reaction yields B_2H_6 as well.[46] A related feature of laser photoreactivity is the effect of varying wavelength. In the series of reactions

$$B(CH_3)_3 + HBr = B(CH_3)_2Br + CH_4$$

$$B(CH_3)_2Br + HBr = BCH_3Br_2 + CH_4$$

$$BCH_3Br_2 + HBr = BBr_3 + CH_4$$

the first and third are accelerated by tuning the laser to $\lambda = 10.30\ \mu m$ while the rate of the last two is increased when $\lambda = 9.62\ \mu m$.[47]

Inasmuch as laser photochemistry is a relatively unexplored area of research, it abounds with ambiguities and contradictions. Two instances suffice. The decomposition of CF_2ClCF_2Cl was studied using a low-power ($\sim$25 W cm^{-2}) laser at high pressure ($\sim$0.4 atm).[48] The decomposition rate was very dependent on the irradiation wavelength; it was 160 times faster when $\lambda = 10.86\ \mu m$ than when $\lambda = 9.51\ \mu m$. Since both wavelengths are strongly absorbed and since, under the experimental conditions, there are $\sim 10^4$ collisions between each absorption event, V–T transfer should lead to rapid laser heating and loss of selectivity which was not observed. Speculation surrounds possible V–V transfer which may further excite one molecule up the vibrational ladder. Frequency matching may be better at the longer wavelength. However, there is presently no clear resolution of the apparent contradictions.

Large-scale utilization of laser techniques would seem to require high pressures ($\sim$0.1 atm) and high intensity ($\sim 10^6$ W cm^{-2}). However, under such conditions one important feature of laser photochemistry would appear to be easily lost—selective vibrational excitation. While thermalization via V–T transfer is slow, collisions still affect atomic motions and work to *redistribute* energy localized in one mode among the molecules' other vibrational degrees of freedom. In CH_3F, $\sim$100 collisions are required to complete the energy redistribution within the molecule; however, energy transfer between nearly resonant modes requires $\lesssim$10 collisions.[49] The consequence for high-pressure laser activation is apparent; at $p \sim 0.1$ atm efficient energy redistribution should require $<$ 50 nsec. However, studies

[46] S. D. Rockwood and J. W. Hudson, *Chem. Phys. Lett.* **34**, 542 (1975).

[47] H. R. Bachmann, H. Nöth, R. Rinck, and K. L. Kompa, *Chem. Phys. Lett.* **33**, 261 (1975).

[48] R. N. Zitter and D. F. Koster, *J. Am. Chem. Soc.* **98**, 1613 (1976).

[49] E. Weitz and G. W. Flynn, *J. Chem. Phys.* **58**, 2679 (1973).

of the decomposition of CCl_3F and $CClF_3$ suggest otherwise.[50] In experiments using a 500-nsec pulse the vibrational energy seems neither to be thermalized by V–T transfer nor to be collisionally redistributed within the molecules. Instead it seems to remain in the vibrational mode that was excited initially. Again no clear resolution of the contradictory evidence has been presented.

6.9. *Laser-Controlled Isotope Separation*

As mentioned in the previous section a CO_2 laser can be tuned to selectively excite the isotopically different species $^{10}BCl_3$ or $^{11}BCl_3$. This phenomenon has been exploited to separate the two isotopes of boron.[51] An isotope of BCl_3 is excited to a high vibrational state by an ultrahigh-power CO_2 laser via multiphoton absorption. It is then exceptionally reactive. In one series of experiments the excited BCl_3 dissociated and the boron atoms reacted with oxygen in the cell.[52] The remaining BCl_3 was thereby enriched in one or another isotope depending upon which species had been photodissociated. The effect is not small. Figure 6.10 shows the chemiluminescence of BO under conditions where the CO_2 laser was tuned to the absorption frequency in $^{10}BCl_3$ or $^{11}BCl_3$. The isotopic composition of the BO formed corresponds to the composition of the BCl_3 excited. Another set of scavenging experiments was done using H_2S. Here reaction appears to occur with vibrationally highly excited BCl_3. Again the remaining BCl_3 was isotopically enriched.[53]

As a matter of practical interest the molecule that one wants to separate isotopically is UF_6. However, its infrared active vibrational frequencies do not fall in the 9–11 μm range and are thus not accessible to excitation by a CO_2 laser. Thus much work has been done on the structurally similar molecule SF_6 which has the infrared active fundamental ν_3 at 10.6 μm; it has been possible to separate $^{32}SF_6$ from $^{34}SF_6$. Here radiation is used to dissociate $^{32}SF_6$, probably to $^{32}SF_4$ and F_2; the $^{32}SF_4$ can be hydrolyzed with water and the $^{32}SOF_2$ separated. By using hundreds of laser pulses it is possible to increase the naturally occurring $^{34}S/^{32}S$ ratio by more than

(50) D. F. Dever and E. Grunwald, *J. Am. Chem. Soc.* **98**, 5055 (1976).

(51) For a recent review see V. S. Letokhov and R. V. Ambartzumian, *Acc. Chem. Res.* **10**, 61 (1977).

(52) R. V. Ambartzumian, N. V. Chekalin, V. S. Doljikov, V. S. Letokhov, and E. A. Ryabov, *Chem. Phys. Lett.* **25**, 515 (1974).

(53) S. M. Freund and J. J. Ritter, *Chem. Phys. Lett.* **32**, 255 (1975).

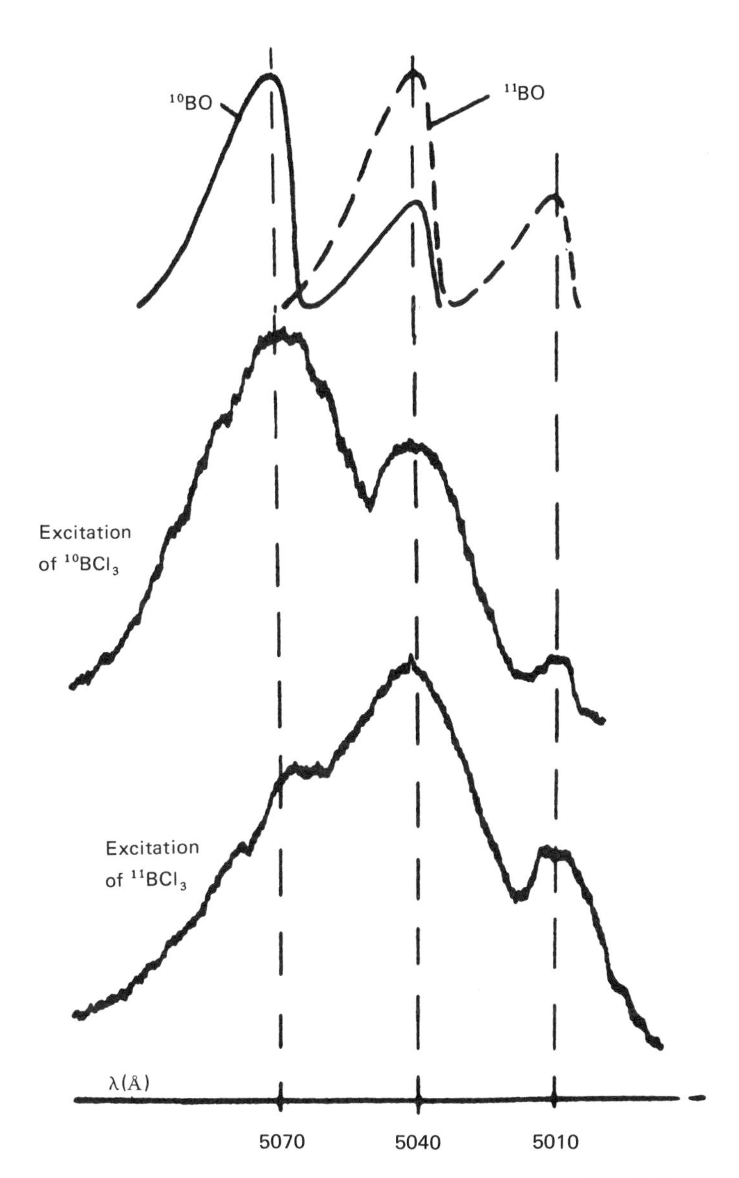

Fig. 6.10. Chemiluminescence spectrum of BO formed by reaction of O_2 with laser-activated BCl_3 when laser is tuned to absorption frequency of $^{10}BCl_3$ or $^{11}BCl_3$. In each case the corresponding isotope of BO is formed preferentially [R. V. Ambartzumian and V. S. Letokhov, *Acc. Chem. Res.* **10**, 61 (1977), reprinted by permission, American Chemical Society].

10^3–10^4.[54] The enrichment ratio is sensitive to a range of variables: laser frequency, pulse energy, pulse duration, pulse number, and gas pressure. Naturally, increase of the pressure is rapidly counterproductive; above $\sim 10^{-3}$ atm V–T transfer is sufficiently rapid so that SF_6 cannot readily be driven into a highly excited state and the enrichment ratio falls.

Efforts to maximize the enrichment ratio led to a variety of experiments, some superficially unlikely to succeed. One of the latter had most unexpected results. The SF_6 was irradiated using two lasers. One, at lower power (50 kW cm^{-2} to 2 MW cm^{-2}), was tuned to the excitation frequency; the ultrahigh-power laser ($\sim$20 MW cm^{-2}) was not tuned. This mode of operation led to a substantial increase in the enrichment ratio.[55]

Theoretical understanding of the laser-driven dissociation is far from complete. Certainly multiphoton absorption is required. However, rotational compensation for vibrational anharmonicity, as illustrated by Fig. 6.9, cannot possibly be operative all the way up the vibrational ladder. In SF_6 more than 30 quanta at 10.6 μm are required to provide the dissociation energy. The upper vibrational levels become so narrowly spaced that most of the photon energy would have to go into rotational excitation in a transition $v \rightarrow v + 1$. If, however, the mechanism of Fig. 6.9 accounts for the initial excitation, the SF_6 molecule may rapidly be driven to a vibrational level where the vibrational–rotational states of one vibrational mode are in the same energy range as those of another mode. The net effect is an essentially continuous distribution of states *all very anharmonic*, the so-called "quasi-continuum." Ordinary selection rules no longer matter greatly and photons of *any* energy may be absorbed to further energize the molecule.[56] By means of this arbitrary division of the vibrational–rotational energy-level manifold, the two laser experiment can be rationalized. The low-power tuned laser selectively excites SF_6 into the quasi-continuum. Then the high-power, untuned one drives only these excited molecules to dissociation. Further study is required to verify this interpretation.

[54] R. V. Ambartzumian, Yu. A. Gorokhov, V. S. Letokhov, and G. N. Makarov, *Pis'ma Zh. Eksp. Teor. Fiz.* **21**, 375 (1975), *JETP Lett.* **21**, 171 (1975) (English translation); J. L. Lyman, R. J. Jensen, J. Rink, C. P. Robinson, and S. D. Rockwood, *Appl. Phys. Lett.* **27**, 86 (1975).

[55] R. V. Ambartzumian, Yu. A. Gorokhov, V. S. Letokhov, G. N. Makarov, A. A. Puretskiǐ, and N. P. Furzikov, *Pis'ma Zh. Eksp. Teor. Fiz.* **23**, 217 (1976), *JETP Lett.* **23**, 194 (1976) (English translation).

[56] R. V. Ambartzumian, Yu. A. Gorokhov, V. S. Letokhov, G. N. Makarov, and A. A. Puretskiǐ, *Pis'ma Zh. Eksp. Teor. Fiz.* **23**, 26 (1976); *JETP Lett.* **23**, 22 (1976) (English translation).

The impetus for laser isotope separation studies is given by UF_6. As mentioned in the previous section the presently known infrared lasers emit in the vicinity of 2.7 μm (HF), 5 μm (CO), and 10 μm (CO_2). Thus to employ the laser approach in uranium isotope separation requires the development of a new high-power, low-frequency, laser source since the ν_3 excitation mode used in the SF_6 experiments appears at 16.0 μm in UF_6.[(57)]

Problems

6.1. Quenching reactions with photoexcited species are subject to diffusion limitation as discussed in Section 5.7. A more general mechanism than (6.16) is

$$A^* + B \underset{k_{\text{diss}}}{\overset{k_{\text{diff}}}{\rightleftharpoons}} A^* \cdot B \xrightarrow{k_2} \text{products}$$

(a) Modify the arguments leading to (6.18) and show that $\varphi_0/\varphi = 1 + \gamma k_{\text{diff}}\tau_0 c$, where $\gamma = k_2/(k_2 + k_{\text{diss}})$.
(b) The phosphorescence lifetime of benzophenone is ~160 sec. If a compound quenches benzophenone triplets at a diffusion-controlled rate ($k_{\text{diff}} \sim 10^{10}\ M^{-1}\ \text{sec}^{-1}$) what concentration is required to quench 95% of the phosphorescence?

6.2. In addition to diffusion of A* and B the species A* · B may also be formed by direct excitation of an A · B complex. Assume molar extinction coefficients ϵ and ϵ' for A and A · B, respectively, and the equilibrium

$$A + B = A \cdot B, \qquad K$$

(a) Show that the concentration of A · B is $[A \cdot B] = (\sigma K + 1 - R)/2K$, where

$$\sigma = A_0 + B_0, \quad \delta = B_0 - A_0, \quad \text{and} \quad R = (\delta^2 K^2 + 2\sigma K + 1)^{1/2}$$

(b) Show that the *total* rate of light absorption is proportional to

$$A_0(\epsilon + \epsilon' K[B])/(1 + K[B])$$

where $[B] = (\delta K - 1 + R)/2K$.
(c) Now show that the fluorescence efficiency of A* is

$$\frac{\varphi}{\varphi_0} = \frac{1}{\epsilon + \epsilon' K[B]} \frac{\epsilon + (1 - \gamma)\epsilon' K[B]}{1 + k_{\text{diff}}\tau_0[B]} = \frac{1 - \gamma\alpha}{1 + k_{\text{diff}}\tau_0[B]}$$

(57) Infrared sources being developed are discussed by J. J. Ewing, in *Chemical and Biochemical Applications of Laws*, C. B. Moore, ed. (New York: Dekker, 1977), Vol. 2, Chapter 6.

where $\gamma \equiv (1 + k_2\tau_{AB})/[1 + (k_2 + k_{diss})\tau_{AB}]$, $\alpha = (1 + \epsilon/\epsilon' K[B])^{-1}$, and τ_{AB} is the radiative lifetime of $A^* \cdot B$.

(d) Under what conditions will the Stern–Volmer plot be a linear function of B_0?

6.3. In acridine quenching by various amines in 0.03 M NaOH the experimental values of γk_{diff} vary from 2.8×10^7 M^{-1} sec^{-1} (NH_3 quencher) to 3.0×10^9 M^{-1} sec^{-1} (ethyl$_3$N quencher). Assuming that meaningful measurements require that $0.95 \geq \varphi/\varphi_0 \geq 0.05$, what quencher concentrations are needed?

6.4. Which of the following pairs of molecules will be more likely to undergo intersystem crossing?

(i) H_2O or H_2Te.

(ii) Fluorene ($E_{S_1} - E_{T_1} = 113$ kJ mol^{-1}) or biacetyl ($E_{S_1} - E_{T_1} =$ 25 kJ mol^{-1}).

6.5. The gas-phase fluorescence efficiency and spectral structure of benzene are invariant from 5×10^{-6} atm to 10^{-4} atm.

(a) What conclusion does this suggest about the collisional efficiency of intersystem crossing?

(b) Assuming $T \sim 300$ K and a collision diameter of 0.4 nm estimate the number of collisions required to induce an $S_1 \rightarrow T_1$ transition.

6.6. Given the information on triplet and singlet energies shown in Table 6.4, which compounds in the table will sensitize or quench chrysene (a) fluorescence, (b) phosphorescence?

Table 6.4

Compound	$E(S_1)$ (kJ mol^{-1})	$E(T_1)$ (kJ mol^{-1})
Anthracene	310	176
Benzene	481	356
Benzophenone	310	289
Biacetyl	255	230
Chrysene	331	238
Fluorene	397	285
Naphthalene	377	255
Phenanthrene	339	259
Quinoline	381	259
Triphenylene	343	280

6.7. Which of the following compounds will quench naphthalene fluorescence at a diffusion-controlled rate: benzene, benzophenone, biacetyl, fluorene, quinoline, or triphenylene?

6.8. Describe the general shape of the vibronic spectrum observed for excitation from the $v = 0$ level of the ground state for the potential functions of Figs. 6.4a, b, and d.

6.9. The phosphorescence decay of benzophenone in benzene solution is consistent with the rate law

$$d[T_1]/dt = -k_1[T_1] - k_2[T_1]^2$$

where, in 1 M benzophenone, k_2 is close to the diffusion-controlled limit, $k_2 = 1.1 \times 10^{10}\ M^{-1}\ \text{sec}^{-1}$.

(a) Estimate the distance between benzophenone molecules in a 1 M solution.

(b) At intermolecular distances $\lesssim 1.5$ nm, "hopping" excitation transfer

$$T_1 + S_0' \xrightarrow{k_H} S_0 + T_1'$$

may be important. At a separation R the probability of "hopping" is proportional to $e^{-R/L}$, where L is a constant. Show that the "hopping" rate $\propto [T_1][S_0]L^3$.

(c) If "hopping" excitation transfer significantly increases the effective mobility of triplet benzophenone, show that it contributes a term proportional to $k_H[S_0]$ to the second-order rate constant k_2.

(d) What conclusions can you draw from the following measurements of k_1 and k_2 [T.-S. Fang, R. Fukuda, R. E. Brown, and L. A. Singer, *J. Phys. Chem.* **82**, 246 (1978)]?

[Benzophenone] (M):	0.001	0.005	0.01	0.015	0.5	1.0
$10^{-5}k_1$ (sec^{-1}):	1.3	1.4	1.3	1.6	2.9	5.0
$10^{-10}k_2$ ($M^{-1}\ \text{sec}^{-1}$):	0.9	0.7	0.9	0.7	1.1	1.1

(e) Suggest experimental conditions that will quench diffusion but permit hopping.

6.10. What is the likely product when the substituted cyclohexadiene

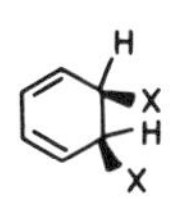

undergoes rearrangement to form a 1,3,5-hexatriene via (a) a thermal pathway; (b) a photochemical pathway?

Table 6.5

$CClF_3$ at 0.079 atm		CCl_3F at 0.079 atm	
E_{abs} (kJ mol^{-1})	CPF × 100	E_{abs} (kJ mol^{-1})	CPF × 100
77.8	1.58	55.6	2.8
56.9	0.374	46.9	0.8
50.2	0.120	41.4	0.4
35.1	0.0060	28.9	0.009

6.11. In laser-activated reactions progress is defined in terms of conversion per flash (CPF) and temperature corresponds to the energy absorbed per mole of reactants, E_{abs}.

(a) Show that in a first-order process $-d \ln [A]/dt \propto$ (CPF).
(b) Show that activation energy corresponds to the slope of a plot of ln(CPF) vs. $(-1/E_{abs})$. What assumptions are required to make this correspondence?
(c) Experiments on laser-induced decomposition of $CClF_3$ and CCl_3F [D. F. Dever and E. Grunwald, *J. Am. Chem. Soc.* **98**, 5055 (1976)] yielded the results shown in Table 6.5. Estimate E_{act} and its variance for the two decompositions.
(d) To what temperatures do the values of E_{abs} correspond?

6.12. The fluorescence efficiency of benzene solutions containing 5 : 1 mixtures of 1-C(hloroanthracene) and P(erylene) was studied for values of $[P] \lesssim 0.006\ M$ [E. T. Bowen and B. Brockelhurst, *Trans. Faraday Soc.* **49**, 1131 (1953)]. The fluorescence efficiency of P *increases* as solute concentration increases.

(a) Show that this phenomenon is rationalized if there is excitation transfer with C acting as donor and P acting as acceptor.
(b) Assume that excitation is transferred or decays by the following reactions:

(1) $C^* \rightarrow C + h\nu$
(2) $C^* \rightarrow\rightarrow C$
(3) $C^* + C \rightarrow 2C$
(4) $C^* + P \rightarrow C + P$
(5) $P^* \rightarrow P + h\nu$
(6) $P^* \rightarrow\rightarrow P$
(7) $P^* + P \rightarrow 2P$
(8) $P^* + C \rightarrow P + C$
(9) $C^* + P \rightarrow C + P^*$

Show that

$$\tilde{\varphi}^0/\tilde{\varphi}_C = \{1 + (k_3[C] + (k_4 + k_9)[P])\tau_C\}(1 + \gamma)$$
$$\tilde{\varphi}^0/\tilde{\varphi}_P = \{1 + (k_7[P] + k_8[C])\tau_P\}(1 + \gamma)/(f_9 + \gamma)$$

where τ_P and τ_C are the radiative lifetimes; γ determines the fraction of radiation absorbed by C and P, $\gamma \equiv \epsilon_P[P]/\epsilon_C[C]$; $\tilde{\varphi}^0$ is the total fluorescence efficiency of the mixture at infinite dilution; $\tilde{\varphi}_C$ and $\tilde{\varphi}_P$ are the concentration-dependent fluorescence efficiencies; f_9 is the fraction of excitation transfer,

$$f_9 \equiv k_9\tau_C[P]/[1 + (k_3[C] + (k_4 + k_9)[P])\tau_C]$$

(c) When $[P] = 0.002\ M$, $\tilde{\varphi}_P/\tilde{\varphi}_P{}^0 \sim 2$; estimate the range of the C–P interaction.
(d) At infinite dilution $\tilde{\varphi}_P{}^0/\tilde{\varphi}^0 \sim \frac{2}{3}$. Estimate γ and the extinction coefficient ratio ϵ_P/ϵ_C.
(e) Measured fluorescence efficiencies are

$[P]\times10^3\ (M)$:	1	2	3.4	4.5	5.5	6.6
$\tilde{\varphi}_P/\tilde{\varphi}_C$:	4.0	5.7	8.0	9.0	12	12
$\tilde{\varphi}^0/\tilde{\varphi}_C$:	3.5	3.8	4.2	4.6	5.7	5.1

Assume k_7 and k_8 are negligible and estimate k_9; τ_C is 1.2×10^{-8} sec. (The diffusion limit in benzene is $\sim10^{10}\ M^{-1}\ \text{sec}^{-1}$.)

General References

Photochemistry

P. G. Ashmore, F. S. Dainton, and T. M. Sugden, eds., *Photochemistry and Reaction Kinetics* (Cambridge: University Press, 1967).
A. Cox and T. J. Kemp, *Introductory Photochemistry* (London: McGraw–Hill, 1971).
J. P. Simons, *Photochemistry and Spectroscopy* (London: Wiley-Interscience, 1971).

Energy Transfer

L. Krause, *Adv. Chem. Phys.* **28**, 267 (1975).
D. Secrest, *Ann. Rev. Phys. Chem.* **24**, 379 (1973).
I. W. M. Smith, in *Gas Kinetics and Energy Transfer*, P. G. Ashmore and R. J. Donovan, eds. (London: Chemical Society, 1977), pp. 1–56.

Chemiluminescence

T. Carrington and J. C. Polanyi, in *MTP International Review of Science, Physical Chemistry*, Series 1, Vol. 9, J. C. Polanyi, ed. (London: Butterworths, 1972), pp. 135–171.

Orbital Symmetry Correlations

T. L. Gilchrist and R. C. Storr, *Organic Reactions and Orbital Symmetry* (Cambridge: University Press, 1972).

R. G. Pearson, *Symmetry Rules for Chemical Applications* (New York: John Wiley, 1976).

Lasers in Chemistry

M. J. Berry, *Ann. Rev. Phys. Chem.* **26**, 259 (1975).

S. Kimel and S. Speiser, *Chem. Revs.* **77**, 437 (1977).

J. T. Knudtson and E. M. Eyring, *Ann. Rev. Phys. Chem.* **25**, 255 (1974).

C. B. Moore, ed., *Chemical and Biochemical Applications of Lasers* (New York: Academic Press, 1974), Vol. 1, Chapters 6 and 7.

7

Nonstationary State Mechanisms

7.1. Introduction

In the last two chapters chemical mechanisms were studied under simplified conditions. An attempt was made to identify each elementary step individually. Naturally this was not always possible. Competition, reverse reaction, consecutive reaction, and inhibition all complicated the analyses. However, by limiting consideration to systems for which the steady-state assumption is appropriate, reasonably simple portraits of reaction could be drawn.

Problems of practical interest such as those that arise in combustion, explosion, metabolism, and manufacturing processes represent a more formidable challenge. Here the system is far from equilibrium. Various forms of feedback may operate so that a wide range of behavior can be found. The most familiar form of feedback leads to explosion. An exothermic reaction evolves heat thus raising the system's temperature. At the higher temperature the reaction velocity is accelerated, the rate of heat evolution is also accelerated, and therefore so is the rate of temperature increase. Unless enough heat can be lost to the surroundings, explosion is inevitable.

A less familiar form of coupling involves chemical feedback. Imagine two parallel competing processes. As we have seen, each retards the other since both deplete the same pool of reactants. If, in addition, an intermediate produced in one pathway is an intermediate reactant in the other, the two paths are no longer parallel and independent. Instead the intermediate's concentration can control whichever of the two pathways is dominant. In a system far from equilibrium there may be periodic switching from one pathway's dominance to the other's. As a consequence the concentration of some components can oscillate in homogeneous systems. Such behavior is particularly evident in biological systems where numerous periodic processes occur in nature.

The oscillations resulting from coupled competing processes need not just involve a temporal change of concentration. Diffusion, temperature, and voltage also couple with chemical reaction resulting in concentration waves, temperature oscillations, or voltage oscillations. Under special conditions a system may even be stable in more than one stationary state and, as a system evolves, a chemical "hysteresis" can be observed.

7.2. Thermal Explosion

In the treatment of chain reactions (Section 5.6) it was clear that under suitable conditions the system could become unstable. In an exothermic reaction, the temperature rises thus accelerating the reaction. Positive feedback causes another increase in temperature. Unless depletion of the reactants or heat loss to the surroundings is sufficiently rapid, the coupling of temperature and reaction rate sets the stage for an explosion.

There are three thermal effects: thermal conduction within the reaction vessel, described by (2.27); heating due to reaction exothermicity; and heat loss due to interaction with the surroundings maintained at a temperature T_0. The local temperature is the solution of the differential equation

$$\frac{\partial T}{\partial t} = \left(\varkappa \nabla^2 T + Q \frac{d\xi}{dt}\right) \frac{V}{C_v} - \gamma(T - T_0) \tag{7.1}$$

The second term is the chemical contribution and the last one accounts for coupling with the surroundings; ξ is the progress variable for chemical reaction, Q is the *enthalpy release* per mole of reaction,[1] and γ accounts for the effectiveness of thermal coupling between system and surroundings.[2] Solution of (7.1) requires knowing how the rate of reaction, $d\xi/dt$, depends upon T. Assume a simple situation and consider a single, irreversible

[1] In general more than one chemical process contributes to (7.1) and $Q\xi$ is generalized to $\sum Q_i \xi_i$. The resulting equations are extraordinarily complex.

[2] If the system were adiabatic γ would be zero.

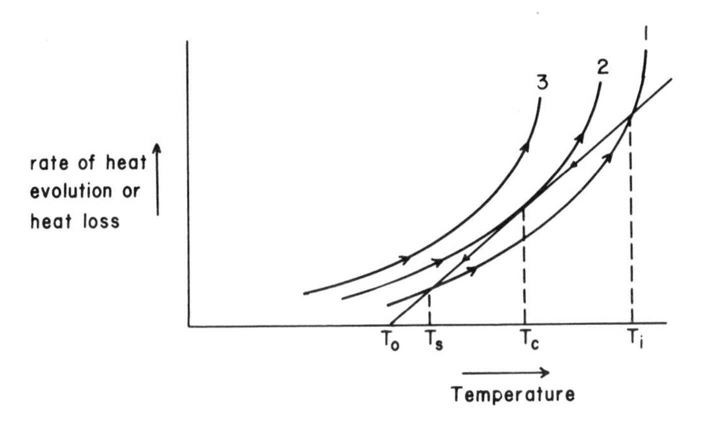

Fig. 7.1. Heat evolution curves (1, 2, 3) and heat loss curve (straight line) as a function of temperature.

bimolecular reaction

$$A + B \xrightarrow{k} \text{products}$$

for which the rate of progress is

$$\frac{d\xi}{dt} = k(T)(A_0 - \xi)(B_0 - \xi) \tag{7.2}$$

where A_0 and B_0 are the initial reactant concentrations. Depending upon initial conditions (7.1) and (7.2) may or may not permit an explosion.

Analysis of (7.1) and (7.2) is difficult; however, some limiting cases are tractable. Ignoring depletion ($A_0 \gg \xi$, $B_0 \gg \xi$) and assuming homogeneity ($\nabla^2 T = 0$), combining (7.2) with (7.1) yields

$$\frac{dT}{dt} \approx k(T)A_0B_0\frac{QV}{C_v} - \gamma(T - T_0) \tag{7.3}$$

The first (heat evolution) term accounts for chemical heating and the second (heat loss) term accounts for conduction to the surroundings. Their separate temperature dependences are shown in Fig. 7.1, where the effect of temperature on Q, C_v, and γ is not considered. Only the rapid increase of rate constant is accounted for. Due to activation effects heat evolution increases exponentially as T increases; heat loss is but a linear function of T.[3]

[3] For an extended treatment with examples see N. N. Semenov, *Some Problems in Chemical Kinetics and Reactivity*, translation by M. Boudart (Princeton, N.J.: Princeton University Press, 1958), Vol. 2, Chapter 8.

Consider low initial reactant density (curve 1). At low temperature heat evolution exceeds heat loss. The temperature increases until T_s is reached when the two rates balance and there is a steady state. Any fluctuations around T_s, possibly caused by local inhomogeneities, are rapidly damped—for $T > T_s$ the cooling rate exceeds the heating rate while for $T < T_s$ the reverse is true. In either case the system is stable; the temperature returns to T_s. Another steady state appears to occur when $T = T_i$; however, this is unstable. A fluctuation for which $T < T_i$ leads to rapid cooling while one for which $T > T_i$ leads to unchecked heating. At T_i the system is on the brink of exploding and may be driven in that direction by local temperature fluctuations. Since any process which allows T to exceed T_i, no matter how briefly, leads to thermal instability the temperature T_i is an ignition temperature. At this temperature a thermal explosion occurs even though the walls of the system remain at T_0; heat dissipation is too slow to forestall explosion.

As reactant concentration increases, heat evolution at fixed temperature also increases. Critical conditions finally arise such that explosion is inevitable. At the critical concentration (curve 2) T_s and T_i coalesce; the steady state is unstable to local temperature fluctuations and ignition occurs. The external temperature is too high to permit sufficient heat loss to the surroundings. At yet higher initial concentration (curve 3) no steady state is possible when the external temperature is T_0; explosion always occurs.

Uncritical consideration of Fig. 7.1 suggests that any exothermic reaction has an ignition point; this is a consequence of our simplification of (7.1) and (7.2). We assumed no significant depletion occurs during the preignition stages. However, if A_0 and/or B_0 are too small depletion keeps the rate of progress (7.2) under control. As a consequence the unchecked temperature increases suggested by (7.3) and illustrated in Fig. 7.1 need not occur.

Coupling between system and surroundings greatly affects the possibility of thermal explosion. With adiabatic walls, explosion is only avoided if there is rapid depletion. The stronger the diathermal coupling, the greater the heat loss and the less likelihood there is of an explosion. Thus the size and shape of the reaction vessel are significant variables; increasing the surface-to-volume ratio decreases the possibility of ignition.

The qualitative features which control thermal explosion do not depend upon the specific chemical process taking place. Whether the reaction is a simple one such as (7.2) or a multistep chain reaction, the principles remain the same. Chemical heat evolution accelerates the reaction which then produces more heating.

7.3. Population Explosion—Autocatalysis

In Section 5.6 we noted that, for a branching-chain reaction, the *population* of chain carriers may grow without limit. As a consequence reaction velocity accelerates independent of chemical heating. At low reactant concentrations the rate law for carrier formation is, from (5.35),

$$\frac{d[\mathrm{R}]}{dt} = nr_I + (\alpha - 1)r_B[\mathrm{R}] - r_T[\mathrm{R}], \qquad \alpha > 1 \tag{7.4}$$

Here n chain carriers are formed in the initiation step for which the rate per carrier is r_I. In the branching step (or steps) α carriers are formed for each one that is consumed; the rate is $(\alpha - 1)r_B[\mathrm{R}]$. The termination rate is $r_T[\mathrm{R}]$.[4]

To illustrate the general considerations consider the hydrogen–oxygen system for which the important elementary steps at low pressure appear to be the following,[5] where S is used for surface and the values in parentheses represent ΔH (kJ mol^{-1}):

$$\begin{array}{lll}
\textit{Initiation:} & \mathrm{H_2 + S \xrightarrow{k_1} H \cdot S + H} & \\
\textit{Branching:} & \left\{\begin{array}{l} \mathrm{H + O_2 \xrightarrow{k_2} HO + O} \\ \mathrm{O + H_2 \xrightarrow{k_3} HO + H} \end{array}\right. & \begin{array}{l} (+64) \\ (-\ 4) \end{array} \\
\textit{Propagation:} & \mathrm{HO + H_2 \xrightarrow{k_4} HOH + H} & (-58) \\
\textit{Termination:} & \left\{\begin{array}{l} \mathrm{H + S \xrightarrow{k_5} stable\ species} \\ \mathrm{HO + S \xrightarrow{k_6} stable\ species} \end{array}\right. &
\end{array} \tag{7.5}$$

Chain initiation occurs by surface reaction, as in this example, or homogeneously, as in the $CO + O_2$ chain.[6] Since heterogeneous processes are very significant in the kinetics of branching chains great care in surface preparation is required to achieve reproducible results.[7] The propagation

[4] See N. N. Semenov, *Some Problems in Chemical Kinetics and Reactivity*, translation by M. Boudart (Princeton, N.J.: Princeton University Press, 1958), Vol. 2, Chapter 9 for a general discussion of ignition in branching-chain systems.

[5] S. W. Benson, *The Foundations of Chemical Kinetics* (New York: McGraw–Hill, 1960), p. 454.

[6] A. S. Gordon and R. H. Knipes, *J. Phys. Chem.* **59**, 1160 (1955).

[7] N. N. Semenov, *Some Problems in Chemical Kinetics and Reactivity*, translation by M. Boudart (Princeton, N.J.: Princeton University Press, 1958), Vol. 2, Chapter 8.

and branching steps of (7.5) are typical; each is bimolecular. In a mixture of fixed composition we expect that

$$r_B \propto p_r \exp(-E_B/RT) \tag{7.6}$$

where p_r is the *reactant* pressure and E_B an activation energy for branching. At low pressure termination is always heterogeneous; any probable homogeneous bimolecular reaction of R must, at the very least, propagate the chain. Termination is diffusion controlled. In Section 2.5 we showed that the characteristic time to diffuse a distance d in a linear system is $\sim d^2/D$, where D is the diffusion coefficient. A similar result is obtained in three-dimensional systems.[8] Using (2.41), the termination rate is

$$r_T \propto D/d^2 \propto T^{3/2}/pd^2 \tag{7.7}$$

where p is the *total* pressure and d is now the diameter of a spherical or a long cylindrical reaction vessel.

Ignoring the effect of reaction on the various r,[9] (7.4) can be integrated

$$[\mathrm{R}] = nr_I(e^{\lambda t} - 1)/\lambda, \qquad \lambda = (\alpha - 1)r_B - r_T \tag{7.8}$$

It is assumed that initially $[\mathrm{R}] = 0$. Now, if $(\alpha - 1)r_B < r_T$, λ is negative and, after sufficient time, [R] reaches the steady-state limit (5.36). However, if $\lambda > 0$ there is positive feedback. The system is unstable, branching dominates, and [R] grows exponentially. There is a population explosion of chain carriers. In the $H_2 + O_2$ system, which is typical, the branching and propagation sequence, for which the stoichiometry is $3H_2 + O_2 = 2H_2O + 2H$, is only slightly exothermic. Here 28 kJ are released per mole of H_2O while $\Delta H_f \sim -240$ kJ mol^{-1}. The bulk of the chemical heating occurs as a consequence of termination. Since the population explosion leads to more termination the system heats rapidly and ignition occurs.

From (7.8) the critical condition for explosion is $\lambda = 0$ or with (7.6) and (7.7)

$$p_r \exp(-E_B/RT) \propto T^{3/2}/pd^2$$

For a buffer-gas pressure of p_0 we find a critical pressure, p_1,

$$\begin{aligned} p_1 &\propto T^{3/4} \exp(E_B/2RT)/d, \qquad p_0 = 0 \\ p_1 &\propto T^{3/2} \exp(E_B/RT)/p_0 d^2, \qquad p_0 \gg p_r \end{aligned} \tag{7.9}$$

[8] L. D. Landau and E. M. Lifschitz, *Fluid Mechanics* (London: Pergamon, 1959), Chapter 6.

[9] This is equivalent to ignoring depletion of reactants and thermal heating.

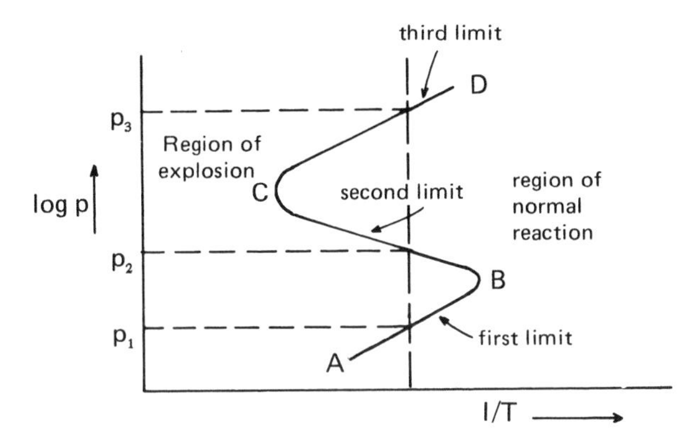

Fig. 7.2. Explosion limits for a typical branched-chain reaction. The region ABC is known as the explosion peninsula. The curve ABCD is the boundary between explosive and bounded reaction rates.

Ignition is expected when $p > p_1$. Reality is more complicated. The explosion limits characteristic of branching chains are illustrated in Fig. 7.2. Increasing pressure with fixed temperature and composition leads to ignition at p_1. The unexpected feature is the region of stability at high pressure, $p_2 < p < p_3$. The reaction velocity at constant temperature and composition is plotted in Fig. 7.3. For $p < p_1$ the rate is very slow, if not imperceptible. Above p_2 the rate is again very slow but here it is measurable, increasing gradually until p_3 is reached. The behavior near p_1 and p_2 is a significant indicator of a population explosion. A system on the verge of thermal explosion does not have an imperceptible reaction velocity.

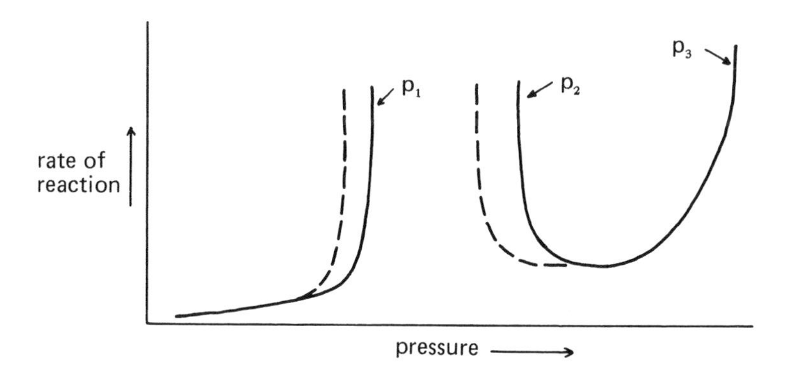

Fig. 7.3. Rate of a branching-chain reaction as a function of pressure for fixed temperature and composition. The dashed lines illustrate the effect of inert buffer gas on the first and second explosion limits.

Consider p_1; its temperature dependence corresponds to (7.9). Increasing the volume-to-surface ratio (increasing d) or increasing buffer-gas pressure is predicted to lower p_1. Precisely such behavior is found in the H_2–O_2, or other branched-chain systems as indicated in Fig. 7.3. Note that were the explosion at p_1 thermal in nature a buffer gas would stabilize the system as C_v would increase and as a result the heating rate would drop (Fig. 7.1).

In order for the system to be stable at high pressure, a process that *consumes* chain carriers must become significant, otherwise autocatalysis would remain dominant. Chain-trapping generally involves a three-body step

$$R + A + M \rightharpoonup AR + M \tag{7.10}$$

where M is buffer gas and A is reactant. As a result adding inert buffer gas to a mixture of fixed composition shifts p_2 to lower values, as shown in Fig. 7.3. The trapping reaction in the H_2–O_2 system is[10]

$$H + O_2 + M \xrightarrow{k_7} HO_2 + M, \qquad \Delta H \sim -167 \text{ kJ mol}^{-1} \tag{7.11}$$

HO_2 is stabilized by heterogeneous reaction or by combining with chain carriers.[11] In general, quenching reduces the carrier formation rate (7.4) by adding a term of the form $-r_Q[R]$. Assuming an elementary step like (7.11) is typical

$$r_Q \propto pp_r \exp(-E_Q/RT) \tag{7.12}$$

where E_Q is the activation energy of the quenching reaction.[12] Modifying (7.8), the critical condition is

$$\lambda \equiv -r_Q + (\alpha - 1)r_B - r_T = 0$$

which, if p_2 is fixed by the quenching-to-branching balance, yields

$$p_2 + p_0 \propto \exp[-(E_B - E_Q)/RT] \tag{7.13}$$

[10] S. W. Benson, *The Foundations of Chemical Kinetcs* (New York: McGraw–Hill, 1960), p. 454.

[11] Possible trapping reactions are $H + HO_2 \rightharpoonup H_2 + O_2$, $HO + HO_2 \rightharpoonup H_2O + O_2$; reactions prejudicial to branching are $H + HO_2 \rightharpoonup 2OH$, $O + HO_2 \rightharpoonup OH + O_2$. All are highly exothermic.

[12] In the CO–O_2 system a rather different reaction sequence is significant: initiation $CO + O_2 \rightharpoonup CO_2 + O$; branching and propagation $O + CO \rightharpoonup CO_2^*$, $CO_2^* + O_2 \rightharpoonup CO_2 + 2O$; termination $CO_2^* + S \rightharpoonup CO_2$, $CO_2^* + M \rightharpoonup CO_2 + M$. Modification of (7.12) is needed in this case; see A. S. Gordon and R. H. Knipes, *J. Phys. Chem.* **59**, 1160 (1955). The chain carrier CO_2^* is probably *electronically* excited CO_2; see B. H. Mahan and R. B. Solo, *J. Chem. Phys.* **37**, 2669 (1962).

Since quenching reactions require little activation, $E_B \gg E_Q$, and p_2 should increase with increasing T (Fig. 7.2). As already mentioned adding buffer decreases p_2.

No general theory of the third explosion limit has been advanced; however, it is generally assumed that such explosions are thermal in nature and reflect the system's inability to dissipate heat. However, in the H_2–O_2 system a new kinetic pathway may become important. As diffusion of HO_2 is retarded surface reaction is less important. The reaction

$$HO_2 + H_2 \rightarrow H_2O + OH$$

may be significant; HO_2 then acts as a propagator rather than a chain trap.[13]

7.4. The Lotka Problem: Chemical Oscillations

We have just seen that positive feedback can lead to explosive reaction, either by unchecked heating or unchecked population growth. We now consider a different example of instability. Consider the hypothetical mechanism

$$A + X \xrightarrow{k_1} 2X, \qquad X + Y \xrightarrow{k_2} 2Y, \qquad Y \xrightarrow{k_3} Z \tag{7.14}$$

which does not describe any known chemical process but has a crude ecological analog. If A is grass, X are deer, Y are wolves, and Z are dead wolves (or fertilizer) then deer eat grass and reproduce, wolves eat deer and reproduce, and wolves die. In wilderness management it is well known that animal populations oscillate rather than settling into a steady state. As we shall see (7.14) can account for such behavior.

Before analyzing this mechanism consider the related problem, where reverse reaction is possible,

$$A + X \underset{k_{-1}}{\overset{k_1}{\rightleftharpoons}} 2X, \qquad X + Y \underset{k_{-2}}{\overset{k_2}{\rightleftharpoons}} 2Y, \qquad Y \underset{k_{-3}}{\overset{k_3}{\rightleftharpoons}} Z \tag{7.15}$$

For simplicity assume that [A] has a constant value A_0 which may be arranged if, initially, A is in great excess or if the system is open and A is

[13] N. N. Semenov, *Some Problems in Chemical Kinetics and Reactivity*, translation by M. Boudart (Princeton, N. J.: Princeton University Press, 1958), Vol. 2, p. 196; J. Nicholas, *Chemical Kinetics* (New York: Halstead, 1976), p. 148.

continually replenished. Since reverse reaction is possible (7.15) admits of a true equilibrium state. The equilibrium constraints are

$$k_1 A_0 = k_{-1} X_0, \qquad k_2 X_0 = k_{-2} Y_0, \qquad k_3 Y_0 = k_{-3} Z_0 \tag{7.16}$$

and the coupled rate equations are

$$\begin{aligned} \frac{d[\mathrm{X}]}{dt} &= k_1 A_0 [\mathrm{X}] - k_{-1}[\mathrm{X}]^2 - k_2[\mathrm{X}][\mathrm{Y}] + k_{-2}[\mathrm{Y}]^2 \\ \frac{d[\mathrm{Y}]}{dt} &= k_2[\mathrm{X}][\mathrm{Y}] - k_{-2}[\mathrm{Y}]^2 - k_3[\mathrm{Y}] + k_{-3}[\mathrm{Z}] \\ \frac{d[\mathrm{Z}]}{dt} &= k_3[\mathrm{Y}] - k_{-3}[\mathrm{Z}] \end{aligned} \tag{7.17}$$

If there are small perturbations about the equilibrium state the system, as discussed in Section 4.9, relaxes to equilibrium (see Problem 7.4). The displacement variables $[\mathrm{X}] - X_0$, etc., are linear combinations of decaying exponential functions. Direct negative feedback, due to reverse reactions like $2\mathrm{X} \rightharpoonup \mathrm{X} + \mathrm{A}$, keep [X] and [Y] under control no matter what the initial conditions.

If reverse reaction is forbidden no true equilibrium exists. No direct feedback controls [X]; instead it is kept in check by reaction with [Y]. From (7.14) or (7.17) the coupled rate equations, maintaining $[\mathrm{A}] = A_0$, are[14]

$$\begin{aligned} \frac{d[\mathrm{X}]}{dt} &= k_1 A_0 [\mathrm{X}] - k_2[\mathrm{X}][\mathrm{Y}] \\ \frac{d[\mathrm{Y}]}{dt} &= k_2[\mathrm{X}][\mathrm{Y}] - k_3[\mathrm{Y}] \end{aligned} \tag{7.18}$$

There appears to be a steady state when $d[\mathrm{X}]/dt = d[\mathrm{Y}]/dt = 0$ in which case

$$k_2 Y_0 = k_1 A_0, \qquad k_2 X_0 = k_3 \tag{7.19}$$

which are substantially different constraints than (7.16). Defining displacement variables

$$[\mathrm{X}] = \xi(t) + X_0, \qquad [\mathrm{Y}] = \eta(t) + Y_0 \tag{7.20}$$

[14] There will be a steady increase in Z but its growth is of no immediate interest. The system (7.18) was first analyzed by A. J. Lotka, *J. Am. Chem. Soc.* **42**, 1595 (1920); *Proc. Nat. Acad. Sci.* (*U.S.A.*) **6**, 410 (1920).

and substituting in (7.18) yields, with (7.19),

$$\begin{aligned}\dot{\xi} &= -k_2X_0\eta - k_2\xi\eta = -k_2\eta(\xi + X_0)\\ \dot{\eta} &= k_2Y_0\xi + k_2\xi\eta = k_2\xi(\eta + Y_0)\end{aligned} \tag{7.21}$$

If the displacements are small the quadratic terms can be dropped. Using the matrix method of Chapter 5 and assuming that ξ and η are linear combinations of decaying exponential functions $e^{-\lambda t}$, the decay constants λ are solutions of the determinantal equation

$$\begin{vmatrix} \lambda & -k_2X_0 \\ k_2Y_0 & \lambda \end{vmatrix} = 0 \tag{7.22}$$

from which we find $\lambda^2 + k_2{}^2X_0Y_0 = 0$ or, with (7.19),

$$\lambda = \pm i\omega = \pm ik_2(X_0Y_0)^{1/2} = \pm i(k_1k_3A_0)^{1/2} \tag{7.23}$$

The "decay constants" are imaginary! The displacement variables, being linear combinations of the functions $e^{i\omega t}$ and $e^{-i\omega t}$ (or equivalently $\sin \omega t$ and $\cos \omega t$), do not describe relaxation to a steady state. Instead a system perturbed from the steady state undergoes perpetual oscillation. The origin in the ξ–η plane (i.e., X $= X_0$, Y $= Y_0$) is not a steady state at all; rather it is a *singular point which is not accessible*. Even if the system were prepared at the steady state, it could not remain there. Due to the fluctuations which constantly arise, a system in which initially $\xi = \eta = 0$ is unstable. Fluctuations create displacements and oscillations commence.

The analysis of the displacement equation (7.21) has been limited to small values of ξ and η. To generalize it is easier to determine the functional relationship between the displacement variables ξ and η than to find the separate solutions $\xi(t)$ and $\eta(t)$. Divide $\dot{\xi}$ by $\dot{\eta}$; the result is

$$\frac{\dot{\xi}}{\dot{\eta}} = \frac{d\xi}{d\eta} = -\frac{\eta}{\xi}\frac{X_0 + \xi}{Y_0 + \eta} \tag{7.24}$$

which may be rewritten

$$d\xi\left(1 - \frac{X_0}{\xi + X_0}\right) = -d\eta\left(1 - \frac{Y_0}{\eta + Y_0}\right)$$

and integrated

$$\xi - X_0 \ln(\xi + X_0) + \eta - Y_0 \ln(\eta + Y_0) = \text{const} \tag{7.25}$$

The equation determines a set of curves that describe the relationship between $\xi(t)$ and $\eta(t)$. The curves are characterized by certain properties:

(i) they *cannot cross* since the slope, $d\xi/d\eta$, is uniquely specified at each point in the ξ–η plane [from (7.24)];
(ii) as [X] > 0 and [Y] > 0, $\xi > -X_0$ and $\eta > -Y_0$ [from (7.20)];
(iii) unbounded increase of ξ or η is impossible [from (7.25)];
(iv) for [X] > 0 and [Y] > 0, to each η there corresponds one value of $\xi > 0$ and one value of $\xi < 0$; similarly there are two η for each ξ [from (7.25)];
(v) when $\xi > 0$ and $\eta > 0$, $\dot{\xi} < 0$ and $\dot{\eta} > 0$, etc. [from (7.21)].

The general features of such curves are indicated in Fig. 7.4a. They form a set of nonintersecting closed curves as shown in Fig. 7.4b. The singularity is the origin in the ξ–η plane. Trajectories do *not* terminate at this point. Instead a system displaced from the singularity oscillates in perpetuity. Furthermore the larger the initial perturbation the larger the amplitude of the oscillations. No relaxation occurs, even for large amplitude perturbations.[15] What is not apparent from Fig. 7.4 is the oscillation period. However, by expanding $\xi(t)$ and $\eta(t)$ in Fourier series the oscillation frequency (7.23) may be shown to be a general feature of the rate law (7.18); ω does not depend upon the perturbation amplitude.[16]

The model mechanism predicts undamped oscillations in the concentrations of intermediates X and Y. The concentrations of reactant A and product Z do not oscillate; however, their *rates* of consumption and formation do. Three conditions were needed to generate such behavior:

(i) continuous flow of matter into the system ([A] is constant);
(ii) a mechanism with coupled indirect feedback steps;
(iii) a system far from equilibrium (some steps are irreversible).

The constraints appear to be general.[17] Unlike explosion where feedback leads to direct acceleration of reactant consumption, oscillatory mechanisms are more subtle. According to (7.18) increasing [X] when [Y] is small

[15] The approach outlined, considering the trajectories in the ξ–η plane instead of $\xi(t)$ and $\eta(t)$ individually, has numerous applications. For an introduction to the theory of trajectory classification see D. A. Sanchez, *Ordinary Differential Equations and Stability Theory* (San Francisco: Freeman, 1968).

[16] A. J. Lotka, *J. Am. Chem. Soc.* **42**, 1595 (1920); *Proc. Nat. Acad. Sci.* (*U.S.A.*) **6**, 410 (1920).

[17] A discussion of the requirements for oscillatory behavior is given by G. Nicolis and J. Portnow, *Chem. Rev.* **73**, 365 (1973); see particularly Sections I. D. and III.A. A more qualitative treatment is given by H. Degn, *J. Chem. Ed.* **49**, 302 (1972).

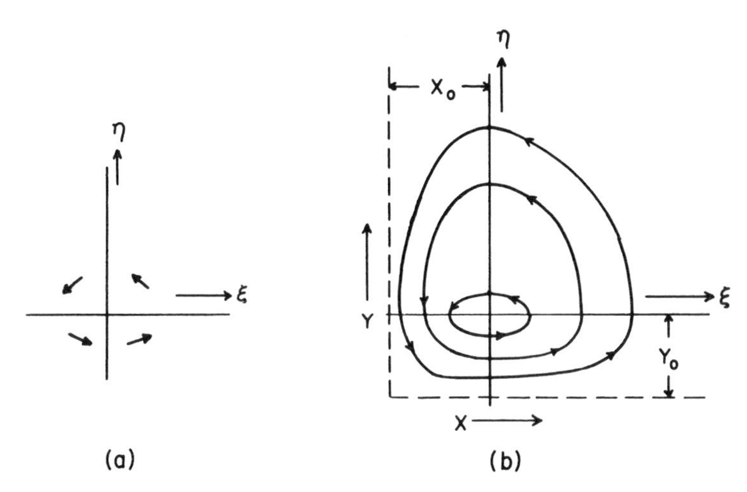

Fig. 7.4. (a) General features of trajectories (7.25). (b) Detailed curves showing three trajectories of the concentration displacement variables ξ and η for different initial displacements. The steady state is the origin of the ξ–η plane. It is inaccessible. [Figure 7.4b adapted from A. J. Lotka, *J. Am. Chem. Soc.* **42**, 1595 (1920).]

leads to acceleration in the rate of production of X. Finally enough Y has formed to consume X with sufficient rapidity so that [X] starts to drop and the first step in (7.14) slows down. As Y is consumed by the third reaction its concentration drops and [X] can build up again resuming the cycle. The net effect on [X] is immediate autocatalysis followed by delayed inhibition by Y. It is *delayed* feedback that characterizes chemical oscillation.

In a closed system reactant concentration must decrease. Thus perpetual oscillations in the concentration of intermediates are not possible. However, if initially A is in great excess and conditions (ii) and (iii) are met, very slowly damped oscillations may be observed. In the remainder of this chapter we discuss some specific examples.[18] Figure 7.5 illustrates allowed and forbidden chemical oscillation. Detailed balance forecloses oscillations about equilibrium, as shown in Fig. 7.5a. Oscillations about the steady state, as shown in Fig. 7.5b, violate no thermodynamic principle.

While the Lotka mechanism exhibits many features found in chemical oscillators, it is misleading in one important qualitative respect. According to Fig. 7.4b the amplitude of the concentration oscillations is determined by the magnitude of the initial perturbation. Such behavior is apparently not characteristic of real systems. Rather, *independent of the initial values of*

[18] For a catalog of chemical, thermochemical, electrochemical, and biochemical oscillators see G. Nicolis and J. Portnow, *Chem. Rev.* **73**, 365 (1973), Section II.

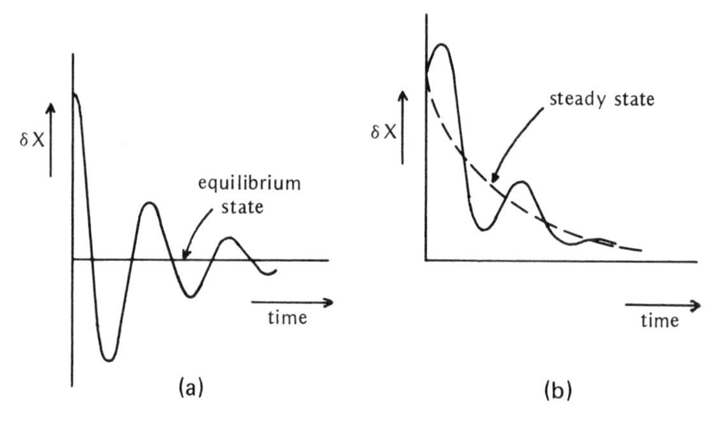

Fig. 7.5. (a) Forbidden and (b) allowed oscillations in the relaxation of a concentration displacement, δX, as a function of time.

intermediate concentrations, chemical oscillators evolve to periodic behavior determined by the rate constants of the elementary steps and by the concentrations of the major reactants. This phenomenon, known as *limit cycle* behavior, has its analog in the evolution of systems displaced from equilibrium; the size of the initial displacement has no effect on the approach to equilibrium. The study of *limit cycle* mechanisms is a complex problem of trajectory analysis and classification.[19]

7.5. The Belousov–Zhabotinskii Reaction

While concentration oscillations have been observed in a variety of systems, only two homogeneous chemical oscillators have been unambiguously established. When malonic acid (HMa) is oxidized by BrO_3^- in sulfuric acid solution using a Ce(IV)/Ce(III) couple as catalyst, Belousov observed that the [Ce(IV)]/[Ce(III)] ratio oscillated periodically.[20] Further studies by Zhabotinskii demonstrated oscillatory behavior when Ce(IV)/Ce(III) was replaced by Fe(III)/Fe(II) or Mn(III)/Mn(II) and

[19] Two such mechanisms, presumed to describe specific oscillating systems, have been formulated. For details see P. Glansdorff and I. Prigogine, *Thermodynamic Theory of Structure, Stability, and Fluctuations* (New York: Wiley-Interscience, 1971), Chapter 14; I. Prigogine and G. Nicolis, *Quart. Rev. Biophys.* **4**, 107 (1971); R. J. Field and R. M. Noyes, *J. Chem. Phys.* **60**, 1877 (1974).

[20] B. P. Belousov, *Sb. Ref. Radiats. Med.* (*Medgiz, Moscow*) **1958**, 145 (1959).

Table 7.1. Range of Reactant Concentrations Permitting Concentration Oscillations in the Belousov–Zhabotinskii Reaction[a]

Species	Concentration (M)
Malonic acid	0.013–0.5
$KBrO_3$	0.013–0.063
$Ce(NH_4)_2(NO_3)_5$	0.0001–0.01
H_2SO_4	0.5–2.5
KBr	Trace ($\lesssim 2 \times 10^{-5}$)

[a] R. F. Field, E. Körös, and R. M. Noyes, *J. Am. Chem. Soc.* **94**, 8649 (1972).

when malonic acid was replaced by other organic acids.[21] Another example of oscillation is the IO_3^-/I_2 catalyzed decomposition of H_2O_2. The production of O_2 is not continuous but proceeds in bursts that are correlated with oscillations in $[I_2]$.[22] Since oxygen is produced, considerable care was required to demonstrate that the oscillations reflect homogeneous processes.[23]

Both the Belousov–Zhabotinskii and the Bray–Liebhafsky systems have been characterized mechanistically.[24] We consider the Belousov reaction in detail. Oscillatory behavior is *not* limited to a narrow range of concentrations. Initial conditions which lead to periodic concentration variation are given in Table 7.1. The period of oscillation is quite dependent upon initial concentrations, as suggested by the Lotka result (7.23); it varies from ~15 to ~200 sec in different experiments. Not only are there oscillations in the [Ce(IV)]/[Ce(III)] ratio but $[Br^-]$, which is present in trace quantities, oscillates as well. Typical results are shown in Fig. 7.6; the amplitudes diminish only very gradually over dozens of cycles. For the experiment il-

(21) A. M. Zhabotinskii, *Dokl. Akad. Nauk. SSSR* **157**, 392 (1964); *Biofizika* **9**, 306 (1964).

(22) W. C. Bray, *J. Am. Chem. Soc.* **43**, 1262 (1921).

(23) W. C. Bray and H. T. Liebhafsky, *J. Am. Chem. Soc.* **53**, 38 (1931); H. T. Liebhafsky, *J. Am. Chem. Soc.* **53**, 896, 2074 (1931); H. T. Liebhafsky and L. S. Wu, *J. Am. Chem. Soc.* **96**, 7180 (1974). Liebhafsky studied the reaction at two widely separated periods in his career; the papers cited represent only a fraction of his work.

(24) Belousov reaction: R. J. Field, E. Körös, and R. M. Noyes, *J. Am. Chem. Soc.* **94**, 8649 (1972); Bray–Liebhafsky reaction: K. R. Sharma and R. M. Noyes, *J. Am. Chem. Soc.* **98**, 4345 (1976).

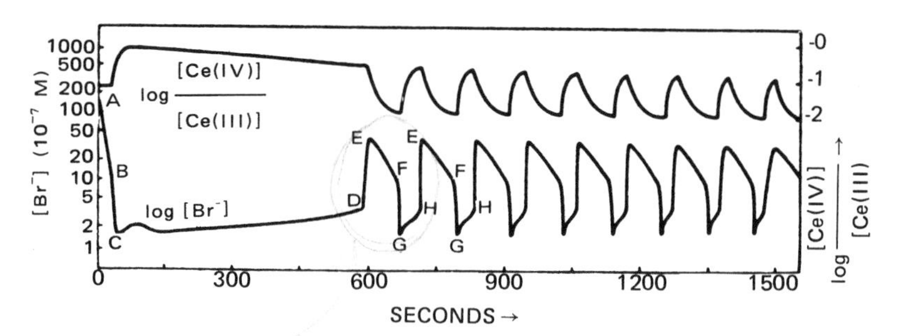

Fig. 7.6. Periodic behavior of $[Br^-]$ and $[Ce(IV)]/[Ce(III)]$ in the Belousov reaction [R. J. Field, E. Körös, and R. M. Noyes, *J. Am. Chem. Soc.* **94**, 8649 (1972), reprinted by permission of the American Chemical Society].

lustrated in Fig. 7.6 cyclical behavior can be characterized as follows:

(i) $[Br^-]$ relatively large and decreasing, $[Ce(IV)]/[Ce(III)]$ ratio decreasing;
(ii) $[Br^-]$ plummets sharply;
(iii) $[Br^-]$ relatively small and increasing, $[Ce(IV)]/[Ce(III)]$ ratio increasing;
(iv) $[Br^-]$ grows abruptly.

The overall stoichiometry, in which none of the species $[Br^-]$, $[Ce(IV)]$, or $[Ce(III)]$ appear, is

$$3BrO_3^- + 3H^+ + 5CH_2(COOH)_2 = 3BrCH(COOH)_2 + 4CO_2 + 2HCOOH + 5H_2O \tag{7.26}$$

The key to understanding the phenomena is knowledge of the chemistry of BrO_3^- which, while a powerful oxidizing agent, does not react directly with organic acids. The brominating agent in (7.26) is Br_2 while production of CO_2 and formic acid is due to attack by Ce(IV) on either HMa or BrMa. As is so often the case, kinetically important species do not appear in the overall stoichiometry.

Two separate sequences account for (7.26). Schematically they may be summarized as

$$BrO_3^- + HMa \xrightarrow{(+Br^-)} BrMa \qquad (A)$$

$$BrO_3^- + HMa \xrightarrow[Br^- \text{ inhibited}]{Ce(IV)/Ce(III)\text{ catalyzed}} HCOOH + CO_2 + Br^- \qquad (B) \tag{7.27}$$

Sequence (A) requires Br^- as a reactant; Br^- also acts indirectly to inhibit sequence (B). When Br^- is depleted reaction switches from (A) to (B). However, (B) replenishes Br^-, becoming self-inhibitory and reinitiating (A). Oscillations arise because of the switching.

The individual steps are complex and will only be outlined.[25] Reduction of BrO_3^- to Br_2 proceeds by consecutive removal of oxygen (two-electron transfer)

$$\begin{aligned}&\text{(a)}\ Br^- + BrO_3^- + 2H^+ \rightharpoonup HBrO_2 + HOBr\\&\text{(b)}\ Br^- + HBrO_2 + H^+ \rightharpoonup 2HOBr\\&\text{(c)}\ Br^- + HOBr + H^+ \rightharpoonup Br_2 + H_2O\end{aligned} \tag{7.28}$$

The Br_2 reacts with HMa

$$\text{(d)}\ Br_2 + HMa \rightharpoonup BrMa + H^+ + Br^- \tag{7.29}$$

The overall process (a) + (b) + 3(c) + 3(d) is sequence (A)

$$\text{(A)}\ 2Br^- + BrO_3^- + 3H^+ + 3HMa \rightharpoonup 3BrMa + 3H_2O \tag{7.30}$$

The slow step in the process is (a). None of the oxybromine intermediates are thermodynamically plausible reactants in one-electron steps such as are required to oxidize Ce(III).

A way to reduce BrO_3^- in one-electron transfers postulates the $BrO_2\cdot$ radical as intermediate,

$$\begin{aligned}&\text{(e)}\ BrO_3^- + HBrO_2 + H^+ \rightharpoonup 2BrO_2\cdot + H_2O\\&\text{(f)}\ Ce(III) + BrO_2\cdot + H^+ \rightharpoonup Ce(IV) + HBrO_2\end{aligned} \tag{7.31}$$

where the net reaction (e) + 2(f) is *autocatalytic*

$$\text{(C)}\ 2Ce(III) + BrO_3^- + 3H^+ + HBrO_2 \rightharpoonup 2Ce(IV) + H_2O + 2HBrO_2 \tag{7.32}$$

It might appear that (C) would rapidly deplete BrO_3^-, however, as $HBrO_2$ disproportionates

$$\text{(g)}\ 2HBrO_2 \rightharpoonup HOBr + H^+ + BrO_3^- \tag{7.33}$$

[25] The analysis is due to R. J. Field, E. Körös, and R. M. Noyes, *J. Am. Chem. Soc.* **94**, 8649 (1972), where corroboratory details are given.

$[HBrO_2]$ is always small and (C) cannot lead to a population explosion. Here the slow step is (e). As long as $[Br^-]$ is sufficiently large (b) competes favorably with (e) and the one-electron sequence cannot proceed. However, when $[Br^-]$ drops sufficiently, as it must in sequence (A), the autocatalytic pathway (C) takes over and $[Br^-]$ drops precipitously since (b) continues. As HOBr is produced in the disproportionation (g) it can react with HMa via (c) and (d). Thus the net reaction of Ce(III), BrO_3^-, and HMa is 2(C) + (g) + (c) + (d)

$$\text{(D)}\ 4\text{Ce(III)} + \text{BrO}_3^- + \text{HMa} + 5\text{H}^+ \rightarrow 4\text{Ce(IV)} + \text{BrMa} + 3\text{H}_2\text{O} \tag{7.34}$$

The individual steps which regenerate Br^- and Ce(III) and complete the cycle have not been as precisely characterized. However, Ce(IV) can oxidize HMa or BrMa in a series of rapid one-electron transfers; the net result is

$$\begin{aligned} &\text{(E)}\ 6\text{Ce(IV)} + \text{CH}_2(\text{COOH})_2 + 2\text{H}_2\text{O} \rightarrow 6\text{Ce(III)} + \text{HCOOH} \\ &\qquad\qquad + 2\text{CO}_2 + 6\text{H}^+ \\ &\text{(F)}\ 4\text{Ce(IV)} + \text{BrCH}(\text{COOH})_2 + 2\text{H}_2\text{O} \rightarrow 4\text{Ce(III)} + \text{HCOOH} \\ &\qquad\qquad + 2\text{CO}_2 + 5\text{H}^+ + \text{Br}^- \end{aligned} \tag{7.35}$$

The set of processes (D), (E), and (F) comprise sequence (B).

Now consider a complete cycle in the oscillatory regime as illustrated schematically in Fig. 7.7. At α, $[Br^-]$ is large; the major pathway is (A) so $[Br^-]$ decreases. The ratio [Ce(IV)]/[Ce(III)] drops as reactions (E) and (F) continue. At β, $[Br^-]$ has become too small to sustain sequence (A); reaction switches to sequence (B) and $[Br^-]$ plummets due to reaction (b) reaching point γ. Then the [Ce(IV)]/[Ce(III)] ratio grows due to (C) and, as more Ce(IV) forms it replenishes $[Br^-]$ via (F). At δ, $[Br^-]$ is large enough to allow switching back to (A); sequence (B) stops abruptly; the remaining Ce(IV) continues to oxidize HMa via (F) and $[Br^-]$ rises sharply. The system has returned to α.

Our analysis of the Belousov reaction has just considered the oscillatory domain. Whether sequence (A) or sequence (B) is followed, initially there must be an induction period before oscillation commences since periodic behavior is triggered by replenishing $[Br^-]$, which cannot occur until [BrMa] is large enough so that (F) becomes important. The behavior illustrated in Fig. 7.6 is consistent with this interpretation. After initiation

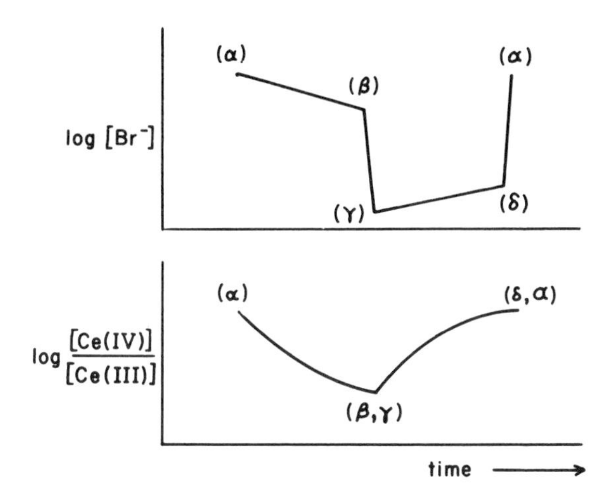

Fig. 7.7. Schematic diagram of one cycle in the Belousov reaction.

$[Br^-]$ drops rapidly and remains small for a considerable time until it rises abruptly signaling the onset of oscillations.

The Belousov reaction satisfies the prerequisite conditions for oscillation. The consumption of reactants is slow; there are irreversible steps so the system is far from equilibrium; there is delayed feedback, in this case both catalytic and inhibitory. The concentrations of intermediates oscillate around the steady state, which is a singular point that is never attained. There are no oscillations in the concentrations of reactants or products, only in their rates of consumption and formation. Evolution seems characteristic of a limit cycle; the initial concentrations of the intermediates appear not to affect the limiting amplitudes of the stable concentration oscillations.(26) The effect of perturbing the oscillations by adding extra Br^- or Ce(IV) has been studied; after a short lag time the oscillations revert to their preperturbation behavior, again indicative of limit cycle behavior.(27) A simplified mechanistic scheme, based upon the reactions outlined in this section, has been analyzed using numerical methods such as those outlined in Section 5.3. The time dependence of the various concentrations calculated from the model is in good agreement with experiment; the numerical solutions appear consistent with limit cycle behavior.(28)

(26) R. J. Field, E. Körös, and R. M. Noyes, *J. Am. Chem. Soc.* **94**, 8649 (1972).

(27) V. A. Vavilin, A. M. Zhabotinskii, and A. N. Zaikin, in *Biological and Biochemical Oscillators*, B. Chance, E. K. Pye, A. K. Ghosh, and B. Hess, eds. (New York: Academic Press, 1973), pp. 71–79.

(28) R. J. Field and R. M. Noyes, *J. Chem. Phys.* **60**, 1877 (1974).

7.6. Other Oscillating Systems

As may well be apparent periodicity in chemical systems requires coupled, competing processes. Without coupling via bromide ion the Belousov–Zhabotinskii reaction would just provide parallel pathways for the consumption of bromate ion and malonic acid. The coupling compels feedback which creates the possibility of switching from one path to the other. Since the time scale of the processes is comparable switching and oscillation are observable. In a similar way thermochemical and electrochemical oscillators may be designed. Furthermore spatial, rather than temporal, periodicity may be observed. We consider a few examples.

7.6.1. Concentration Waves in the Belousov Reaction

Zhabotinskii observed that when the Belousov system was not stirred the [Ce(IV)]/[Ce(III)] ratio, instead of oscillating homogeneously, evolved as traveling waves.[29] A geometry and set of initial concentrations which dramatize the effect has been developed by Winfree.[30] An explanation requires considering diffusion. In an unstirred system local fluctuations ensure that all regions do not reach the end of the induction period simultaneously. Consider a domain in which [Br^-] has reached the triggering concentration, as shown in Fig. 7.8a. According to the model of the previous section [Br^-] rises abruptly and the [Ce(IV)]/[Ce(III)] ratio is high. Thus Br^- and Ce(IV) tend to flow out of the area and Ce(III) tends to flow into it. As a consequence sequence (A) is triggered in the neighboring regions; as [Ce(III)] and [Br^-] are dropping in the central area it switches to sequence (B), as shown in Fig. 7.8b. Diffusive triggering of sequence (A) propagates outward. Subsequently diffusion and depletion cause a shift to sequence (B). The interchange continues with resulting concentration waves. Their wavelength is determined by the diffusion coefficients of the intermediates in addition to the rate constants and the concentrations of major reactants.[31]

(29) A. M. Zhabotinskii, *Dokl. Akad. Nauk SSSR* **157**, 392 (1964); *Biofizika* **9**, 306, (1964).

(30) A. T. Winfree, *Science* **175**, 634 (1972); *Sci. Amer.* **230**(6), 82 (1974). The latter article has particularly pretty photographs.

(31) A. M. Zhabotinskii, in *Biological and Biochemical Oscillators*, B. Chance, E. K. Pye, A. K. Ghosh, and B. Hess eds. (New York: Academic Press, 1973), pp. 89–95.

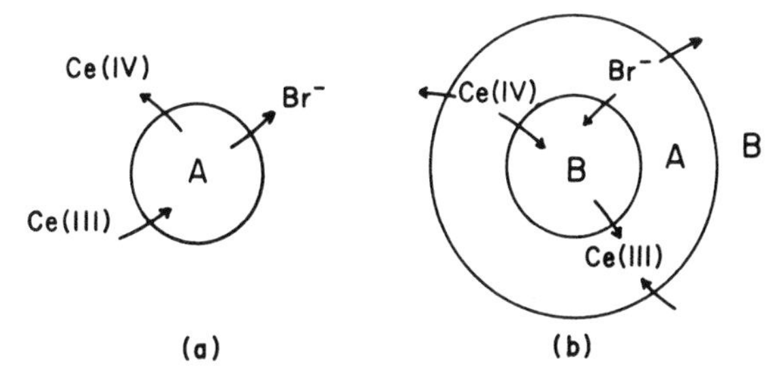

Fig. 7.8. Schematic diagram showing development of concentration waves in the Belousov reaction. (a) Sequence A is triggered in central domain; diffusion flows are indicated. (b) At a later time sequence B dominates in central region; rings of concentration variations propagate outward.

7.6.2. Liesegang Bands: Periodic Precipitation

When a concentrated salt solution of, e.g., $AgNO_3$, is allowed to diffuse into a sheet of gel impregnated with, e.g., K_2CrO_4, the resulting Ag_2CrO_4 does not precipitate uniformly. Instead it forms in bands (or rings) parallel to the diffusion front.[32] The phenomenon is quite general; similar structures are found with many sparingly soluble salts. Numerous explanations have been proposed. The simplest presumes that some degree of supersaturation is necessary before precipitation occurs.[33] As the $AgNO_3$ diffuses into the gel it forms a diffusion front parallel to the edge of the gel sheet. Crystallization occurs when the leading area is supersaturated with Ag_2CrO_4. As the Ag_2CrO_4 precipitates, the $[CrO_4^{2-}]$ concentration in that region drops and CrO_4^{2-} diffuses to the growing crystals thus depleting the neighboring area. As $AgNO_3$ continues to diffuse into the gel it passes through a region of low $[CrO_4^{2-}]$; precipitation recommences when a normal area is reached. The phenomenon then repeats itself leading to alternating bands.

A somewhat different interpretation postulates that precipitation is preceded by the formation of colloidal particles of the insoluble material.[34] These then become charged by preferential adsorption of either anion or cation which, due to electrostatic interactions, is autocatalytic. The larger

[32] The effect was discovered by R. E. Liesegang, *Naturw. Wochschr.* **11**, 353 (1896). A voluminous literature exists on the subject; see K. H. Stern, *Chem. Revs.* **54**, 79 (1954).

[33] Originally due to Ostwald, a modern treatment is due to S. Prager, *J. Chem. Phys.* **25**, 279 (1956).

[34] M. Flicker and J. Ross, *J. Chem. Phys.* **60**, 3458 (1974).

the colloidal particle the more charge it may carry and the more counterions it will attract thus accelerating its growth and ultimate precipitation. Then, due to diffusive coupling, the characteristic rings are formed. The significant feature of this model is that spatial inhomogeneities form *before* precipitation, as seems to occur. The supersaturation explanation requires both to be simultaneous.

7.6.3. The "Beating Mercury Heart": Electrocapillarity

When $\sim 1\ cm^3$ of mercury is placed in a watch glass and covered with a dilute solution of strong acid (e.g., HNO_3, HCl, H_2SO_4) to which a few crystals of a strong oxidizing agent (e.g., $K_2Cr_2O_7$, $Na_2S_2O_8$, or $KMnO_4$) are added, the solution is complex and a film forms on the surface of the mercury. Then, if an iron wire just touches the Hg globule, the "heart" pulsates 2 or 3 times per second.[35] Recent experiments have shown that the phenomenon is not arcane.[36] The system forms an electrochemical cell which may be crudely described as

$$Fe(s) \mid H^+(aq), Cr_2O_7^{2-}(aq) \mid Hg(liq) \tag{7.36}$$

with iron being the negative electrode. Iron is not required; similar results are obtained using aluminum or magnesium. The oxidizing agent may also be discarded although the "heartbeat" is then less powerful.

When there is Fe–Hg contact the iron is oxidized, $Fe \rightharpoonup Fe^{2+} + 2e^-$, and solvated Fe^{2+} diffuses into solution. Electrons are transferred to the Hg surface where a cathode reaction occurs; as a result negative charge, due to ions adsorbed on the surface, builds up. As long as there is contact between the iron and the mercury the surface charge continues to increase; it is apparently only limited by the rate of ion desorption from the surface. However, the surface charge, being negative, is repelled by the negatively charged iron electrode. If the wire does not penetrate the liquid too deeply the Hg drop can deform sufficiently for electrical contact to break. The adsorbed negative ions desorb into solution; the liquid resumes its spherical shape and the circuit is reestablished. Continuous repetition leads to pulsation.

[35] This curious phenomenon was discovered by G. Lippman, *Ann. Phys.* **2S 149**, 546 (1873).

[36] S.-W. Lin, J. Keizer, P. A. Rock, and H. Stenschke, *Proc. Nat. Acad. Sci.* (*U.S.A.*) **71**, 4477 (1974).

7.6.4. Hydrocarbon Combustion: Thermal Oscillations

Hydrocarbon combustion proceeds via complicated chain mechanisms.[37] A simplified low-temperature scheme is

$$\begin{array}{lll} \textit{Initiation:} & RH + O_2 & \rightharpoonup R\cdot + HO_2\cdot \\ \textit{Propagation:} & \left\{ \begin{array}{l} R\cdot + O_2 \\ ROO\cdot + RH \end{array} \right. & \begin{array}{l} \rightharpoonup ROO\cdot \\ \rightharpoonup ROOH + R\cdot \end{array} \\ \textit{Branching:} & \left\{ \begin{array}{l} ROOH \\ HO\cdot + RH \end{array} \right. & \begin{array}{l} \rightharpoonup RO\cdot + HO\cdot \\ \rightharpoonup R\cdot + H\ O \end{array} \\ \textit{Termination:} & R\cdot + R\cdot & \rightharpoonup R_2 \end{array} \tag{7.37}$$

$HO_2\cdot$ is unreactive (see Section 7.3) and the chain carriers are $R\cdot$, $ROO\cdot$, $RO\cdot$, and $HO\cdot$. The effect of buffer gas on energy transfer is kinetically unimportant since the experimental conditions conform to the high-pressure regions for unimolecular decomposition and bimolecular recombination.

As temperature is increased an unusual effect is observed; the reaction rate drops. Thus explosion can be quenched by *raising* the temperature. A new pathway that reduces the chain-carrier concentration must become significant. A plausible candidate is the decomposition of the weakly bound peroxy radical

$$ROO\cdot \rightharpoonup R\cdot + O_2$$

followed by olefin formation, e.g.,

$$CH_3CHCH_3 + O_2 \rightharpoonup CH_3CH = CH_2 + HO_2\cdot \tag{7.38}$$

This sequence both reduces branching by forming less $ROO\cdot$ and increases termination by forming more $HO_2\cdot$. At still higher temperatures $HO_2\cdot$ itself becomes reactive and may lead to propagation and branching. Thus the reaction rate will again rise.

In the temperature domain where reaction rate falls as temperature increases, thermal oscillations are observed for systems at pressures below the explosion limit. The temperature of the burning gas is a periodic function of time.[38] The temperature may increase by as much as 200 K in a

[37] See S. W. Benson, *The Foundations of Chemical Kinetics* (New York: McGraw–Hill, 1960), pp. 479–489 or J. Nicholas, *Chemical Kinetics* (New York: Halstead, 1976), pp. 150–157 for details.

[38] For a discussion of propane combustion see P. Gray, J. F. Griffiths, and R. J. Moule, *Faraday Symp. Chem. Soc.* **9**, 103 (1974).

pulse. While it might seem that the phenomenon is an indication of inhomogeneity in the reactor, in fact oscillation is more regular and less damped in a well-stirred system. In closed reactors only a few pulses are observed before depletion damps the oscillations; in a continuous-flow reactor pulsation continues indefinitely.[39]

From (7.1) the temperature–time profile in a homogeneous reactor is

$$\frac{dT}{dt} = \frac{QV}{C_v}\frac{d\xi}{dt} - \gamma(T - T_0)$$

In the domain of falling reaction rates, at lower temperatures dT/dt is large and positive since heat evolution is large and heat loss is small. As a consequence the temperature in the reactor increases abruptly after which the reaction rate drops sharply. As a result heat loss greatly exceeds heat evolution and the temperature plummets. As long as reactants are not depleted or in a continuous-flow system the process may repeat itself. There is obviously an intermediate temperature at which the heat evolution and the heat loss balance; one might therefore expect a steady state instead of limit cycle oscillations. However, as in all the other examples considered there is delayed feedback. The reaction rate does not instantaneously reflect the ambient temperature; thus the temperature overshoots the steady-state value before a correction occurs.

7.6.5. The Glycolytic Pathway: A Biochemical Oscillator

Whereas oscillatory phenomena have been, until recently, considered chemical curiosities, they had been of great interest to biochemists for years. Metabolic pathways conform to all requirements for the observation of oscillations: the systems are open—reactants constantly enter across cell walls and products are removed; they are far from equilibrium; there are feedback loops to maintain and reassert control even when the cell is drastically perturbed. Most biochemical oscillators are so complex that a detailed mechanistic analysis, such as that given for the Belousov reaction, is not available. The mechanism of glycolytic oscillation in yeast is the most well established.

The overall yeast metabolism is

$$\text{glucose} \rightarrow \rightarrow \rightarrow \text{ethanol, glycerol, } CO_2 \tag{7.39}$$

[39] B. F. Gray and P. G. Felton, *Comb. and Flame* **23**, 295 (1974).

As well as being of biochemical interest glycolysis is commercially important; it is used in the production of bread, beer, and wine. For a constant rate of glucose input, oscillation is observed in the rate of CO_2 production as well as in the concentrations of many reaction intermediates.[40] There are multiple feedback loops; the control mechanism is far more elaborate than in any of the physicochemical oscillators discussed previously. We shall sketch some of the results to indicate the complexity of even a rather simple biochemical pathway.[41]

The overall reaction (7.39) involves at least 10 separate intermediates. The various steps, requiring C—C bond cleavage or dehydrogenation, require mediation by a minimum of six enzymes working together. Feedback is affected by variation of pH and of the concentration of the reduced dinucleotide, NADH.[42] Oscillations in concentration have been found for at least 16 separate species involved in glycolysis. The oscillators fall into two categories. In each group the maxima and minima in the concentrations occur at the same time. However, the two groups oscillate out of phase, an observation that has been used to provide insight as to some features of the feedback loop. Furthermore, the character of the oscillations changes as the rate of glucose input is varied. At very low input level no oscillations are found. Above a critical input level doubly periodic oscillations of small amplitude and frequency develop. As the input rate increases further the multiple periodicities wash out. The amplitude and frequency of the stable oscillations increase. Finally at very high input levels oscillations are again damped.

Of these phenomena, three have simple rationalizations. At very low input levels the biochemical system is not sufficiently displaced from equilibrium for stable oscillations to form. Increasing the input level increases the concentration of the fundamental reactant which, in terms of the Lotka model, requires an increase in frequency from (7.23). Too great an input rate finally saturates the intermediate enzymes in a fashion similar to that outlined in Section 5.10. As a result feedback controls are no longer effective and oscillations damp out.

[40] There is an enormous literature on the subject. For representative work see *Biological and Biochemical Oscillators*, B. Chance, E. K. Pye, A. K. Ghosh, and B. Hess, eds. (New York: Academic Press, 1973), Part III.

[41] For an analysis of many of the features of the metabolic pathway see A. Boiteux and B. Hess, in *Biological and Biochemical Oscillators*, B. Chance, E. K. Pye, A. K. Ghosh, and B. Hess, eds. (New York: Academic Press, 1973), pp. 243–252.

[42] NADH is the reduced form of nicotinamide–adenine–dinucleotide.

7.7. Multiple Stationary States: The NO_2–N_2O_4 Reaction

Systems far from equilibrium exhibit unusual kinetic behavior. Feedback may lead to instability and ultimately explosion or to undamped (or weakly damped) oscillations about an inaccessible steady state. Here we treat another example of chemical instability—the abrupt switching between steady states with an attendant chemical hysteresis.[43] The coupled variables are temperature and the progress variable for a chemical reaction, the same coupling that creates thermal explosion (Section 7.2) and thermal oscillation (Section 7.6.4). The difference is that here the chemical reaction is endothermic so explosion is ruled out. Instability is induced by heat flow into the system. Unlike the examples considered previously, the mathematical description of this coupled system is simple. The resulting equations may be solved and detailed theoretical predictions verified.

Consider the qualitative aspects of the NO_2–N_2O_4 equilibrium; the reaction

$$N_2O_4 \underset{k_-}{\overset{k_+}{\rightleftharpoons}} 2NO_2 \tag{7.40}$$

is endothermic ($\Delta H = 58.1$ kJ mol^{-1}). When an NO_2–N_2O_4 mixture is irradiated with visible light only the NO_2 absorbs energy which, by various photochemical relaxation processes, ultimately serves to heat the mixture thus raising the temperature. At higher temperature the equilibrium is shifted to favor NO_2; thus a greater fraction of the incident radiation may be absorbed and the temperature is further raised. The feedback is positive.[44]

In terms of concentration variables

$$[NO_2] = \xi, \qquad [N_2O_4] = C - \xi/2 \tag{7.41}$$

where C would be $[N_2O_4]$ if no NO_2 were present, extension of (7.1) to account for radiation heating yields

$$\frac{\partial T}{\partial t} = \left(\frac{\epsilon I_0 l \xi}{V} + \varkappa \nabla^2 T + Q \frac{\partial \xi}{\partial t} \right) \frac{V}{C_v} - \gamma(T - T_0) \tag{7.42}$$

Here I_0 is the radiation intensity, ϵ the molar extinction coefficient of NO_2, and l the path length; Q, the heat evolved per mole of NO_2 formed, is now negative. The rate law for NO_2–N_2O_4 interconversion is, from (7.40) and

[43] The phenomenon was predicted on theoretical grounds by A. Nitzan and J. Ross, *J. Chem. Phys.* **59**, 241 (1973).

[44] See C. L. Creel and J. Ross, *J. Chem. Phys.* **65**, 3779 (1976).

(7.41),

$$\frac{\partial \xi}{\partial t} = 2k_+(T)(C - \xi/2) - 2k_-(T)\xi^2 \tag{7.43}$$

In (7.42) and (7.43) the quantities under experimental control are I_0, T_0, and C. Increasing the radiation dose has two direct consequences—radiation heating and increased heat loss to the surroundings—and one indirect one—heat absorption as more NO_2 forms.

In a uniform steady state $\nabla^2 T$, $\partial T/\partial t$, and $\partial \xi/\partial t$ are all zero. Since the parameters in (7.42) and (7.43) are known the equations, which are non-linearly coupled due to the temperature dependence of the rate constants k_+ and k_-, may be solved for ξ (or T) as functions of I_0, T_0, and C. Typical results, for the same value of C, are presented in Fig. 7.9. As the surroundings cool from 250 K to 230 K the structure of the solutions changes dramatically. At higher temperatures there is a unique NO_2 concentration for each value of I_0. However, as T_0 decreases a critical point is reached below which there are three possible values of $\xi(I_0, T_0)$. Two represent stable steady states; one is unstable.

Consider the 230 K curve in detail for a system initially in steady state A. As I_0 increases ξ increases until the marginal stability point B is reached where $d\xi/dI_0 \rightarrow \infty$. Further increase of I_0 requires an abrupt jump in ξ to point C if a steady state is to be maintained. Yet higher-power levels again lead to a gradual increase of ξ. Now consider reducing the radiation

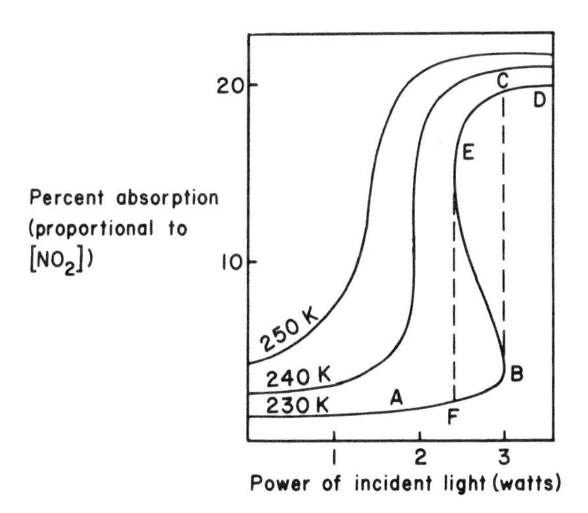

Fig. 7.9. Steady-state solutions of (7.42) and (7.43) for various choices of external temperature T_0. Note the hysteresis in the 230 K curve. [Adapted from C. L. Creel and J. Ross, *J. Chem. Phys.* **65**, 3779 (1976).]

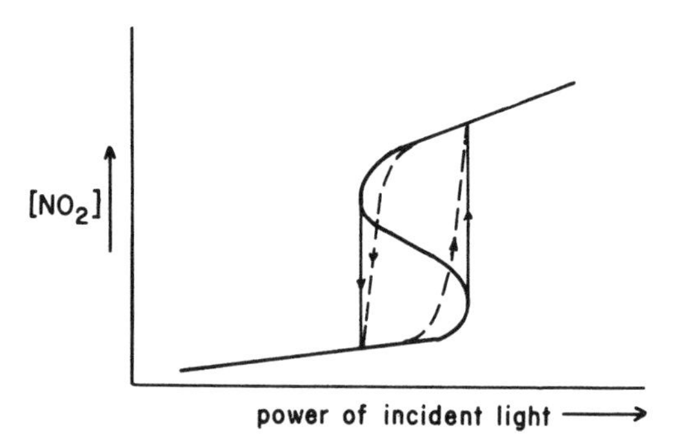

Fig. 7.10. Comparison of ideal (solid line) and observed (dashed line) hysteresis in the NO_2–N_2O_4 reaction.

intensity. The system traces a path DCEFA since it is only at E that marginal stability occurs. The net result is a hysteresis loop as I_0 is varied. Above 240 K there is no hysteresis; ξ varies continuously with I_0. The set of curves for different values of T_0 look just like isotherms of a van der Waals gas (turned on their side) above and below a critical point. Hysteresis below the critical temperature has its analog in supercooling and superheating.[45]

Experiment confirms the predictions of Fig. 7.9. Above the critical value T_0 there is no hysteresis. Below T_0 hysteresis is observed; however, the jump from one branch to the other is not as abrupt as predicted.[46] A comparison of observed and predicted hysteresis loops is shown in Fig. 7.10. Transition occurs before the marginal stability point is reached. Under the experimental conditions there is sufficient inhomogeneity to account for this effect. The temperature of the gas is probably not uniform throughout the sample so that observation is of a set of systems undergoing sequential transitions between the branches.

If uniformity were established one would still expect deviations from the ideal hysteresis loop. Fluctuations about the steady state could induce transitions from one branch to the other before the marginal stability point is reached. Such behavior is analogous to nucleation of a supercooled liquid; freezing then occurs abruptly before the limit of metastability is reached. Referring to Fig. 7.9, one might wonder whether there could be an analog to the Maxwell construction in the van der Waals fluid. While a definitive answer is not yet available, the tentative answer is yes. Since

[45] For extension of the thermodynamic analogy see A. Nitzan, P. Ortoleva, J. Deutch, and J. Ross, *J. Chem. Phys.* **61**, 1056 (1974).

[46] C. L. Creel and J. Ross, *J. Chem. Phys.* **65**, 3779 (1976).

power dissipation in the mixture depends on ξ and T, it is multiply valued in the domain where there are multiple solutions. After taking proper account for heat flow into (by radiation) and out of (by conduction at the walls) the system, the state of lower power loss could be found. In this sense point F is more stable than point E; similarly point C is more stable than point B. At some intermediate power level the states on each branch should be equally stable, thus defining a Maxwell-like construction.[47]

Problems

7.1. (a) From Fig. 7.1 it is clear that T_c is the temperature at which thermal explosion is inevitable. Use (7.3), assume an Arrhenius form for $k(T)$, and show that $T_c - T_0 = RT^2/E_a$.
(b) Compute T_c for a reaction where $E_a = 300$ kJ mol^{-1} when the system is in a thermostatted furnace at (a) 600 K, (b) 800 K, (c) 1000 K.
(c) Why does $T_c - T_0$ increase as T_0 increases?

7.2. The seven elementary steps of (7.5) and (7.11) account for kinetics in the H_2–O_2 system below the third explosion limit.
(a) Show that at the first limit $k_5 = 2k_2[O_2]/(1 + k_6/k_4[H_2])$ and thus, as volume-to-surface ratio increases, $k_5 = 2k_2[O_2]$.
(b) At the second limit show that $k_7[M] = 2k_2/(1 + k_6/k_4[H_2])$ which, for a large reaction vessel, reduces to $k_7[M] = 2k_2$.

7.3. In a stoichiometric mixture of H_2 and O_2 the temperature dependence of the second explosion limit is

p_2 (atm):	0.043	0.066	0.104	0.156
T (K):	750	775	800	825

Estimate E_B; the quenching reaction (7.11) has a small negative activation energy, $E_Q \sim -7$ kJ mol^{-1}. To which of the branching reactions does the computed E_B correspond?

7.4. To demonstrate that chemical oscillations are a far-from-equilibrium phenomenon consider the mechanism (7.15) with the equilibrium constraints (7.16).
(a) Define displacement variables $\xi = [X] - X_0$, $\eta = [Y] - Y_0$, and $\zeta = [Z] - Z_0$ and use (7.17) to show that, near equilibrium,

$$\dot{\xi} = -(k_{-1}X_0 + k_2Y_0)\xi + k_{-2}Y_0\eta$$
$$\dot{\eta} = k_2Y_0\xi - (k_{-2}Y_0 + k_3)\eta + k_{-3}\zeta$$
$$\dot{\zeta} = k_3\eta - k_{-3}\zeta$$

[47] I. Procaccia and J. Ross, *J. Chem. Phys.* **67**, 5558 (1977).

(b) Assume $Z = [Z_0]$, i.e., that product is always drawn off so that $\zeta = 0$. Use the matrix method of Chapter 5 and show that the decay constants are solutions of the equation

$$\lambda^2 - [k_{-1}X_0 + (k_{-2} + k_2)Y_0 + k_3]\lambda + k_{-1}X_0(k_3 + k_{-2}Y_0) + k_2k_3Y_0 = 0$$

(c) Show that both roots are positive and therefore that the system relaxes to equilibrium without oscillation. (The roots of the equation $\lambda^2 - B\lambda + C = 0$ satisfy the relation $\lambda_1\lambda_2 = C$.)

7.5. Show that when ξ and η are small (7.25) becomes $\xi^2/X_0 + \eta^2/Y_0 = \text{const}$, i.e., the ξ–η trajectory is an ellipse.

7.6. Mechanistic details strongly influence the possibility of chemical oscillation. Consider the autocatalytic mechanism

$$A \xrightarrow{k_1} X, \qquad X + Y \xrightarrow{k_2} 2Y, \qquad Y \xrightarrow{k_3} Z$$

[A. J. Lotka, *J. Phys. Chem.* **14**, 271 (1910)].

(a) Assume [A] is constant and formulate the rate laws. Show that in the steady state $X_0 = k_3/k_2$ and $Y_0 = k_1A_0/k_3$.

(b) Define displacement variables $\xi \equiv [X] - X_0$ and $\eta \equiv [Y] - Y_0$ and show that close to the steady state

$$\dot{\xi} = -k_2Y_0\xi - k_2X_0\eta, \qquad \dot{\eta} = k_2Y_0\xi$$

(c) Show that the decay constants are

$$\lambda_\pm = k_2Y_0[1 \pm (1 - 4X_0/Y_0)^{1/2}]/2$$

and thus that the system exhibits damped oscillations as it approaches the steady state if $Y_0 < 4X_0$, i.e., if $4k_3^2 > k_1k_2A_0$.

(d) Can you give a physical rationalization for the condition governing the onset of damped oscillation?

7.7. Spatial inhomogeneities couple with chemical relaxation. One set of possible consequences is illustrated when the Lotka rate law (7.18) is coupled with Fick's law (2.25)

$$[\dot{X}] = k_1A_0[X] - k_2[X][Y] + D_x \nabla^2[X]$$
$$[\dot{Y}] = k_2[X][Y] - k_3[Y] + D_y \nabla^2[Y]$$

(a) Show that the linearized rate equations for the displacement functions $\xi(t, \mathbf{r})$ and $\eta(t, \mathbf{r})$ are

$$\dot{\xi} = D_x \nabla^2\xi - k_2X_0\eta, \qquad \dot{\eta} = k_2Y_0\xi + D_y \nabla^2\eta$$

(b) Assume a perturbation of wave vector $\mathbf{q}$, displacement functions of the form $\xi = \alpha(t) \cos \mathbf{q} \cdot \mathbf{r}$ and $\eta = \beta(t) \cos \mathbf{q} \cdot \mathbf{r}$, and show that

$$\dot{\alpha} = -D_xq^2\alpha - k_2X_0\beta, \qquad \dot{\beta} = k_2Y_0\alpha - D_yq^2\beta$$

(c) Now show that the decay constants are

$$\lambda_\pm = (D_x + D_y)q^2/2 \pm [(D_x - D_y)^2 q^4/4 - \omega_0{}^2]^{1/2}$$

where ω_0 is the Lotka frequency (7.23).
(d) Do inhomogeneities in the Lotka system persist? What is the effect of increasing q?

7.8. A proposed model for the Belousov reaction is

$$A + Y \xrightarrow{k_1} X, \qquad X + Y \xrightarrow{k_2} P, \qquad B + X \xrightarrow{k_3} 2X + Z,$$
$$2X \xrightarrow{k_4} Q, \qquad Z \xrightarrow{k_5} fY$$

with the identifications $X = HBrO_2$, $Y = Br^-$, $Z = Ce(IV)$, and $A = B = BrO_3^-$ [R. J. Field and R. M. Noyes, *J. Chem. Phys.* **60**, 1877 (1974)].
(a) Identify the steps in the Belousov system which correspond to reactions 1–5.
(b) Show that an overall stoichiometry is $fA + 2B = fP + Q$.
(c) From the discussion in Section 7.5 determine the permissible range of values for the stoichiometric factor f.
(d) Formulate the rate laws and show that in the steady state, assuming [A] and [B] are constant: $Z_0 = k_3 B_0 X_0/k_5$, $Y_0 = fk_3 B_0 X_0/(k_1 A_0 + k_2 X_0)$, and X_0 is the solution to the equation $k_2 Y_0 + k_4 X_0 = (1 + f)k_3 B_0/2$.
(e) If $A_0 = B_0$ show that $X_0 = k_3 B_0\{b - a - c + [(a+b)^2 + c^2 + 2c(a-b)]^{1/2}\}/2$ where $a = k_1/k_2 k_3$, $b = (1 + f)/2k_4$, and $c = f/k_2 k_4$.
(f) Define displacement variables and show that the decay constants, assuming small deviations from the steady state, are solutions of the cubic equation

$$0 = \begin{vmatrix} k_3 B_0 - k_2 Y_0 - 4k_4 Y_0 + \lambda & k_1 A_0 - k_2 X_0 & 0 \\ -k_2 Y_0 & -k_1 A_0 - k_2 X_0 + \lambda & fk_5 \\ k_3 B_0 & 0 & -k_5 + \lambda \end{vmatrix}$$

(g) The critical condition for onset of oscillation occurs when $\lambda = 0$. Show that, assuming $A_0 = B_0$, the constraint is $X_0 \geq k_3 B_0 a/[k_4(a - b)]$.

7.9. Consider the effect of inhomogeneity in the NO_2–N_2O_4 system discussed in Section 7.7. The modified form of (7.43) is

$$\dot{\xi} = k_+(2C - \xi) - 2k_-\xi^2 + D\,\nabla^2\xi$$

(a) If the system is very slightly perturbed from the steady state so that $T = T_{ss} + \tau$ and $\xi = \xi_{ss} + x$ show that

$$\dot{\tau} = (\epsilon I_0 l x/V + \varkappa\,\nabla^2\tau + Q\dot{x})V/C_v - \gamma\tau$$
$$\dot{x} = -k_+ x - 4k_-\xi_{ss} x + D\,\nabla^2 x + (2C - \xi_{ss})k_+'\tau - 2\xi_{ss}^2 k_-'\tau$$

where $k_+' \equiv (dk_+/dT)_{T_{ss}}$, etc.

(b) Show that $k_+' = E_a^{(+)} k_+(T_{ss})/RT_{ss}^2$, etc.

(c) Define displacements of wave vector $\mathbf{q}$: $\tau = f(t) \cos \mathbf{q} \cdot \mathbf{r}$ and $x = g(t) \cos \mathbf{q} \cdot \mathbf{r}$; show that $\dot{f} = -af + bg$ and $\dot{g} = cg - df$, where $a = \gamma + \varkappa q^2 V/C_v$, $b = (\epsilon I_0 l + QV)/C_v$, $c = (2C - \xi_{ss})k_+ \Delta E/RT_{ss}^2$, and $d = k_+ + 4k_- \xi_{ss} + Dq^2$ with $\Delta E = E_a^{(+)} - E_a^{(-)}$.

(d) Show that the decay constants are $\lambda_\pm = \{(a + d) \pm [(a - d)^2 + 4bc]^{1/2}\}/2$.

(e) The parameters a, c, and d are always positive. However, as $Q < 0$ (the reaction is endothermic), $b < 0$ at low radiation intensity. Are there values of q which permit *undamped* oscillation about the steady state? What conditions permit *damped* oscillations? Is it possible that $\lambda_- < 0$? What is the physical significance of a negative λ?

7.10. An interesting ecological problem (with chemical analogs) occurs when two similar predators compete for the same food.

(a) If the predator species P_i have mean lifetimes τ_i and reproduce according to the mechanism $P_i + F \xrightarrow{k_i} 2P_i$ $(i = 1, 2)$ show that the rate laws are

$$[\dot{P}_i] = -\frac{1}{\tau_i}[P_i] + k_i[P_i][F], \qquad i = 1, 2$$

(b) Now show that

$$k_2 \frac{[\dot{P}_1]}{[P_1]} - k_1 \frac{[\dot{P}_2]}{[P_2]} = \left(\frac{k_1}{\tau_2} - \frac{k_2}{\tau_1}\right)$$

and thus that $k_2 \ln[P_1] - k_1 \ln[P_2] = Kt + \text{const}$, with $K = k_1/\tau_2 - k_2/\tau_1$. It is generally possible for both species to persist or will one become extinct?

General References

Explosion and Ignition

P. G. Ashmore, in *Photochemistry and Reaction Kinetics*, P. G. Ashmore, F. S. Dainton, and T. M. Sugden, eds. (Cambridge: Cambridge University Press, 1967), pp. 309–386.

N. N. Semenov, *Some Problems in Chemical Kinetics and Reactivity*, translation by M. Boudart (Princeton, N.J.: Princeton University Press, 1959), Vol. 2.

Oscillatory and Multiple Steady-State Phenomena

B. F. Gray, in *Reaction Kinetics*, P. G. Ashmore, ed. (London: Chemical Society, 1975), Vol. 1, pp. 309–386.

R. J. Field and R. M. Noyes, *Acc. Chem. Res.* **10**, 214 (1977).

R. M. Noyes and R. J. Field, *Acc. Chem. Res.* **10**, 273 (1977).

A. Pacault, P. Hanusse, P. De Kepper, C. Vidal, and J. Boissonade, *Acc. Chem. Res.* **9**, 438 (1976).

I. Prigogine, in *Fast Reactions and Primary Processes in Chemical Kinetics*, S. Claesson, ed. (New York: Interscience, 1967), pp. 371–382.

J. Ross, *Ber. Buns. Gesell. Phys. Chem.* **80**, 1112 (1976).

8

Single-Collision Chemistry

8.1. Introduction

Even if the rate law and the mechanism of a reaction have been determined there are still many unanswered questions of chemical interest. The rate law only accounts for the overall phenomenology. The mechanism provides a model for the basic steps in the reaction. The molecular properties which determine whether or not reaction occurs can only be inferred from other chemical knowledge. Our previous discussion leads to only two very general, and hardly informative, conditions for reaction: molecules must interact (collide) and must have sufficient energy.

There are many other questions which can be posed. How is a theory for computing a rate constant to be constructed? How may the orientational requirements for a reactive collision (such as the familiar Walden inversion) be studied directly? Can reaction take place in a glancing collision or must molecules always collide head-on? Does reaction require the formation of a long-lived complex or can reaction take place as molecules fly past one another? If molecules have sufficient energy, will they invariably react? What molecular properties promote catalysis?

These and other questions can only be investigated by isolating each elementary step and studying the dynamics of the interacting molecules. For gas-phase reactions, which usually proceed via binary collisions, the chemistry of single collisions can be studied using molecular beams thus providing new perspectives. For solution kinetics the problem is more complicated since the basic reaction step is not the collision of two molecules but the trapping of the reactants in a cage of solvent. Gas-phase studies in which a molecule is surrounded by a few solvent molecules provide a way to isolate a solvated molecule from the bulk solvent. Some molecular beam investigations of this type have been carried out.

The inherent distinction between gas and solution kinetics must be part of any successful theory of chemical kinetics. In particular the mechanics of gas-phase reaction depend upon forces involving two, or at most three, molecules at a time. In solution the complicating effects due to interaction with a number of surrounding solvent molecules must also be considered.

In this chapter we focus upon the dynamics of the reactive collision process in the gas phase. To this end we first show how the basic measurable parameter in collision chemistry, the reaction cross section, may be related to its thermal counterpart, the rate constant. We then discuss the experimental approach to measurement of reaction cross section as well as a naive model. The remainder of the chapter is devoted to the interpretation of a number of experiments in terms of simple dynamical models for reactive collisions. We make no attempt to develop a theory of reaction cross sections.

8.2. Reaction Cross Section: Relation to the Rate Constant

To obtain a basis for computing the rate constant for an elementary step consider the simple bimolecular gas-phase transformation

$$\mathrm{A} + \mathrm{BC} \rightarrow \mathrm{AB} + \mathrm{C}$$

The A and BC molecules are constantly colliding with one another, but not all such collisions can lead to reaction. If A collides with the C end of BC, AB bond formation is probably much less likely than if collision is with the B end. If the reaction is endothermic, energy must be supplied in order for reaction to take place. The necessary energy is supplied by collision and then redistributed among the reactants. Thus, when molecules collide, we can expect that there is a probability of reaction (less than 1) which depends upon their relative kinetic energy, their relative orientation, and possibly other factors which have not yet been considered.

The rate of reaction of A atoms per unit volume can be expressed as

$$-\frac{dn_\mathrm{A}}{dt} = \int_0^\infty dc \begin{pmatrix} \text{number of A–BC collisions} \\ \text{per unit volume} \\ \text{with relative speed} \\ \text{between } c \text{ and } c + dc \end{pmatrix} \begin{pmatrix} \text{fraction of such} \\ \text{collisions} \\ \text{which lead to} \\ \text{reaction} \end{pmatrix} \tag{8.1}$$

where $n_A = N_A/V$ is the concentration of A. The first factor in the integral states that as more collisions occur, more molecules can react. The second factor states that each collision need not lead to reaction; there is a reaction probability, $p(\epsilon)$, which may vary with ϵ, the relative kinetic energy of the colliding molecules.[1] From (2.21) we know that the total frequency of A–BC collisions per unit volume is

$$Z = n_A n_{BC} \pi d^2 \bar{c}_{A,BC} \tag{8.2}$$

where $\bar{c}_{A,BC}$ is the mean *relative speed* of A and BC, and d is a hard-core diameter. Naturally, real molecules are not hard spheres and d, as noted in Chapter 2, is to be interpreted as a phenomenological parameter. There, the collision cross section was related to transport coefficients. Here, by introducing suitable modifications we shall be able to define a reaction cross section.

The same arguments which led to (2.21) can be modified to show that the collision frequency per unit volume of A and BC molecules with relative speed between c and $c + dc$ is

$$Z(c)\,dc = n_A n_{BC} \pi d^2 \left[4\pi \left(\frac{\mu}{2\pi kT} \right)^{3/2} \exp(-\mu c^2/2kT) c^3\, dc \right] \tag{8.3}$$

where μ is the reduced mass.[2] The bracketed quantity is c times the probability that a pair of molecules will collide with a relative speed between c and $c + dc$. Integrating (8.3) over all speeds leads immediately to (8.2). Substituting (8.3) into (8.1) the reaction rate is

$$-\frac{dn_A}{dt} = n_A n_{BC} \pi d^2 \left[4\pi \left(\frac{\mu}{2\pi kT} \right)^{3/2} \int_0^\infty \exp(-\mu c^2/2kT) c^3 p(\epsilon)\, dc \right]$$

Since $\frac{1}{2}\mu c^2$ is the relative kinetic energy, ϵ, the integral can be rewritten with energy as a variable

$$-\frac{dn_A}{dt} = n_A n_{BC} \left[\pi d^2 \left(\frac{8}{\pi \mu (kT)^3} \right)^{1/2} \int_0^\infty p(\epsilon) \epsilon e^{-\epsilon/kT}\, d\epsilon \right] \tag{8.4}$$

which is precisely the form of the rate law for the elementary step being studied; the bracketed term is the rate constant. Here πd^2 is the collision

(1) Molecular energy is denoted as ϵ; molar energy is denoted as E.

(2) For a derivation of (8.3) see W. J. Moore, *Physical Chemistry*, 4th ed. (Englewood Cliffs, N. J.: Prentice-Hall, 1972), pp. 150–152.

cross section which measures how close the centers of mass of A and BC must pass in order for collision to occur. The factor $p(\epsilon)$ accounts for the fraction of the collisions that lead to reaction. Both πd^2 and $p(\epsilon)$ are phenomenological parameters. Rather than treating each separately we define a cross section for reaction[3]

$$Q(\epsilon) = \pi d^2 p(\epsilon)$$

and rewrite (8.4)

$$-\frac{dn_A}{dt} = n_A n_{BC}\left[\frac{8}{\pi\mu(kT)^3}\right]^{1/2} \int_0^\infty \epsilon Q(\epsilon) e^{-\epsilon/kT}\, d\epsilon \tag{8.5}$$

The reaction cross section is equivalent to the collision cross section for hypothetical spherical molecules which react with each collision; in the next section we indicate how $Q(\epsilon)$ can be measured directly. The rate constant is thus identified from (8.5)

$$k(T) = \left[\frac{8}{\pi\mu(kT)^3}\right]^{1/2} \int_0^\infty \epsilon Q(\epsilon) e^{-\epsilon/kT}\, d\epsilon \tag{8.6}$$

Since concentrations, in the rate law (8.5), are measured in molecules cm^{-3}, $k(T)$ has units of $cm^3\ sec^{-1}$ molecule^{-1}. The task of calculating a binary rate constant is thus reduced to the problem of measuring the reaction cross section, $Q(\epsilon)$.

The parameters in the Arrhenius expression (4.42) can be related to the rate constant determined from the cross section. The activation energy is *defined* as the negative slope of $\ln k(T)$ vs. $(RT)^{-1}$. This indicates E_a is to be computed as

$$E_a = -R\frac{d \ln k}{d(1/T)} \tag{8.7}$$

In general, values of E_a calculated from (8.7) are temperature dependent (Problem 8.1). There is, in fact, only one functional form of $Q(\epsilon)$ that yields the Arrhenius form (4.42) (Problem 8.17).

[3] The d that measures the reaction cross section may be substantially different from a d determined from gas-phase transport experiments. Ion–molecule and alkali metal–halogen reactions often have cross sections substantially larger than what would be expected from analysis of transport data for the corresponding species. As we have defined $Q(\epsilon)$, it is a function of molecular energy; however, experimental data usually present $Q(E)$, a function of molar energy. At 300 K the mean collision per colliding pair is $\sim 6\times 10^{-21}$ J (equivalent to 0.04 eV or, in molar units, to 4 kJ mol^{-1}).

8.3. Scattering Measurements—Molecular Beams

The study of individual molecular collisions is a topic of great current interest. The experimental methods are variants of the model experiment outlined in Section 2.2 to illustrate the idea of mean free path. While that experiment determines a *total* cross section for loss of reactant, with more sophisticated apparatus the nature of the product can be determined, i.e., partial cross sections for elastic, inelastic, and reactive events can be distinguished. Let us reconsider the reaction

$$\mathrm{A} + \mathrm{BC} \rightarrow \mathrm{AB} + \mathrm{C}$$

To measure the reaction cross section as a function of relative kinetic energy requires forming beams of A molecules and BC molecules with specified velocities. The beams are oriented so that they collide; collision products are monitored as a function of energy and orientation. A representative apparatus is sketched in Fig. 8.1.

Beam experiments are rather difficult to carry out since they require intense molecular beams of particles with a well-defined velocity; the technical problems are greatest for neutral species.[4] To form an ionic beam electric fields are used to accelerate the ions to the desired speed. To form a beam of neutral molecules various techniques have been used. The most intense beams are formed when the gas is driven, under high pressure, through a small nozzle; the gas emerges at supersonic speeds. The result, on the molecular scale, is similar to running water through a hose with a nozzle. As the nozzle radius diminishes (within limits) the stream of water emerges at higher speeds; all regions in the stream move with the same speed.

In a molecular beam experiment the rate of product formation is measured as a function of the energy of the products and the direction in which they are scattered. The immediate problem is to interpret these data and deduce the reaction cross section. Consider crossed beams such as those of Fig. 8.1. The beams have fluxes I_{A} and I_{BC}; they collide with a relative energy ϵ for which the reaction cross section is $Q(\epsilon)$. Reaction occurs in the small collision volume V_{coll}. As the beam fluxes are kept constant, the number of BC molecules in the collision volume is also a constant, N_{BC}. The quantity $I_{\mathrm{A}}Q(\epsilon)$ is the number of A molecules which

[4] For a description of the apparatus necessary for production, detection, and analysis see J. M. Parson and Y. T. Lee, *J. Chem. Phys.* **56**, 4658 (1972); Y. T. Lee, J. D. McDonald, P. R. Le Breton, and D. R. Herschbach, *Rev. Sci. Instr.* **40**, 1402 (1969).

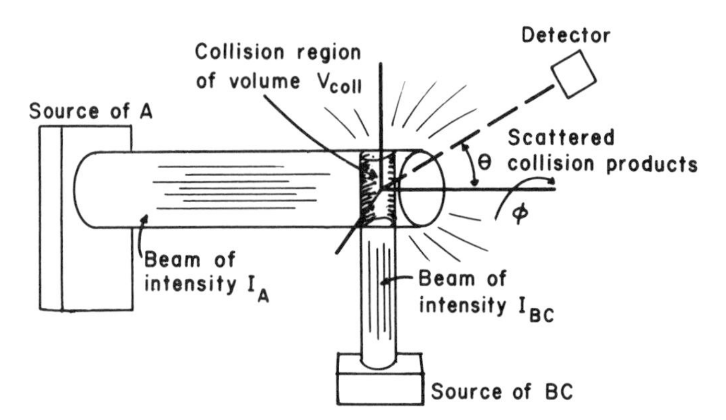

Fig. 8.1. Schematic diagram of a molecular beam experiment. The detector is movable; it subtends the solid angle $d\Omega = \sin\theta \, d\theta \, d\varphi$.

react per unit time *per BC molecule* in the collision volume so that the *total* rate of product formation is

$$\frac{d[\text{Prod}]}{dt} = [I_A Q(\epsilon)] N_{BC} \tag{8.8}$$

To compute the number of BC molecules in the collision zone requires adapting the molecular flux calculations of Chapters 1 and 2. From the perspective of a viewer in the collision volume, A and BC molecules approach one another with the relative speed c_{rel}. The larger the relative speed, the fewer the molecules that are in the collision zone at any instant.

In Section 1.5 we showed that, for particles moving in the $+x$ direction, the flux of molecules with x component of velocity between u and $u + du$ is

$$G(u)\, du = \frac{N}{V} u \varrho_x(u)\, du$$

Here the fraction of molecules in the specified velocity domain, the analog to $\varrho_x(u)\, du$, is 1 since the beam is velocity-selected. The approach speed is now c_{rel}, not u, and the beam intensity is I_{BC}, not $G(u)\, du$. Thus, the flux is

$$I_{BC} = \frac{N_{BC}}{V_{coll}} c_{rel}$$

from which, with (8.8), we obtain

$$\frac{d[\text{Prod}]}{dt} = \frac{V_{coll}}{c_{rel}} Q(\epsilon) I_{BC} I_A \tag{8.9}$$

The result is symmetric in A and BC as is mandatory. It is somewhat similar to a binary rate law with beam flux corresponding to concentration and the quantity $Q(\epsilon)/c_{\text{rel}}$ corresponding to the rate constant. The rate constant provides a measure of reaction probability for molecules at a specific temperature while the cross section (or equivalently $Q(\epsilon)/c_{\text{rel}}$) is a measure of reaction probability for molecules with specific relative kinetic energy. The units of Q/c, length $\times$ time, do not immediately suggest its similarity to a rate constant. However, the differences arise because fluxes, not concentrations, appear in the rate law (8.9). The parallel is, of course, implicit in the expression (8.6). The beam intensities, relative speed, and collision volume are the experimental parameters. The rate of product formation is measured from which, with (8.9), $Q(\epsilon)$ can be determined.

The detector shown in Fig. 8.1 may be moved to measure the rate of product formation as a function of position. At an orientation specified by θ and φ the detector subtends a solid angle $d\Omega = \sin\theta \, d\theta \, d\varphi$. Generalizing (8.9) to account for angular variations in the rate of product formation we have

$$\left\{\frac{d[\text{Prod}]}{dt}\right\}_{\theta,\varphi} = \frac{V_{\text{coll}}}{c_{\text{rel}}} I_{\text{BC}} I_{\text{A}} \frac{\partial^2 Q(\epsilon;\theta,\varphi)}{\partial\theta\,\partial\varphi}\,d\Omega \tag{8.10}$$

where $\{d[\text{Prod}]/dt\}_{\theta,\varphi}$ is the rate at which product arrives at the detector. The quantity $\partial^2 Q/\partial\theta\,\partial\varphi$ is the *differential cross section* for molecules with collision energy ϵ.

The detector must be able to distinguish different products as well as determine the amount of each that is formed. A mass spectrometer is generally used for this task. When coupled with a time-of-flight analyzer it is possible to determine the distribution of product velocities as well. Such data provide information about the dynamics of a reactive collision (Section 8.7).

An even more detailed picture can be constructed if the reactant molecules can be prepared in specific excited states. For the example considered, A may be formed in a variety of electronic states and BC may be formed with a specific amount of electronic, vibrational, and rotational energy. If i and j denote the internal states of A and BC, respectively, the reaction cross section is found to be a function of both i and j.

Thus $Q(\epsilon)$, as it is used in (8.5) and (8.6), is really a thermal average since the population of the internal states of the reactants is temperature dependent. A more complete specification requires defining the reaction cross section for molecules in particular internal states colliding with a specified collision energy, $Q(\epsilon, i, j)$. The mean cross section in a thermally equilibrated system is a function of temperature and energy; it is

$$Q(\epsilon; T) = \sum_i \sum_j f_A{}^{(i)}(T) f_{BC}^{(j)}(T) Q(\epsilon, i, j) \tag{8.11}$$

where $f_A{}^{(i)}(T)$ is the fraction of A molecules in the internal state i at temperature T. From statistical thermodynamics this fraction is $e^{-\beta\epsilon_i}/q_A$, where q_A is the partition function for the internal degrees of freedom of A.[5] With this result and its analog for the fraction of BC molecules, the average cross section can be computed from (8.11); it is this average which is substituted in (8.6) to calculate a rate constant.

If the colliding molecules can produce a variety of collision products, there are separate reaction cross sections for each chemical transformation. In the terminology of the field the individual pathways are called *channels*. A channel is open if the total energy available, defined as the energy of the separated reactants plus the collision energy, is greater than that of the separated products. Even though a channel is open there may be no detectable product yield if the reaction cross section is sufficiently small. To each channel there corresponds a rate constant for reaction in a system at thermal equilibrium to be calculated using (8.11) and (8.6); the activation energy is found from (8.7).

8.4. Reaction Cross Section: Hard-Sphere Model

Consider a particularly simple model of chemical reaction. The molecules are imagined to be hard spheres of mean diameter σ and the reaction has an energy barrier of height ϵ_0. For reaction to occur two conditions must be met: the molecules must collide and they must have sufficient collision energy to overcome the barrier.

Case A: *All collision energy available for surmounting barrier*. In these circumstances the energy dependence of the reaction cross section is particularly simple

$$Q(\epsilon) = \begin{cases} 0, & \epsilon < \epsilon_0 \\ \pi\sigma^2, & \epsilon > \epsilon_0 \end{cases} \tag{8.12}$$

Even at this level of simplification the model cannot be correct. Not all the collision energy is available to overcome the barrier. The fraction of

[5] A good elementary discussion of statistical thermodynamics is given by T. L. Hill, *Introduction to Statistical Thermodynamics* (Reading, Mass.: Addison–Wesley, 1960). See especially Chapter 3.

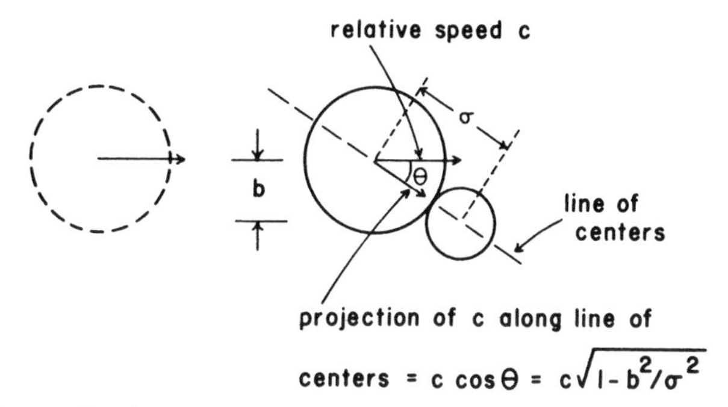

Fig. 8.2. Diagram for determining line-of-centers velocity for hard spheres which do *not* collide head-on; the impact parameter is b.

energy accessible depends upon the impact parameter, which, as illustrated in Fig. 8.2, is the quantitative measure of the extent to which the collision is off center. As b increases from 0 to σ the character of the collision changes from head-on to glancing. If $b > \sigma$ no collision takes place.

Case B: *Line-of-centers collision energy available for surmounting barrier*. For the collision geometry of Fig. 8.2 the velocity directed along the line of centers is $c(1 - b^2/\sigma^2)^{1/2}$. The energy available for overcoming the barrier is therefore only

$$\epsilon(b) = \tfrac{1}{2}\mu c^2(1 - b^2/\sigma^2) = \epsilon(1 - b^2/\sigma^2)$$

Since $\epsilon(b) \geq \epsilon_0$ for reaction to occur, we can solve for b and find that, for reactive collisions, b is less than a critical value $b_c(\epsilon)$

$$b^2 \leq b_c^2(\epsilon) = \sigma^2(1 - \epsilon_0/\epsilon)$$

Collisions which occur with impact parameter b have their loci on a cylinder of radius b. Thus, the collision cross section for an impact parameter between b and $b + db$ is the annular area $2\pi b\, db$. The fraction of collisions which occur with this range of impact parameters is therefore $2\pi b\, db$ divided by the total collision cross section $\pi\sigma^2$. Since reaction only occurs if $b < b_c(\epsilon)$, the fraction of reactive collisions is

$$\int_0^{b_c(\epsilon)} (2\pi b\, db/\pi\sigma^2) = b_c^2(\epsilon)/\sigma^2 = 1 - \epsilon_0/\epsilon$$

which when combined with (8.1) yields the result

$$Q(\epsilon) = \begin{cases} 0, & \epsilon < \epsilon_0 \\ \pi\sigma^2(1 - \epsilon_0/\epsilon), & \epsilon \geq \epsilon_0 \end{cases} \tag{8.13}$$

For $\epsilon < \epsilon_0$ no reaction occurs; as ϵ increases the cross section increases reaching its limiting value of $\pi\sigma^2$ when $\epsilon \gg \epsilon_0$.

Both models for the reactive cross sections are grossly oversimplified. They depend on substituting a hard-sphere potential for a realistic intermolecular potential. The inadequacy of this replacement is apparent from Fig. 2.9 which compares particle trajectories based upon the hard-sphere and Lennard-Jones potentials.

Since the models are mathematically tractable it is instructive to deduce a form for the rate constant. We only consider the (slightly) more realistic Case B and combine (8.13) with (8.6) to yield, with $\beta = 1/kT$,

$$k(T) = (8\beta^3/\pi\mu)^{1/2} \int_{\epsilon_0}^{\infty} \epsilon e^{-\beta\epsilon} \pi\sigma^2 (1 - \epsilon_0/\epsilon)\, d\epsilon$$

$$= \left(\frac{8}{\pi\beta\mu}\right)^{1/2} \pi\sigma^2 \exp(-\beta\epsilon_0) \tag{8.14}$$

In SI, the units of $k(T)$ are m^3 molecule^{-1} sec^{-1}; to convert to the more commonly used units liter mole^{-1} sec^{-1} multiply (8.14) by $10^3 N_0$. The activation energy can be computed from (8.7); it is

$$E_a = \frac{R}{k}\epsilon_0 + \tfrac{1}{2}RT = N_0\epsilon_0 + \tfrac{1}{2}RT \tag{8.15}$$

which, as expected, varies slowly with temperature. The activation and threshold energies are *not* the same, which is generally the case. However, the change in activation energy predicted by this model is ~2 kJ mol^{-1} for a temperature change of 500 K. Experimental determinations of E_a are rarely both sufficiently accurate and made over a large-enough range of temperature to render such variation observable.

The preexponential factor in the line-of-centers model is proportional to $T^{1/2}$ which is a direct consequence of the cross-section function (8.13). With a different, but less readily rationalized, model for $Q(\epsilon)$, a constant preexponential factor is predicted (Problem 8.18).

Both ϵ_0 and σ are parameters of the model. It is not possible to determine ϵ_0 from the properties of the individual reactants. However, σ can be estimated from the molecular diameters of isolated reactants using transport, spectroscopic, or X-ray diffraction data. Thus, the preexponential factor in the Arrhenius expression can be computed. From (4.42), (8.14), and (8.15) we find

$$A = \left(\frac{8e}{\pi\beta\mu}\right)^{1/2} \pi\sigma^2 \tag{8.16}$$

which is compared with experimental results in Table 8.1. As expected,

Table 8.1. Kinetic Parameters for Some Bimolecular Reactions

Reaction	Activation energy (kJ mol^{-1})	Frequency factor $\times 10^{-12}$ (cm^3 mol^{-1} sec^{-1})		
		Experiment	Calculated from (8.16)	Reference
$NO + O_3 \rightarrow NO_2 + O_2$	10.5	0.80	47	[a]
$NO_2 + F_2 \rightarrow NO_2F + F$	43.5	1.6	59	[a]
$NO_2 + CO \rightarrow NO + CO_2$	132	12	110	[a]
$F_2 + ClO_2 \rightarrow FClO_2 + F$	36	0.035	47	[a]
$2NOCl \rightarrow 2NO + Cl_2$	102.5	9.4	59	[a]
$2ClO \rightarrow Cl_2 + O_2$	0	0.058	26	[a]
$COCl + Cl \rightarrow CO + Cl_2$	3.5	400	65	[a]
$H + I_2 \rightarrow HI + I$	2	200	1070	[b]
$H + N_2H_4 \rightarrow H_2 + N_2H_3$	8	0.35	900	[b]
$H + N_2H_4 \rightarrow NH_2 + NH_3$	29	10	920	[b]
$H + CCl_4 \rightarrow HCl + CCl_3$	16	7	1400	[b]
$Br_2 + Ar \rightarrow 2Br + Ar$	138	14	480	[b]
$Cl + CCl_4 \rightarrow CCl_3 + Cl_2$	84	85	340	[b]
$Cl + H_2 \rightarrow HCl + Cl$	18	12	330	[b]

[a] D. R. Herschbach, H. S. Johnston, K. S. Pitzer, and R. E. Powell, *J. Chem. Phys.* **25**, 736 (1956), where original references are given.

[b] V. N. Kondratiev, *Rate Constants of Gas Phase Reactions*, translation by L. J. Holtschlag (Washington: National Bureau of Standards, 1972), where original references are given.

the model is poor. Calculated values of A all fall in the fiftyfold range, 3×10^{13}–1.5×10^{15} cm^3 mole^{-1} sec^{-1}, in disagreement with experiment, a result that cannot be ascribed to errors in the estimate of σ. A twofold error would only change A by a factor of 4 and the actual discrepancies are usually much larger. In one case ($H + N_2H_4$), two reaction channels with significantly different A-factors exist; collision theory makes little distinction between them. In all but one case the calculated value of A is much larger than that found experimentally, indicating that (8.13) generally overestimates reaction cross sections.

Some of the defects of the hard-sphere model for the reaction cross section are apparent. It assumes that all the line-of-centers kinetic energy can be used for overcoming the threshold. It does not account for the dependence of $Q(\epsilon, i, j)$ on the internal states of the reactants. It does not consider any effect due to molecular structure or to the details of the collision process. Improvement requires either direct experimental measurement of $Q(\epsilon, i, j)$ or equivalently, a calculation which takes into account the interaction potential between colliding molecules in specific internal states.

8.5. Reaction Cross Section: Ion–Molecule Systems

The hard-sphere model, as described in Chapter 2, only accounted in a qualitative fashion for the mass, momentum, and energy transport that underlie the phenomena of diffusion, viscous flow, and heat conduction. For systems other than monatomic gases, the model was of limited utility. Thus, its failure to predict reasonable values of the Arrhenius A-factor is to be expected.

An understanding of reaction cross sections is only possible by considering some aspects of the dynamics of molecular interaction.[6] Naturally not all molecular collisions lead to reaction. There is always the possibility (often quite substantial) of inelastic collision. However, we shall only discuss reactive events.

Simplest to describe are those systems in which there is no barrier to reaction. Exoergic[7] ion–molecule reactions fall into this category. Due both

[6] For an extension of the discussion in this and the next section see R. Wolfgang, *Acc. Chem. Res.* **2**, 248 (1969).

[7] This term denotes that electronic energy is released in the reaction, i.e., the products are more stable than the reactants. It is the equivalent, in discussions of chemical dynamics, to the familiar word exothermic which is used to indicate that heat is evolved in a reaction at constant temperature.

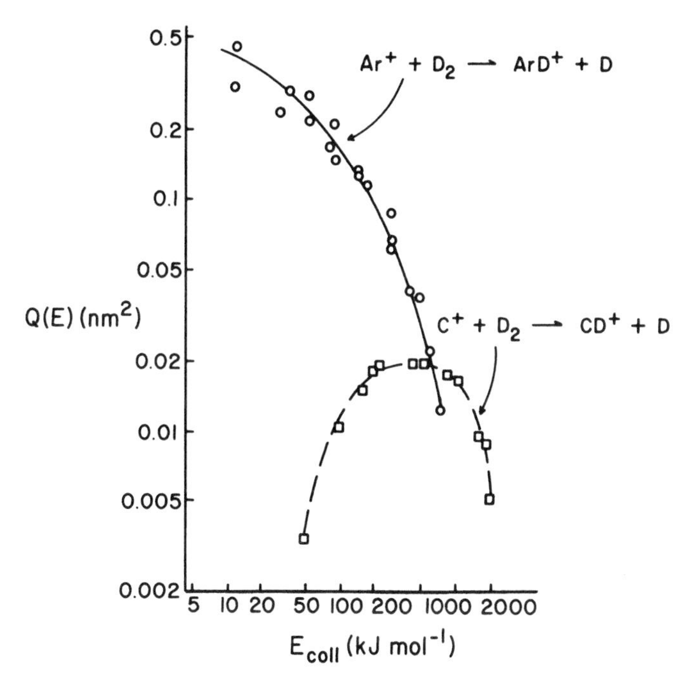

Fig. 8.3. Reaction cross section, $Q(E)$, for two ion–molecule systems as a function of collision energy. [Data for the $Ar^+ + D_2$ reaction from A. J. Masson as cited by M. J. Henchman, in *Ion–Molecule Reactions*, J. L. Franklin, ed. (New York: Plenum Press, 1972), p. 180. Data for the $C^+ + D_2$ reaction from E. Lindemann, L. C. Frees, R. W. Rozett, and W. S. Koski, *J. Chem. Phys.* **56**, 1003 (1971).]

to electrostatic forces and to a tendency toward chemical binding, there are long-range attractive interactions between ions and molecules. Reactants with zero initial collision energy are accelerated toward each other.

In Fig. 8.3 the energy dependence of $Q(E)$ for the

$$Ar^+ + D_2 \rightarrow ArD^+ + D \tag{8.17}$$

reaction is shown.[8] The data indicate that $Q(E)$, for this barrierless reaction, drops steadily as the collision energy of the reactants increases, an observation that directly contradicts the prediction of the hard-sphere model (8.13).

To understand the behavior of $Q(E)$ requires considering the effect of the long-range attractive forces on the reactants' trajectories. This is

[8] In the remainder of this chapter we conform to the experimental convention and present data as a function of molar energy, E.

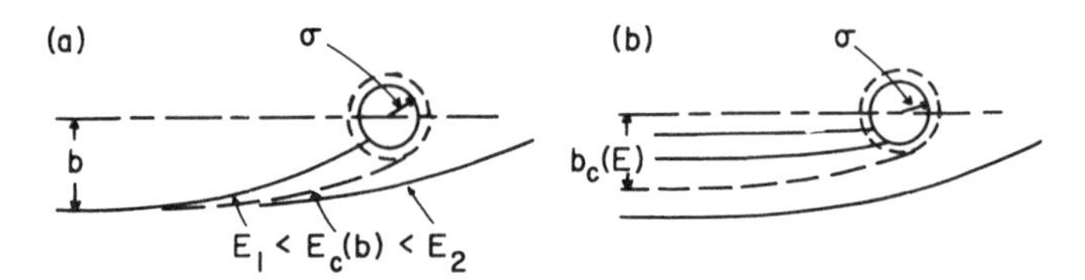

Fig. 8.4. Effect of collision energy and impact parameter on trajectories when there are long-range attractive forces and no threshold for reaction. (a) Variation of energy at constant impact parameter; (b) variation of impact parameter at constant collision energy.

illustrated in Fig. 8.4. When the impact parameter is large, only reactants with low collision energy are deflected sufficiently to spiral in toward one another and react. As the collision energy increases, the relative speed increases and the trajectory is less curved. As E increases further, there is less deflection and the ion and molecule no longer pass close enough for reaction to take place. The qualitative picture is the same as that illustrated in Fig. 2.9 for interaction via a Lennard-Jones potential; at large impact parameters significant deflection only occurs for small collision energies. The effect of varying E at constant b is shown in Fig. 8.4a. There is a critical energy $E_c(b)$ above which reaction cannot occur; as b increases E_c must drop. The effect of varying b at constant E is shown in Fig. 8.4b. It is complementary. At any E there is a critical impact parameter $b_c(E)$ above which no reaction can take place. As E increases, b_c must drop until it becomes comparable with the "hard-core" distance σ which is determined by the short-range repulsive forces between the reactant species. At this and higher energies the repulsive forces are presumably most important for determining reaction dynamics.

The existence of the critical impact parameter is often discussed in terms of the so-called *centrifugal barrier*.[9] As particles spiral in toward one another, the energy available for radial motion is less than the initial collision energy. Conservation of angular momentum of two colliding particles requires that there be nonzero centrifugal kinetic energy (that part of the kinetic energy which accounts for motion perpendicular to the line of centers). As the interparticle separation decreases so does the moment of inertia; the centrifugal kinetic energy increases to conserve the angular momentum.

For a pair with reduced mass μ, collision energy E_0, impact parameter

[9] For a quantitative treatment of the relevant collision dynamics see R. E. Weston, Jr., and H. A. Schwartz, *Chemical Kinetics* (Englewood Cliffs, N.J.: Prentice–Hall, 1972), Chapter 3.

b, and interaction potential $V(R)$, conservation of energy requires that, at any interparticle separation R,

$$E_0 = \mu(dR/dt)^2/2 + E_0 b^2/R^2 + V(R) \tag{8.18}$$

Here $E_0 b^2/R^2$ is the centrifugal kinetic energy. The radial speed dR/dt is therefore governed by an effective potential $E_0 b^2/R^2 + V(R)$. The first term (the centrifugal barrier) is always positive and the second term is negative for R large enough for particles to attract one another. Again referring to Fig. 8.4b, which illustrates the effect of varying b at fixed E_0, we note that as b increases, the distance of closest approach R_0 (where $dR/dt = 0$) also increases. When b is large, particles do not "collide"; instead they bounce off the centrifugal barrier and reaction, involving approach to within distances comparable to chemical bond lengths, may not occur. As b decreases, so does the centrifugal barrier; finally particles may surmount it and penetrate to the repulsive wall. The critical impact parameter $b_c(E_0)$ is the maximum value of b for which such penetration occurs.

For the $Ar^+ + D_2$ reaction the cross section is very large at low collision energies. Even for $b < b_c(E)$, not all collisions are reactive; thus for a given energy an upper bound to $Q(E)$ is $\pi b_c^2(E)$. The maximum measured value of Q is $\sim$0.5 nm^2, which occurs at the lowest energy, $\sim$10 kJ mol^{-1}. Thus, at this energy b_c is *greater* than 0.4 nm while the hard-core diameter, from Table 2.6, is $\sim$0.3 nm. The observed cross sections are consistent with a model in which long-range attractive forces dominate dynamics at low collision energy. Both the data and the model illustrate an important generalization: *the effective range of interaction decreases as the collision energy increases.*

As the collision energy increases further $Q(E)$ continues to drop steadily. Reaction cross sections of $\lesssim$0.03 nm correspond to values of $b \lesssim 0.1$ nm, which is much less than the "hard-core" diameter of 0.3 nm. *Reaction cross sections decrease at high energy* for an obvious reason. If too much energy is transferred to the product it dissociates. Since the strongest chemical bonds are $\sim 10^3$ kJ mol^{-1}, at larger collision energies we may always expect fragmentation to be significant.[10] There is an important corollary. Dissociation only occurs if the collision energy of the reactants is converted into vibrational energy of the products. Thus *the energy disposition between translation and vibration in the products depends*

[10] Naturally if the products which are formed have lower bond energies, dissociation occurs at smaller collision energies.

upon the initial energy. We shall investigate this point when we study a model for the mechanics of reaction dynamics in Chapter 10.

Thus far we have considered the barrierless, exoergic $Ar^+ + D_2$ reaction. To illustrate the effect of a potential barrier the reaction cross section for

$$C^+ + D_2 \rightharpoonup CD^+ + D$$

is included in Fig. 8.3. The threshold for reaction is $\sim$40 kJ mol^{-1}. The cross section is zero if the collision energy is less than the threshold value; it then rises rapidly until the inevitable decrease occurs at high energy. The low-energy cross sections for reactions with barriers are always significantly less than for those without barriers. At higher energies they are comparable. At yet higher energies the cross section is determined by the repulsive, short-range forces which are similar for all molecules.

8.6. Reaction Cross Section: Atom–Molecule Systems

Most gas-phase reactions of chemical interest involve neutral species. Because of the difficulties of producing monoenergetic beams of neutral molecules few systematic studies of cross section vs. energy have been carried out when both reactants are neutral. The available evidence, both experimental and theoretical, indicates that the general features of $Q(E)$ are much the same whether or not the reactants are charged. The important consideration is whether there is a barrier.

Although cross sections have been measured for many reactions, few have been studied over a range of energies. One for which data are available is the reaction

$$CH_3I + K \rightharpoonup KI + CH_3, \qquad \Delta E = -100 \text{ kJ mol}^{-1}$$

The cross section is plotted in Fig. 8.5. The threshold is small ($\lesssim$ 3 kJ) and the cross sections large, comparable to those in the low-energy domain of the $Ar^+ + D_2$ system (Fig. 8.3). The studies of the $K + CH_3I$ reaction were not continued to energies where there is a precipitous drop in $Q(E)$.

Even larger low-energy cross sections have been observed. Studies of reactions of alkali metals with halogens are even more dramatic, as long as there is no barrier. For the $K + Br_2$ system a cross section of $\sim$2 nm^2 has been measured[11]; assuming each collision to be reactive,

[11] J. H. Birely, R. R. Herm, K. R. Wilson, and D. R. Herschbach, *J. Chem. Phys.* **47**, 993 (1967).

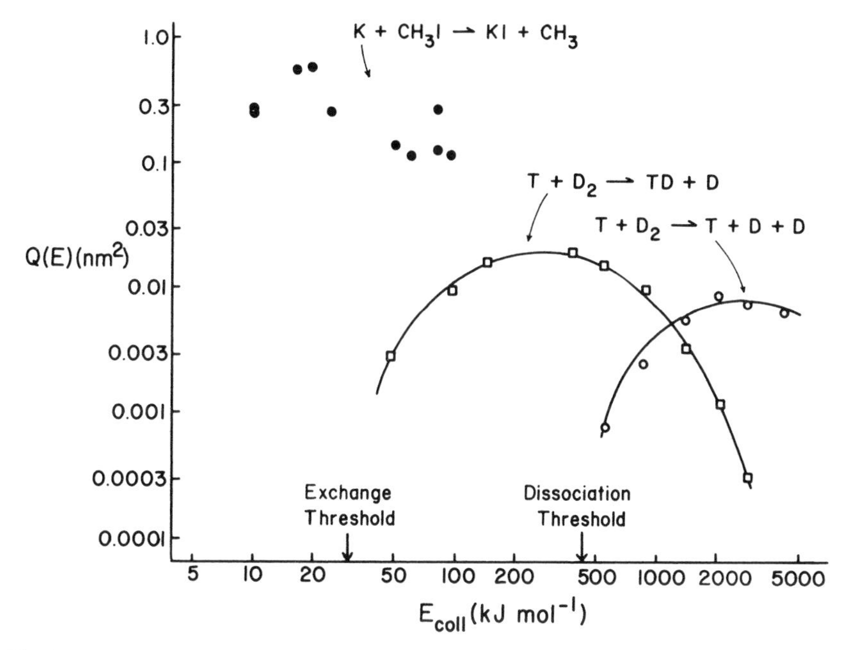

Fig. 8.5. Reaction cross section, $Q(E)$, for three atom–molecule reactions as a function of collision energy. [Data for K + CH_3I reaction from M. E. Gersh and R. B. Bernstein, *J. Chem. Phys.* **55**, 4661 (1971). Data for T + D_2 reactions from M. Karplus, R. N. Porter, and R. D. Sharma, *J. Chem. Phys.* **45**, 3871 (1966).]

reaction occurs at impact parameters of $\gtrsim 0.8$ nm, far larger than any sensible "hard-core" distance. A reasonable chemical explanation of such enormous cross sections may be based on the low ionization potentials of alkali metals and the high electron affinities of halogens. Because of this complementary relationship at distances of $\lesssim 1$ nm electron transfer is highly probable, producing two charged species which are strongly attracted to each other; reaction is then very likely. The mechanism has been dubbed "harpooning" because of the following picturesque analogy. At a fairly large distance the alkali metal "tosses out" an electron which acts as a harpoon and attaches itself to the halogen; the ionic attraction then provides the line for the whaler (alkali metal) to haul in the whale (halogen).[12]

[12] As is the case in small-boat whaling the distinction between the hauler and the hauled is fuzzy. In addition, if the whale is powerful enough (large collision energies) it can break the line and escape. In that event the whale carries the harpoon with it; here the electron is transferred back (unless the collision energies are huge).

The available experimental data on reactions with a significant threshold suggest that cross sections are $\lesssim 0.05 \text{ nm}^2$, as found in the $C^+ + D_2$ system. A theoretical study of the tritium–deuterium reaction

$$T + D_2 \rightharpoonup TD + D, \qquad \Delta E = -1.6 \text{ kJ mol}^{-1}$$

is available; the results are included in Fig. 8.5. This cross-section function, when used to compute rate constants via (8.6), led to results in good agreement with experimentally determined Arrhenius parameters.[13] The threshold is $\sim 30 \text{ kJ mol}^{-1}$ and the qualitative similarity with the $C^+ + D_2$ system (Fig. 8.3) is apparent. The decomposition cross section for the reaction

$$T + D_2 \rightharpoonup T + D + D, \qquad \Delta E = +437 \text{ kJ mol}^{-1}$$

was also computed and is included in Fig. 8.5. Note that it does *not* show any significant decrease at high energy, which is hardly surprising. If the collision energy is large enough fragmentation and/or ionization is the almost-certain outcome of any collision.

In the examples considered so far collision led to one of three results: inelastic energy transfer, reaction, or decomposition. In more complicated systems a bewildering variety of reactions becomes possible at large collision energy.[14] The reactions of hydrogen atoms with propane provide another good example.[15] These experiments were done using tritium produced by neutron bombardment of ^{3}He via the nuclear reaction

$$^3\text{He} + n \rightharpoonup p + T$$

The tritium, being radioactive, provided a convenient analytical label. At low collision energies under conditions where the tritium reacts, only hydrogen abstraction occurs[16]:

$$T + C_3H_8 \rightharpoonup C_3H_7 + HT, \qquad E_0 \lesssim 50 \text{ kJ mol}^{-1} \tag{8.19}$$

[13] M. Karplus, R. N. Porter, and R. D. Sharma, *J. Chem. Phys.* **43**, 3259 (1965); **45**, 3871 (1966).

[14] The $K + CH_3I$ system shows such behavior at higher collision energies. Possible reactions have been cataloged by M. E. Gersh and R. B. Bernstein, *J. Chem. Phys.* **56**, 6131 (1972). The channel forming $K^+ + CH_3I^-$ followed by dissociation of the negative ion has been observed by A. M. C. Moutinho, J. A. Aten, and J. Los, *Chem. Phys.* **5**, 84 (1974).

[15] A. H. Rosenberg and R. Wolfgang, *J. Chem. Phys.* **41**, 2159 (1964).

[16] For a discussion of hot-atom chemistry see R. Wolfgang, *Sci. Amer.* **214**(1), 82 (1966).

Table 8.2. Relative Rate Constants for the Abstraction, Replacement, and Double Replacement Reactions in the $T + C_3H_8$ *System*[a]

Temperature (K)	Abstraction (8.19)	Replacement (8.20)	Double replacement (8.21)
300	8×10^{-8}	7×10^{-16}	6×10^{-74}
600	6×10^{-4}	8×10^{-8}	1×10^{-36}
1000	2×10^{-2}	1×10^{-4}	4×10^{-22}

[a] Estimated as proportional to $\beta\epsilon\, e^{-\beta\epsilon}$ at threshold.

As the energy range is increased replacement also becomes important:

$$\left.\begin{aligned} T + C_3H_8 &\rightarrow C_3H_7T + H \\ &\rightarrow C_2H_5T + CH_3 \\ &\rightarrow CH_3T + C_2H_5 \end{aligned}\right\} E_0 \sim 100 \text{ kJ mol}^{-1} \qquad (8.20)$$

At still higher energies double replacement reactions, of which

$$T + C_3H_8 \rightarrow C_2H_5 + CH_3 + T, \qquad E_0 \sim 450 \text{ kJ mol}^{-1} \qquad (8.21)$$

is one example, occur. The collision energy is then so large that there is partial fragmentation. The generalization is obvious—*as collision energy increases more reaction channels are open.* Reaction (8.21) need not occur in a single step. If vibrationally excited C_2H_5T or CH_3T, formed by replacement as in (8.20), has sufficient vibrational energy, it will undergo secondary decomposition.

At thermal energies only abstraction is observable. From (8.6) the quantity $\beta\epsilon_0 \exp(-\beta\epsilon_0)$ provides a measure of relative rate constants if we assume that the shapes of the cross-section functions near threshold are not too vastly different. Values of $\beta\epsilon_0 \exp(-\beta\epsilon_0)$ are given in Table 8.2 for the reactions (8.19)–(8.21). Only at 1000 K might replacement be detectable; double replacement can never be an important thermal pathway. Thus collision chemistry permits the study of reactions inaccessible by means of ordinary thermal kinetic techniques.

8.7. Product Distribution Analysis

By studying the scattering pattern (the velocity and angular distribution) of the products formed in a crossed molecular beam experiment, much can be deduced about mechanistic details of an elementary reaction. The effect of varying impact parameter, collision geometry, and internal states of the reactants can all be investigated.

We shall consider three systems which exemplify distinct types of collision dynamics. Two are *direct* reactions; collision, reaction, and separation take place in a time comparable to molecular vibration periods, about 10^{-13} sec. The reaction

$$Ar^+ + D_2 \rightharpoonup ArD^+ + D$$

which we discussed as a prototype barrierless ion–molecule system, occurs at large impact parameter and proceeds via a direct, forward-peaked mechanism. The reaction

$$D + Cl_2 \rightharpoonup DCl + Cl$$

is favored by small impact parameters; it proceeds via a *rebound* mechanism. The third example

$$O + Br_2 \rightharpoonup BrO + Br$$

is an *indirect* reaction; a Br_2O complex forms and survives for at least a full rotational period ($\gtrsim 5\times 10^{-12}$ sec) before decomposing. The scattering pattern is quite different in each case and provides a distinctive signature.

Molecular beam experiments are performed in a laboratory frame of reference but the chemically interesting events take place with respect to the center of mass of the colliding species. In order to interpret the data, differential cross sections measured in the laboratory (LAB) coordinate system must be transformed to reflect events which took place in the center-of-mass (CM) coordinate system. To effect this transformation the invariant motion of the center of mass must be subtracted from the scattering data obtained in the LAB system. A simple example which illustrates the difference between LAB and CM kinematics is shown, for an elastic collision, in Fig. 8.6. In CM the particles always move directly toward one another before interaction and directly apart afterwards. This condition is a consequence of momentum conservation in a system with a stationary center of mass. The interaction causes each particle to be deflected through

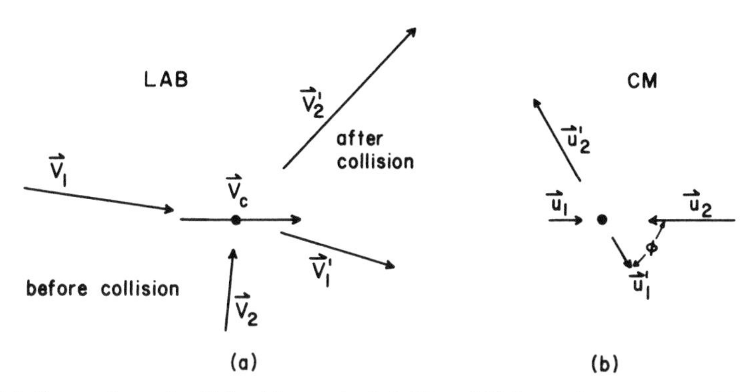

Fig. 8.6. Comparison of collision kinematics in LAB and CM coordinate systems. Unprimed and primed vectors refer to velocities before and after collision. In this example $M_1 = 2.7M_2$. (a) In LAB the center of mass, marked •, moves with constant velocity $\mathbf{V}_c$ throughout the collision. The collision geometry is for crossed beams at 90°. (b) In CM the center of mass is stationary. Conservation of momentum requires that relative motion be linear before and after interaction. Each particle is deflected through an angle φ.

an angle φ. In LAB the situation is vastly different. The heavy particle, which carries most of the center-of-mass momentum, is only slightly deflected while the trajectory of the lighter particle is more significantly perturbed. Similar diagrams can be constructed if collisions are inelastic or reactive. The relationship between LAB and CM is more complicated but, as long as the internal states of both reactants and products are known, the correspondence can be made.[17]

A number of experimental limitations introduce ambiguities into the determination of CM scattering patterns from LAB data. Molecules in the incident beams have a range of velocities. There is therefore a distribution in initial center-of-mass and relative velocities. Thus a particular deflection angle in LAB is correlated with a range of deflection angles in CM. The uncertainty can be reduced by improving velocity selection in the beams. Another source of ambiguity is that, in general, the scattering pattern of only one collision product is determined which, unless the internal energy of the product is measured, is insufficient to precisely characterize the kinematic problem. A final difficulty is that molecules in the incident beams are usually produced in a range of rotational states, which further blurs the construction of the CM scattering pattern from LAB data.

[17] This problem is analyzed by R. D. Levine and R. B. Bernstein, *Molecular Reaction Dynamics* (Oxford: Clarendon Press, 1974), pp. 60–63, 192–196.

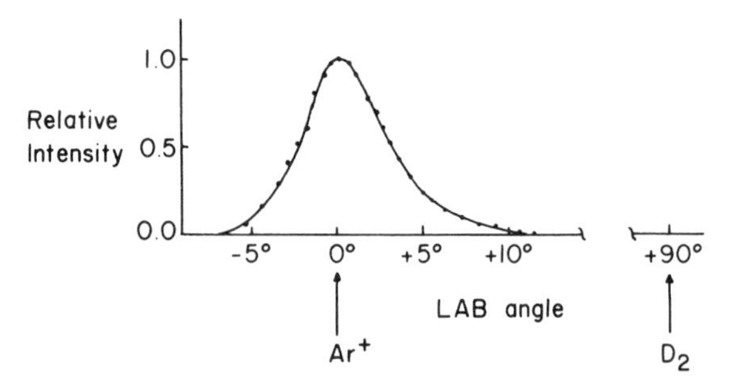

Fig. 8.7. The angular distribution of ArD^+ scattered in LAB due to the reaction $Ar^+ + D_2 \rightarrow ArD^+ + D$. The Ar^+ and D_2 are incident at 0° and 90°. The collision energy is 7.8 kJ mol^{-1}. [Adapted from Z. Herman, J. Kerstetter, T. Rose, and R. Wolfgang, *Disc. Faraday Soc.* **44**, 123 (1967).]

8.8. Direct, Forward-Peaked—The $Ar^+ + D_2$ System

The reaction

$$Ar^+ + D_2 \rightarrow ArD^+ + D$$

has been extensively studied using molecular beams. The scattering pattern has been deduced by monitoring the ionic product ArD^+. Measurements have been carried out for collision energies from 7.8 to 880 kJ mol^{-1}, an energy range over which the reaction cross section drops by a factor of about 50 (Fig. 8.3).[18]

The LAB angular distribution of ArD^+ at the lowest collision energy is shown in Fig. 8.7. All the ArD^+ is scattered at angles close to the direction of the incident Ar^+ beam, which is expected since the center-of-mass momentum does not differ greatly from the momentum of the heavy incident (Ar^+) or scattered (ArD^+) particles. By transforming these data to CM a portrait of the reaction dynamics can be constructed.

The CM angular distributions of ArD^+ determined for collision energies of 7.8 and 440 kJ mol^{-1} are compared in Fig. 8.8. In both cases the product is predominantly forward scattered, i.e., it is moving in roughly the same direction as the Ar^+ incident beam, which reflects the *chemical* process since the scattering pattern refers to CM, not LAB. The results reflect

[18] Z. Herman, J. Kerstetter, T. Rose, and R. Wolfgang, *Disc. Faraday Soc.* **44**, 123 (1967); M. Chiang, E. A. Gislason, B. H. Mahan, C. W. Tsao, and A. S. Werner, *J. Chem. Phys.* **52**, 2698 (1970).

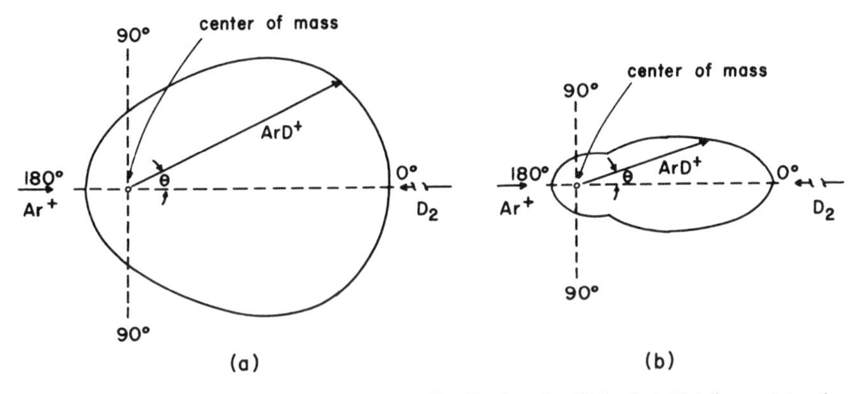

Fig. 8.8. Polar diagrams showing angular distribution in CM of ArD^+ formed in the $Ar^+ + D_2$ reaction at collision energies of (a) 7.8 kJ mol^{-1} and (b) 440 kJ mol^{-1}. The radius vector is proportional to ArD^+ intensity. The diagrams are *not* drawn to the same scale. [Derived from: (a) Z. Herman, J. Kerstetter, T. Rose, and R. Wolfgang, *Disc. Faraday Soc.* **44**, 123 (1967); (b) M. Chiang, E. A. Gislason, B. H. Mahan, C. W. Tsao, and A. S. Werner, *J. Chem. Phys.* **52**, 2698 (1970).]

collisions that occurred for a range of impact parameters since experimental control of the distance of closest approach for molecules in the crossed beams is, of course, impossible.

The predominant forward scatter and the fact that it is more pronounced at the higher energy can be interpreted in terms of the *spectator stripping* model illustrated in Fig. 8.9a. As the Ar^+ flies past the D_2 at large

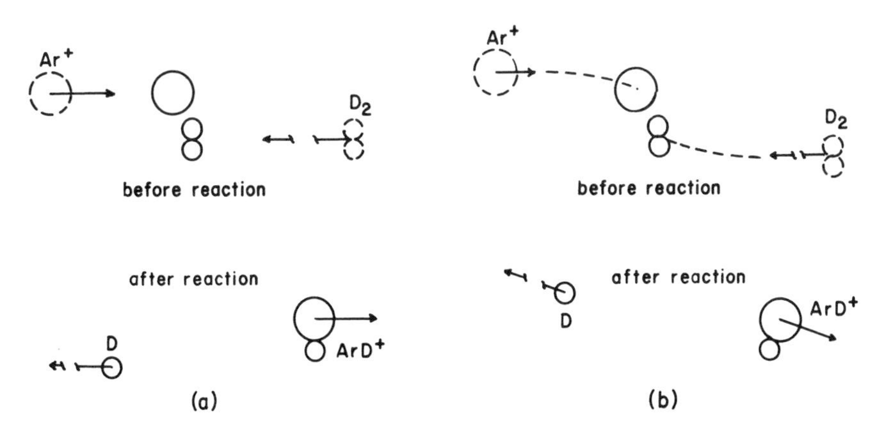

Fig. 8.9. Model for spectator stripping. (a) Idealized: trajectories are not deflected due to intermolecular forces. Product ArD^+ is scattered at 0°. (b) Effect of long-range forces. Product ArD^+ is forward scattered.

impact parameter one D atom is snatched from the molecule while the other is a passive spectator and continues along its original path with velocity unaffected. The process is analogous to the parlor trick of snatching a tablecloth out from underneath a set of dishes. If the cloth is pulled abruptly the dishes do not move; the cloth goes with the prankster and the dishes remain behind.

If the stripping model described the reaction precisely the ArD^+ would appear only at 0°, the direction of the incident Ar^+. Real trajectories are curved due to the forces between Ar^+ and D_2. At low collision energy the curvature is more pronounced (Fig. 8.4a) and so the ArD^+ is considerably deflected as shown in Fig. 8.9b.[19] At higher collision energies kinematics always swamp effects due to longer range, weaker forces, as illustrated in Fig. 2.9 and emphasized in the discussion of the centrifugal barrier (Section 8.5). The experimental consequence for the $Ar^+ + D_2$ system is that it more closely approximates ideal stripping; as a consequence the forward scatter is more pronounced.

Stripping, by its very nature, is a large impact parameter process. At small impact parameters the reactants collide head-on; thus the Ar^+ rebounds and is back scattered in the CM coordinate system either taking a D atom with it or not, depending upon whether the collision is reactive or nonreactive. Thus some back-scattered products are always expected, as is found experimentally (Fig. 8.8).

By considering both scattering and cross-section data a slightly more detailed portrait of the reaction can be drawn. At the lower energy (7.8 kJ mol^{-1}) Q is $\sim$0.5 nm^2. Such a large value implies substantial reaction probability at large impact parameter, conditions favorable for forward scattering. At the higher energy (440 kJ mol^{-1}) Q is only $\sim$0.04 nm^2. Since the scattering indicates that stripping is still the predominant mechanism, most reaction must be due to large impact parameter collisions. The scattering and cross-section data are consistent if two conditions are met. The probability of stripping is much smaller at high collision energy (to account for the lower cross section) and, at all energies, the probability of rebound reaction is far less than that of stripping (to account for the forward scatter).

[19] There is a correspondence in the tablecloth–dish analogy. The less abrupt the tug on the cloth, the more the dishes are perturbed. The reader is cautioned that expertise at the trick comes at the expense of many broken dishes.

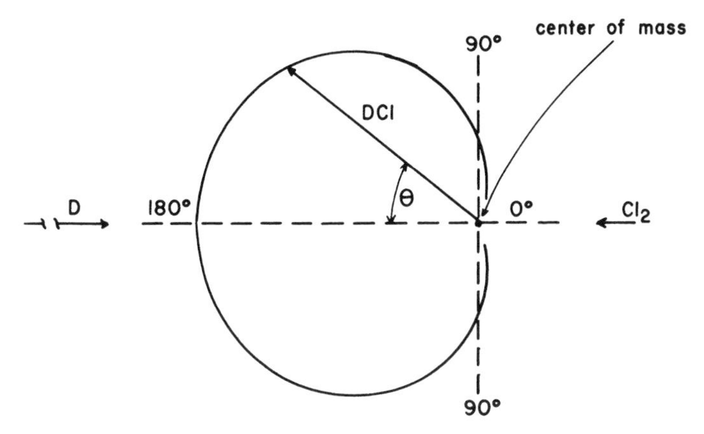

Fig. 8.10. Angular distribution in CM of DCl produced in the D + Cl_2 reaction. The radius vector is proportional to the intensity of DCl. The intensity at angles less than 60° is very small. [Derived from D. R. Herschbach, *Disc. Faraday Soc.* **55**, 233 (1973).]

8.9. Recoil—The D + Cl_2 System

The angular distribution of products in the reaction

$$D + Cl_2 \rightharpoonup DCl + Cl$$

contrasts sharply with that in the $Ar^+ + D_2$ reaction. As shown in Fig. 8.10, the DCl is predominantly back scattered which indicates that small impact parameter recoil collisions favor reaction,[20] an observation consistent with the fact that the reaction cross section is small.

In product analyses both velocity and angular distribution are measured. While knowledge of the angular distribution is usually sufficient to characterize the reaction as a small or large impact parameter process, the velocity distribution can be used to infer more details of the reaction. Figure 8.11 is a contour map representing the angle–velocity flux of DCl produced in the D + Cl_2 reaction; the collision energy is $\sim$35 kJ mol^{-1}. At the peak in the distribution the DCl has a velocity of $\sim$1600 m sec^{-1}. Since the contour map refers to CM, the peak in the distribution of Cl must also occur at roughly the same velocity since Cl and DCl are about equally massive, and in CM, the center of mass of the reacting system is, of course, stationary. Thus the total relative kinetic energy of the collision products at this velocity is $\sim$90 kJ mol^{-1}, nearly three times as large as

[20] J. D. McDonald, P. R. Le Breton, Y. T. Lee, and D. R. Herschbach, *J. Chem. Phys.* **56**, 769 (1972).

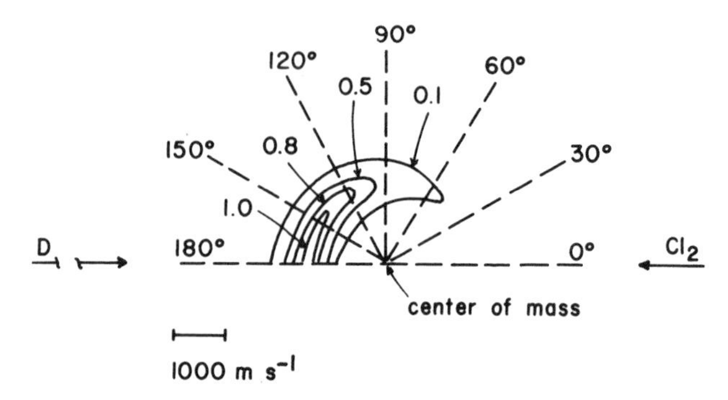

Fig. 8.11. Contour map of the angle–velocity flux distributions for DCl in the D + Cl_2 reaction. The radius vector is proportional to the *velocity* of the Cl. Contours are drawn at relative intensities of 0.1, 0.5, 0.8, and 1.0. [Adapted from D. R. Herschbach, *Disc. Faraday Soc.* **55**, 233 (1973).]

the collision energy of the reactants.[21] Referring to Fig. 8.11, most of the product DCl appears at velocities between 1200 and 2000 m sec^{-1} which corresponds to energies between ~50 and ~150 kJ mol^{-1}. The extra kinetic energy acquired by the products derives from the large exoergicity of the reaction, ~190 kJ mol^{-1}. On the average ~40% of the available energy appears as product recoil energy; the remainder must reside as internal energy (vibration, rotation) of DCl.

These observations can be used to determine the preferred reaction geometry. As the D and Cl_2 approach they are decelerated due to repulsive forces which create an activation barrier. The collision energy being greater than the threshold energy, a D—Cl bond forms with great energy release. The products then recoil. In Fig. 8.12 two extreme geometries are shown. As the D—Cl bond forms the Cl—Cl bond is stretched and broken. Neglecting the mass of the D atom and the initial collision energy (which are both relatively small), the Cl atoms must recoil along the direction of the original Cl—Cl bond in order to conserve momentum (the *D* atom is now attached to one of the Cl atoms). The perpendicular geometry therefore implies sideways scattering (peaks near 90°) while the collinear geometry is consistent with back scattering. Thus, the reaction probability is greatest for collinear (or nearly collinear) collisions.[22]

[21] The final energy in CM is $E = (1/2)(M_{DCl}V_{DCl}^2 + M_{Cl}V_{Cl}^2)$. Since momentum conservation requires that $M_{DCl}\mathbf{V}_{DCl} + M_{Cl}\mathbf{V}_{Cl} = 0$, $\mathbf{V}_{Cl}$ and thereby E can be calculated from a measurement of $\mathbf{V}_{DCl}$.

[22] A rationalization, based upon consideration of the molecular orbitals involved in the rearrangement process, is given by J. D. McDonald, P. R. Le Breton, Y. T. Lee, and D. R. Herschbach, *J. Chem. Phys.* **56**, 769 (1972).

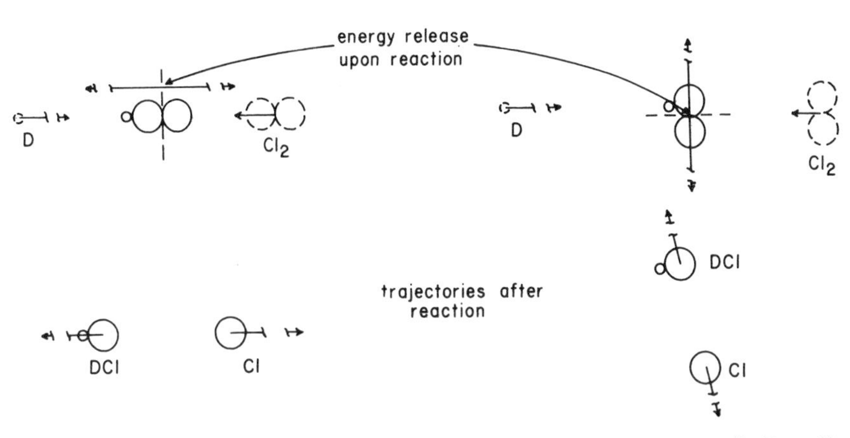

Fig. 8.12. Effect of collision geometry upon scattering of products in exoergic D + Cl_2 reaction.

8.10. *Collision Complexes—The O + Br_2 System*

The angular distribution of product BrO in the reaction

$$O + Br_2 \rightarrow BrO + Br$$

is shown in Fig. 8.13.[23] Unlike the reactions discussed in the two previous sections the scattering pattern is symmetrical; the distribution is strongly peaked at both 0° and 180°, a result that cannot be understood in terms of either the stripping or the recoil mechanism. However, if the reactants form a collision complex a completely different picture emerges as shown in Fig. 8.14. The reactants collide with an impact parameter, b (Fig. 8.14a). If the collision is "sticky" some of the relative momentum is converted into angular momentum of the complex (Fig. 8.14b) so that the three atoms rotate about the center of mass of the system. If the complex survives for at least one full rotation before decomposing, all scattering directions in the collision plane are equally probable, which accounts for the forward–backward symmetry in a CM scattering pattern. The reason that the measured angular distribution is not uniform can be understood from the following considerations. Collisions with impact parameter b are cylindrically distributed about the z-axis of Fig. 8.14. Thus the full scattering pattern is found by superposing the pattern for each collision plane as indicated in Fig. 8.14c. The axis of the resulting sphere is oriented along

[23] D. D. Parish and D. R. Herschbach, *J. Am. Chem. Soc.* **95**, 6133 (1973).

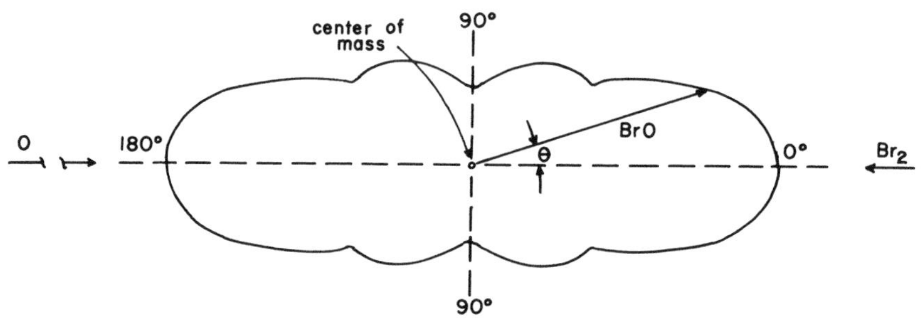

Fig. 8.13. Angular distribution in CM of BrO produced in the Br_2 + O reaction. The radius vector is proportional to the BrO intensity. [Derived from D. D. Parish and D. R. Herschbach, *J. Am. Chem. Soc.* **95**, 6133 (1973).]

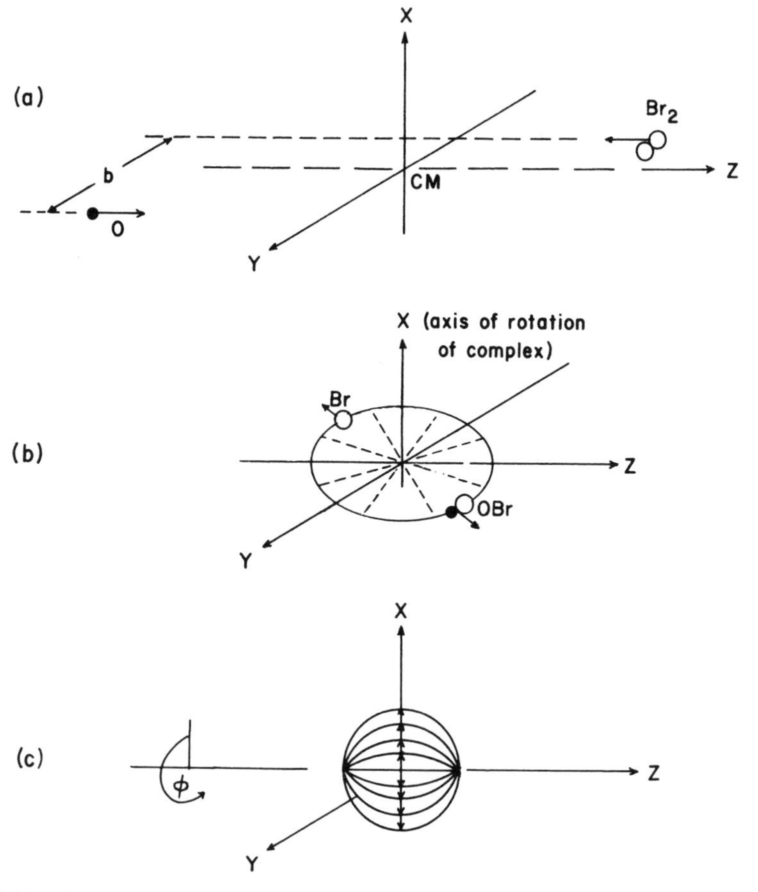

Fig. 8.14. Schematic diagram showing details in a collision typical of complex formation. [Adapted from J. P. Toennies, *Ber. Bunsenges.* **72**, 927 (1968).]

the direction of the relative velocity vector. All 0° and 180° product trajectories are superposed since these angles are the poles of the scattering sphere. At other angles the trajectories are distributed on the parallels of latitude. Thus as a detector is moved from 0° to 180° in a particular plane, observed CM intensity is symmetric with its minimum at the equator (the 90° line).

The angular distribution of BrO shown in Fig. 8.13 is precisely what is expected if complex formation has taken place. It is symmetric and strongly peaked in both forward and backward directions. Such a pattern only occurs if the complex survives at least one full rotational period ($\gtrsim 5 \times 10^{-12}$ sec) before decomposing. Since vibrational periods are typically $\sim 10^{-13}$ sec, the Br_2O complex is long-lived on a molecular time scale. In this example vibrationally unexcited Br_2O is stable relative to Br_2 and O. As is true of all cases in which complex formation has been found to occur, there are strong long-range attractive forces (and thus no activation barrier) which would lead to the formation of a stable molecule, radical, or ion if the excess energy could be removed by collisional deactivation. Energy removal is not possible given the conditions of a beam experiment. The excess energy remains trapped in the vibrational modes of the complex, being transferred between them until enough is concentrated in one mode at which point decomposition occurs. Since the overall reaction

$$O + Br_2 \rightarrow Br + BrO$$

is exoergic by $\sim$40 kJ mol^{-1}, less of the complex's excess energy is required to split off Br than to split off O. Presumably only a small fraction of the complex decomposes to reform the reactants.

8.11. State Selection Experiments

The molecules in an ordinary molecular beam have both rotational and vibrational energy. By suitable modification of the experiment it is possible to produce beams of molecules in a particular vibrational state or with a particular orientation. Therefore one can directly study the effect of vibrational energy or of molecular orientation on chemical reactivity.

By focusing an HCl chemical laser tuned to the $v = 0 \rightarrow 1$ vibrational transition on a beam of HCl molecules, it is possible to excite all the molecules of the beam into the first vibrational state of HCl. The effect on reactivity in the

$$K + HCl \rightarrow KCl + H$$

system is dramatic.[24] The reactive cross section for ground-state HCl molecules is ~0.0015 nm^2, less than 1% of that in the K + CH_3I system (see Section 8.6 and Fig. 8.5) which indicates there is an activation barrier, estimated to be ~10 kJ mol^{-1}. Reaction only occurs at small impact parameters; a recoil mechanism is most likely. One quantum of vibration, in this case ~36 kJ mol^{-1}, provides more than enough energy to surmount the barrier and the cross section increases enormously to ~0.2 nm^2, comparable to that found for the reaction K + CH_3I. Now reaction takes place at much larger impact parameters; the mechanism may well be totally different.

Methyl iodide is a dipolar molecule; it can therefore be aligned in an electric field. By passing a beam of CH_3I molecules through such a field a beam of oriented molecules is produced. This technique was used in a study of the

$$Rb + CH_3I \rightarrow RbI + CH_3$$

system.[25] Reaction was about four times more probable when the methyl iodide molecules in the beam were selected such that the rubidium approached the iodine end of the molecule than if the alignment was such that the rubidium approached the methyl end. The results are consistent with the model in that electron transfer from alkali metal to halide is the primary step.

A similar experiment was performed on the

$$K + CF_3I \rightarrow KI + CF_3$$

system.[26] The results were strikingly different. Reaction is most likely if the molecules in the beam are oriented so that the alkali atom approaches the CF_3 end of the molecule. Presumably electron transfer is still the primary step; since the F atom is far more electronegative than the I atom, it is much more likely to participate in the harpooning step. However, the C—I bond is weaker than the C—F bond (234 vs. 490 kJ mol^{-1}) and the further course of reaction seems governed by energy considerations. The I atom is the one removed which simply requires that the charge in the K^+—CF_3I^- be rapidly redistributed after the electron transfer has taken place.

[24] T. J. Odiorne, P. R. Brooks, and J. V. V. Kasper, *J. Chem. Phys.* **55**, 1980 (1971).

[25] R. J. Beuhler, Jr., and R. B. Bernstein, *J. Chem. Phys.* **51**, 5305 (1969). The experiment is *not* designed to form a beam of parallel dipoles since, at the required field strengths, the molecules would ionize. Instead a field geometry is constructed to reject all molecules which are leading with their iodine (or methyl) end.

[26] P. R. Brooks, *J. Chem. Phys.* **50**, 5031 (1969).

Problems

8.1. Using (8.6) and (8.7) show that

$$E_a = N_0 \langle \epsilon^2 Q(\epsilon) \rangle / \langle \epsilon Q(\epsilon) \rangle - 3RT/2$$

where

$$\langle f(\epsilon) \rangle \equiv \int_0^\infty f(\epsilon) e^{-\beta\epsilon} \, d\epsilon$$

8.2. If the temperature dependence of $Q(\epsilon)$ is considered show that

$$E_a = E_a^{(1)} + RT^2 \langle \epsilon \, \partial Q(\epsilon; T)/\partial T \rangle / \langle \epsilon Q(\epsilon; T) \rangle$$

where $E_a^{(1)}$ is the activation energy calculated in Problem 8.1.

8.3. In Section 8.3 we described the measurement of a reaction cross section. What measurements must be made to determine (a) the *elastic* cross section, (b) the *inelastic* (nonreactive) cross section?

8.4. A molecular beam source contains Cs(liq) in equilibrium with Cs(v) at 600 K. The vapor pressure of Cs is 4.5×10^{-3} atm. The aperture is a circular slit 1 mm in diameter; other apertures maintain the beam as an atomic ribbon. Calculate:

(a) the total cesium atom flux emerging from the source;
(b) the axial intensity in a plane perpendicular to the source and 0.3 m from the last aperture;
(c) the axial intensity assuming the geometry of (b) if only atoms with axial speeds in the range $0.95 \leq v/c_{\mathrm{mp}} \leq 1.05$ are transmitted (c_{mp} is the most probable speed).

8.5. Energy-selected ionic beams are produced by electron bombardment of a thermal distribution of a species A, $A + e^- \rightarrow A^+ + 2e^-$, followed by acceleration of A^+ through a potential drop V_0.

(a) Xe, initially at 300 K, is ionized; a 5-V potential is then applied. What is the mean ionic energy?
(b) What is the mean velocity in the direction of the applied potential?
(c) What is the standard deviation (i) in the speed, (ii) in the energy?

8.6. Another way to produce an energy-selected ionic beam is to pulse the potential for a time, τ, that is much less than that needed to accelerate the ions through the electric field, V/d, where d is the plate separation. Each ion then receives the same *velocity* increment.

(a) What is the velocity increment?
(b) Estimate the maximum pulse duration.
(c) If Xe^+, formed from Xe at 300 K, is accelerated via this procedure to the mean velocity of Problem 8.5b, what is the standard deviation (i) in the speed, (ii) in the energy?

8.7. Total cross section, as determined by measurement of beam attenuation, was discussed in Chapter 2. In one such experiment an atomic beam is passed through a reaction chamber of 2-cm length containing gas at 10^{-7} atm and 300 K. If the attenuation is 50% what is the *total* cross section for atom–gas interaction?

8.8. At 500 K the viscosity of H_2(g) is 1.26×10^{-5} and that of I_2(g) is 2.20×10^{-5} in SI units. Estimate the molecular size and then use collision theory to obtain a value for the A-factor. The experimental value is $\sim 4.3 \times 10^{10}$ M^{-1} sec^{-1}.

8.9. The cross section for the reaction $T + H_2 \rightharpoonup TH + H$ is roughly given by the analytical expressions

$$Q = \begin{cases} 0, & \epsilon < \epsilon_0 \\ By^{0.69} \exp(-0.47y^{1.46}), & \epsilon > \epsilon_0 \end{cases}$$

where

$$y \equiv (\epsilon - \epsilon_0)/\Delta, \quad \epsilon_0 = 4.2 \times 10^{-20}\ \text{J}, \quad \Delta = 6.0 \times 10^{-19}\ \text{J}, \quad B = 0.039\ \text{nm}^2$$

For a temperature 1000 K plot

(a) the Boltzmann function $(\beta\epsilon)^{1/2}e^{-\beta\epsilon}$;
(b) the cross section $Q(\epsilon)$;
(c) the *excitation function* $\beta\epsilon e^{-\beta\epsilon}Q(\epsilon)$.

Note how each of these has its maximum in quite distinct energy domains. From (c) it should be clear why threshold and activation energies are not the same.

8.10. What is your prediction of the scattering pattern for reactive collisions in the $Ar^+ + D_2$ system at high collision energies in CM? Would forward or back scattering predominate? (Ignore fragmentation.)

8.11. Consider the bimolecular reactions of Table 8.1. On the basis of the activation energies and any *chemical* knowledge at your disposal suggest which reactions might proceed (a) via complex formation; (b) via a recoil mechanism.

8.12. In Figs. 8.8, 8.10, and 8.13, CM angular distributions are given for a number of reactions. The polar diagrams plot normalized differential cross sections $\partial^2 Q/\partial\theta\,\partial\varphi$ as a function of θ for fixed φ. As the detector subtends a solid angle $d\Omega = \sin\theta\, d\theta\, d\varphi$, the differential cross section overemphasizes scattering for $\theta \sim 0°$ and $\theta \sim 180°$. For each example estimate the relative value of $\partial^2 Q/\partial\theta\,\partial\varphi$ at 15° intervals. Then plot a normalized scattering pattern as a function of θ by numerically computing

$$\sin\theta \int_0^{2\pi} d\varphi(\partial^2 Q/\partial\theta\,\partial\varphi)$$

Table 8.3

θ (deg)	$\partial^2 Q/\partial\theta\,\partial\varphi$ ($nm^2\ rad^{-2}$) (i)	$\partial^2 Q/\partial\theta\,\partial\varphi$ ($nm^2\ rad^{-2}$) (ii)	θ (deg)	$\partial^2 Q/\partial\theta\,\partial\varphi$ ($nm^2\ rad^{-2}$) (i)	$\partial^2 Q/\partial\theta\,\partial\varphi$ ($nm^2\ rad^{-2}$) (ii)
0	0.16	0	105	0.060	0.0048
15	0.15	0	120	0.049	0.0054
30	0.14	0.0015	135	0.047	0.0061
45	0.13	0.0021	150	0.047	0.0066
60	0.10	0.0026	165	0.044	0.0068
75	0.079	0.0033	180	0.043	0.0074
90	0.071	0.0040			

8.13. For reactive scattering of Na and (i) Br_2, (ii) CH_3I at about the same collision energy the variations of differential cross sections with scattering angles (CM) are [data derived from J. H. Birely, E. A. Entemann, R. R. Herm, and K. R. Wilson, *J. Chem. Phys.* **51**, 5461 (1969)] as given in Table 8.3.

(a) Calculate the reaction cross section for each case.

(b) What are likely mechanisms for each reaction?

8.14. Consider a reactive collision between molecules 1 and 2 with initial masses m_1, m_2 and velocities $\mathbf{v}_1$, $\mathbf{v}_2$. The products 1′ and 2′ have masses m_1', m_2' and the internal energy change is $\Delta\epsilon$ [$\mathbf{c} = \mathbf{v}_1 - \mathbf{v}_2$, $M = m_1 + m_2$].

(a) Show that the relative speed *after* collision is $c' = [(\mu c^2 - 2\Delta\epsilon)/\mu']^{1/2}$; μ and μ' are reduced masses before and after collision, respectively.

(b) Assume a deflection χ during reaction in CM. Show that the relative velocity after collision is $\mathbf{c}' = c'(\mathbf{n}_\parallel \cos\chi + \mathbf{n}_\perp \sin\chi)$; $\mathbf{n}_\parallel$ is a unit vector parallel to $\mathbf{c}$, and $\mathbf{n}_\perp$ a unit vector in the collision plane perpendicular to $\mathbf{c}$.

(c) Show that the product velocities in LAB are

$$\mathbf{v}_1' = \mathbf{v}_1 + (m_2'\mathbf{c}' - m_2\mathbf{c})/M, \qquad \mathbf{v}_2' = \mathbf{v}_2 + (m_1\mathbf{c} - m_1'\mathbf{c}')/M$$

(d) Show that the deflection of 1′ relative to an incident beam of 1 in LAB is

$$\cos\theta = \frac{\mathbf{v}_1' \cdot \mathbf{v}_1}{|\mathbf{v}_1'|\,|\mathbf{v}_1|} = \frac{v_1^2 + \mathbf{v}_1 \cdot \mathbf{q}}{(v_1^2 + 2\mathbf{v}_1 \cdot \mathbf{q} + q^2)^{1/2}}$$

where $\mathbf{q} = (m_2'\mathbf{c}' - m_2\mathbf{c})/M$. If $\mathbf{q}$ is small show that $\cos\theta = 1 - q^2 \sin^2\omega/2v_1^2$, where $\cos\omega = (\mathbf{v}_1 \cdot \mathbf{q})/|\mathbf{v}_1|\,|\mathbf{q}|$.

8.15. In idealized spectator stripping the atom left behind continues with its *velocity* unaltered. The reaction $Ar^+ + D_2 \rightarrow ArD^+ + D$ is 145 kJ mol^{-1} exoergic.

(a) Use conservation of momentum and show that, in ideal stripping, the relative translational energy after reaction is $E' = 10E_0/21$, where E_0 is the collision energy.

(b) Since excess collision energy must appear as internal energy of ArD^+, at what collision energy does the reactive cross section for *stripping* go to zero? The binding energy for ArD^+ is 370 kJ mol^{-1}.

8.16. A "steric factor" is often used to account for discrepancies between observed A-factors and those computed using the hard-sphere model (8.15). The steric factor f accommodates orientational requirements in reaction so that the A-factor is $A = A_0 f$ with A_0 given by (8.15).

(a) What are the steric factors for the reactions of Table 8.1?

(b) Comment, if you can, on possible reasons for differences between f in the following pairs of reactions:

(i) $H + N_2H_4 \rightarrow NH_2 + NH_3$

$H + N_2H_4 \rightarrow H_2 + N_2H_3$

(ii) $H + CCl_4 \rightarrow CCl_3 + HCl$

$Cl + CCl_4 \rightarrow CCl_3 + Cl_2$

(iii) $H + I_2 \rightarrow HI + I$

$Cl + H_2 \rightarrow HCl + Cl$

8.17. Show that the cross section

$$Q(\epsilon) = \begin{cases} 0, & \epsilon < \epsilon_0 \\ A[(\epsilon - \epsilon_0)^{1/2}]/\epsilon, & \epsilon \geq \epsilon_0 \end{cases}$$

yields an Arrhenius form for $k(T)$.

8.18. Plot the line-of-centers cross section (8.13) and the cross section of Problem 8.17, and compare with the data presented in (a) Fig. 8.3, (b) Fig. 8.5.

8.19. Sketch the energy dependence of the reaction cross section for

(a) $D^+ + H_2 \rightarrow DH + H^+$

(b) $D^- + H_2 \rightarrow DH + H^-$

Do not forget the possible effect of a potential barrier.

8.20. Show that (8.18) implies that the distance of closest approach of hard spheres is

$$R_0 = \begin{cases} b, & b > \sigma \\ \sigma, & b < \sigma \end{cases}$$

General References

J. L. Kinsey, in *MTP International Review of Science, Physical Chemistry*, Ser. 1, Vol. 9, J. C. Polanyi, ed. (London: Butterworths, 1972), pp. 173–212.

K. J. Laidler, *Theories of Chemical Reaction Rates* (New York: McGraw–Hill, 1969), pp. 182–204.

R. D. Levine and R. B. Bernstein, *Molecular Reaction Dynamics* (Oxford: Clarendon Press, 1974), Chapter 6.

J. Ross, ed., *Molecular Beams, Advances in Chemical Physics*, Vol. 10 (New York: Interscience, 1966).

Ion–Molecule Reactions

P. F. Knewstubb, *Mass Spectrometry and Ion–Molecule Reactions* (Cambridge: University Press, 1969).

Elastic Scattering

R. J. Cross, Jr., *Acc. Chem. Res.* **8**, 225 (1975).

Inelastic Scattering

G. A. Fish and F. F. Crim, *Acc. Chem. Res.* **10**, 73 (1977).

I. W. M. Smith, *Acc. Chem. Res.* **9**, 161 (1976).

J. P. Toennies, *Ann. Rev. Phys. Chem.* **27**, 225 (1976).

Reactive Scattering

D. R. Herschbach, *Disc. Faraday Soc.* **55**, 223 (1973).

D. A. Micha, *Acc. Chem. Res.* **6**, 138 (1973).

State-to-State Experiments

J. M. Farrar and Y. T. Lee, *Ann. Rev. Phys. Chem.* **25**, 357 (1974).

9

Theories of Reaction Rates

9.1. Introduction

A theoretical model for chemical reaction might take one of two forms. It could be designed to interpret molecular beam experiments which would necessitate detailed consideration of the possible outcomes of molecular collisions; it could yield, e.g., the reaction cross section (Figs. 8.3 and 8.5) and the product angular distribution (Figs. 8.7, 8.8, 8.10, 8.11, and 8.13). Such an undertaking would be exceptionally ambitious. It would essentially require theoretical predictions of all details of the reaction process.

A much more limited objective is to try to calculate the chemical rate constant. Rather than predicting the results of molecular collisions, this approach is designed to determine the Arrhenius parameters E_a and A. Naturally if the reaction cross section had been computed, it could be used to compute the rate constant. On the other hand, in a reaction at constant temperature the probability of high-energy collisions is small. Thus, the reaction pathways that are statistically significant are those with energy barriers near the threshold for reaction. If these chemically significant pathways can be characterized it is possible to apply statistical mechanics and estimate the thermal rate constant.

In either case, it is necessary to know something about the way in which reactants interact with each other. In our discussion of reaction dynamics we have seen that large reaction cross sections suggest that there is long-range attraction between the reactants. Conversely, systems in which there is a barrier to reaction have small reaction cross sections. In this chapter we shall consider how some knowledge of the interparticle interaction can be used to construct theories of the chemical rate constant.

9.2. Potential-Energy Surfaces

In Section 8.2 we found that the reaction cross section provides the link between the rate constant and collision dynamics. However, $Q(\epsilon, i, j)$ is just a function of the collision energy and the internal states of the reactants. The outcome of a single collision also depends upon the impact parameter, as demonstrated for the simple hard-sphere model of reaction cross section discussed in Section 8.4. The probability of a specific result $p(b, \epsilon, i, j)$ is dependent on all these variables. In Section 8.4 we showed that the relative importance of a collision with impact parameter between b and $b + db$ is proportional to the annular area $2\pi b\, db$. The reaction cross section is found by multiplying the weighting factor for a particular b by the probability of a specific result and integrating over all values of the impact parameter

$$Q(\epsilon, i, j) = \int_0^\infty (2\pi b\, db) p(b, \epsilon, i, j) \tag{9.1}$$

The reaction probability measures the outcome of all possible collisions with specified initial values of b, ϵ, i, and j. It is particularly simple for the hard-sphere model of Section 8.4, where

$$p(b, \epsilon) = \begin{cases} 1, & \epsilon > \epsilon_0 \text{ and } b < b_c(\epsilon) \\ 0, & \text{otherwise} \end{cases}$$

and $b_c(\epsilon) = \sigma(1 - \epsilon_0/\epsilon)^{1/2}$. Substituting this expression into (9.1) leads immediately to the result for the reaction cross section obtained previously (8.13). For more realistic models the determination of the reaction probability is a formidable problem which requires, as a preliminary step, a knowledge of the interaction potential (the *potential-energy surface*) of the reactants as a function of the nuclear coordinates. Such a calculation determines the electronic energy of the interacting molecules for all nuclear configurations of interest. Once the potential-energy surface is known the problem of computing the reaction probability can be tackled. This requires applying time-dependent quantum mechanics to treat the problem of nuclear motion in a known potential. The task is exceptionally difficult and is one that has not been accomplished, even for the simplest reactions, at the present time.[1]

[1] Recent calculations on the $H + H_2 \rightarrow H_2 + H$ exchange reaction come close to carrying out this program; see A. Kupperman and G. C. Schatz, *J. Chem. Phys.* **62**, 2502 (1975); A. B. Elkowitz and R. E. Wyatt, *J. Chem. Phys.* **62**, 2504 (1975).

The calculation of a potential-energy surface is itself not easy. To describe the surface for the exchange reaction

$$A + BC \rightharpoonup AB + C \qquad (9.2)$$

three parameters are needed, e.g., the distances A—B and B—C and the A—B—C angle.[2] Correspondingly the surface is four-dimensional. If one of these variables is held fixed, a three-dimensional projection can be constructed. A series of such diagrams permits visualization of the whole potential-energy surface.

Reactions that proceed via an intermediate which is more energetic than either reactants or products, as in Fig. 4.8, have qualitatively similar potential-energy surfaces. A typical slice at fixed A–B–C angle is shown in perspective in Fig. 9.1 and as a contour map in Fig. 9.2. When r_{AB} (the A–B distance) is large, the interaction potential is that of the isolated BC molecule; the point R represents the classical ground state of the reactants. Similarly when r_{BC} is large the potential is that of the AB molecule; P is the classical ground state for products. Large values of both r_{AB} and r_{BC} correspond to total dissociation into A + B + C; this region of the surface is the plateau D.

Reaction is represented by motion from R to P on the surface. As reaction occurs the system starts in the reactant *valley*, travels over a *pass*, and descends into the product valley. Reaction only occurs if the relative energy of the interacting molecules is sufficient to surmount the pass. For thermal reactions the system must move along relatively low-energy paths because, for surfaces like that of Fig. 9.1, the pass height is usually high enough so that collisions which occur with enough relative energy are highly improbable. Reaction thus samples the high-energy tail of the Boltzmann distribution and only collisions with energy close to the pass energy are sufficiently probable to contribute significantly to reaction.[3] The minimum energy path is shown in both Figs. 9.1 and 9.2. The section through this path, known as the *reaction profile*, is shown in Fig. 9.3. It is similar to the schematic diagram of the activation process shown in Fig. 4.8; now, however, progress along the path is correlated with specific changes in geometry. Reactants with collision energy less than the energy of the pass

(2) Many other ways of specifying the ABC triangle are possible but this is the most convenient choice.

(3) If the system strays too far from the pass the collision energy must increase in order for reaction to be possible. At 300 K the fraction of possibly reactive collisions decreases tenfold with each 6 kJ mol^{-1} increase in the barrier.

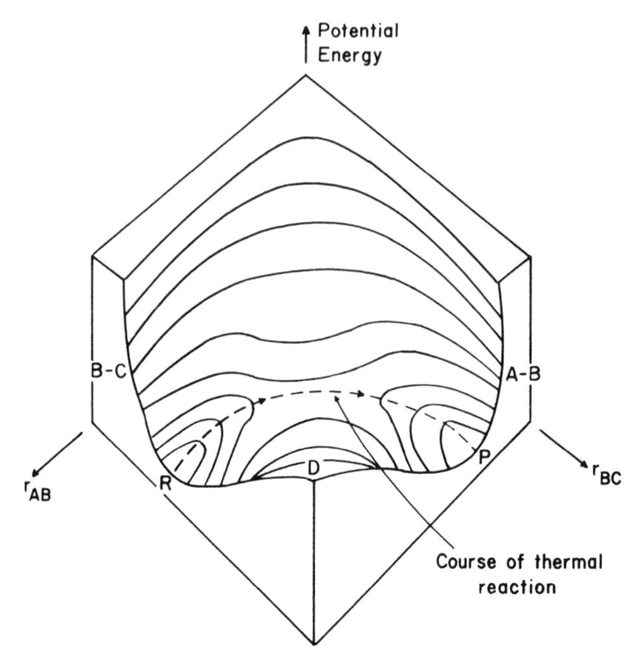

Fig. 9.1. Variation of potential energy with A–B and B–C distances for an ABC system in which the A–B–C angle is held fixed. [Adapted from K. J. Laidler, *Chemical Kinetics* (New York: McGraw–Hill, 1965).]

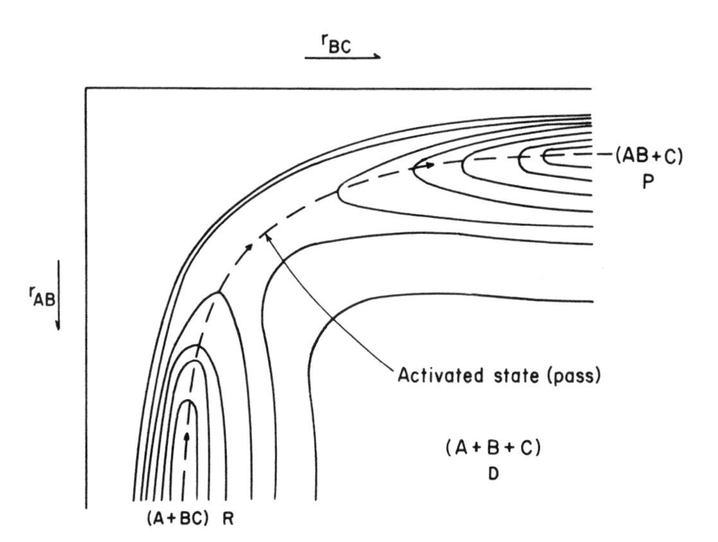

Fig. 9.2. Contour diagram for the potential energy of the ABC system of Fig. 9.1. The dashed line shows the minimum energy path which is followed by a reaction under thermal conditions.

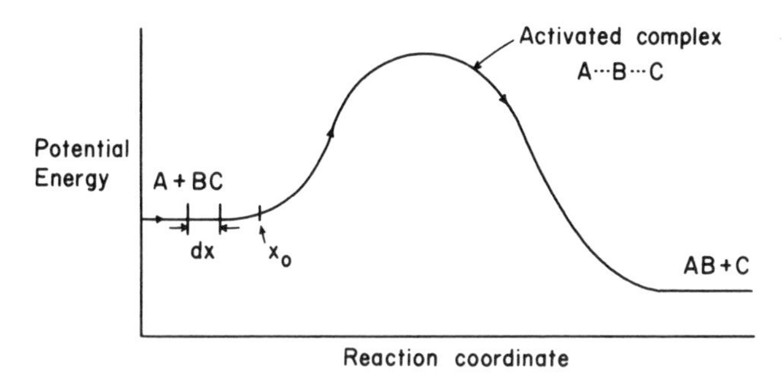

Fig. 9.3. Section through the minimum energy path of Fig. 9.2. There is a barrier to reaction in either direction.

do not react.[4] In *thermal chemical* reaction, only trajectories close to the reaction profile are important; all others require too much collision energy. Thus most reaction paths pass close to the pass on the potential-energy surface. Because of the chemical importance of the relatively low-energy region of the ABC aggregate system, it is specially designated. The pass, which is the peak of the reaction profile, is known as the *transition state* or the *activated complex*. It bears, of course, no relation to the long-lived collision complexes discussed in Section 8.10.

Changing the A–B–C angle affects both the potential-energy surface and the reaction profile but only in those regions where the three particles are close together. The far valleys and the plateau which describe the isolated molecules and the separated atoms are not changed. Thus all reaction profiles are alike at start and finish but their intermediate structure differs. The pass energy depends upon the angle A–B–C as is illustrated in Fig. 9.4, where the reaction profile for the exchange reaction

$$H + H_2 \rightharpoonup H_2 + H$$

is plotted for different H–H–H angles. The lowest pass height in this system occurs in the linear configuration. The activated complex or transition state is identified as the configuration for which the ABC complex has the *minimum pass energy*. It is important to recognize that the most probable reaction paths pass through configurations close to that of the activated

[4] This is not strictly correct. Molecules follow the laws of quantum, not classical, mechanics. As such they have zero point energy; the system does not precisely follow the potential-energy trough which slightly alters the collision energy necessary for surmounting the pass. It is also possible, but not likely, for reactants with collision energy somewhat less than the pass energy to tunnel through the barrier. These quantum effects alter the threshold energy which is usually slightly *less* than the pass energy.

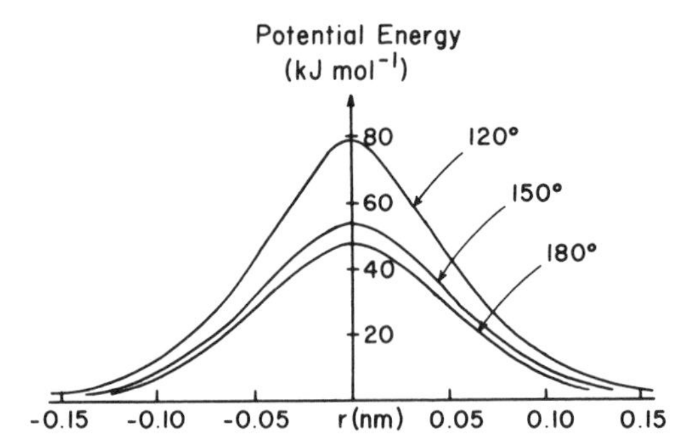

Fig. 9.4. Reaction profiles for the H_3 system as a function of HHH angle. The distance r is the separation of the incoming or leaving H atom from the pass on the potential-energy surface. Energies are measured with respect to the $H + H_2$ system. [Adapted from I. Shavitt, R. M. Stevens, F. L. Minn, and M. Karplus, *J. Chem. Phys.* **48**, 2700 (1968).]

complex. Thus knowledge of the molecular structure of the region at the top of the pass is most important for understanding many aspects of chemical reactivity.

For the model potential-energy surface we have discussed, the lowest energy intermediate (the transition state) is at higher energy than either products or reactants. This is the most common surface; however, other possibilities also occur. Minimum energy paths for reactions in which there is no barrier are illustrated in Fig. 9.5. These are typical of exoergic reactions with large cross sections. The minimum energy path of Fig. 9.5a describes a reaction such as $Ar^+ + D_2$, discussed in Chapter 8. The

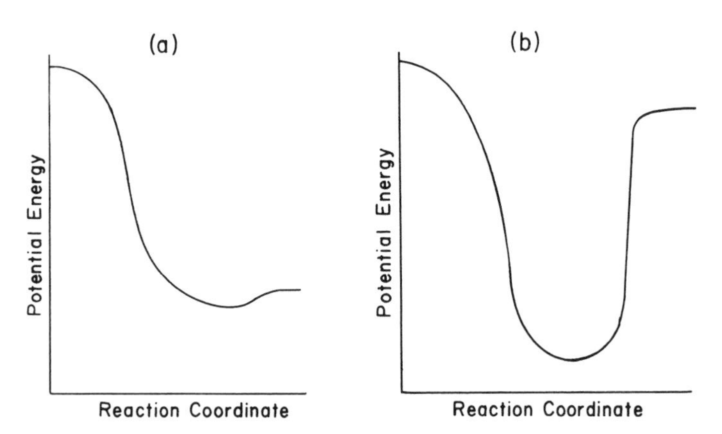

Fig. 9.5. Minimum energy path for systems without barriers. (a) Typical profile for reaction with shallow well when complex formation does not occur. (b) Typical profile for reaction with deep well where intermediate complex may form.

well is too shallow to permit the formation of a collision complex. The minimum energy path of Fig. 9.5b has a deep well; complex formation is probable, e.g., the O + Br_2 system. *All* low-energy paths are now potentially reactive; there is no longer a favored reaction trajectory and the idea of an activated complex is less useful. In these examples the pass energy and the reactant energy are the same. However, this does not imply that the pass geometry is the reactant geometry. In terms of Fig. 9.2 we may imagine a constant energy reactant trough which defines the reaction coordinate as A and BC approach one another. Trajectories that deviate significantly from this coordinate must overcome energy barriers. Thus, while there is no activated complex, there is still a favored intermediate configuration which plays a similar role.

In general, reactions involve more than three atoms, so a complete potential-energy surface depends on more variables. To describe the exchange reaction

$$K + CH_3I \rightarrow KI + CH_3$$

a total of 12 coordinates is needed. Even if a complete calculation could be done, interpreting such a surface would present enormous problems. By considering only reaction channels in which the CH_3 group can be treated as a single particle, the problem is reduced to that of a simple triatomic exchange reaction.

The actual construction of a potential-energy surface is a difficult quantum mechanical problem. It requires calculating the energy of a many-electron system as a function of nuclear positions. For the similar, but simpler, problem of computing the known vibrational and rotational properties of molecules, comparisons of theory and experiment suggest that approximate theories are not too accurate and accurate theories are too difficult to be generally useful. A potential-energy surface calculation involves more electrons (the electrons of all the reactants must be considered simultaneously) and many geometries. It is not surprising that most such calculations are both approximate and of limited accuracy.[5]

[5] Reactions that have been specifically studied include: $H + H_2 \rightarrow H_2 + H$, B. Liu, *J. Chem. Phys.* **58**, 1925 (1973); $K + CH_3I \rightarrow KI + CH_3$, L. M. Raff and M. Karplus, *J. Chem. Phys.* **44**, 1212 (1966); $H_2 + He^+ \rightarrow HeH^+ + H$, P. J. Brown and E. F. Hayes, *J. Chem. Phys.* **55**, 922 (1971); $K + Br_2 \rightarrow KBr + Br$, D. R. Herschbach, in *Molecular Beams*, J. Ross, ed. (New York: Interscience, 1966), p. 376; $F + H_2 \rightarrow H + HF$, C. F. Bender, P. K. Pearson, S. V. O'Neil, and H. F. Schaefer III, *J. Chem. Phys.* **56**, 4626 (1972); $Li + F_2 \rightarrow LiF + F$, G. G. Balint-Kurti, *Mol. Phys.* **25**, 393 (1973); $H_2 + I_2 \rightarrow 2HI$, L. M. Raff, L. Stivers, R. N. Porter, D. L. Thompson, and L. B. Sims, *J. Chem. Phys.* **52**, 3449 (1970).

9.3. Activated Complex Theory

From the viewpoint of a potential-energy surface molecules on one side of the pass are reactants; those on the other side are products. In regions close to the pass such a distinction is hard to maintain since the molecules form an aggregate of strongly interacting particles. Reaction dynamics can be described as molecules *traversing* the pass from reactant side to product side, suggesting that the rate of a chemical reaction may be calculated by determining the number of systems per second which cross the pass in the direction reactant to product, an observation that is the basis for the theory of chemical reaction developed by Eyring.[6]

Consider the bimolecular gas-phase reaction

$$\mathrm{A} + \mathrm{B} \rightharpoonup \text{products} \tag{9.3}$$

with a reaction profile like that of Fig. 9.3. Motion along the reaction coordinate is specified by a position x and a reaction momentum p. By $p > 0$ we mean that the reactants are moving toward each other; as they approach the pass they may be considered to form an aggregate or a composite system (sometimes described as an AB complex). The transition from independent reactants to aggregate system cannot be recognized to have occurred at any particular point along the reaction coordinate. At small separations ($\lesssim$0.5 nm) interaction is strong and the composite description is appropriate. At large separations ($\gtrsim$10 nm) interaction is weak and describing the reactants as independent species is more reasonable.

The basic assumption of the theory is to consider motion along the reaction coordinate separately from other motions on the potential-energy surface. In addition to its translational, rotational, and vibrational modes the AB aggregate has a reaction mode which describes the chemical transformation in terms of progress along the reaction coordinate. The larger the value of the reaction momentum p, the more collision energy is available for surmounting the pass. If the energy in the reaction mode is large enough, the molecules traverse the pass and react; if not, they are reflected.[7]

Far away from the pass, the number of A–B composite systems, $C(p, x)$, with reaction coordinate between x and $x + dx$ and reaction momentum

[6] The development which follows is adapted from H. Eyring, J. Walter, and G. E. Kimball, *Quantum Chemistry* (New York: John Wiley, 1944), pp. 299–311. For a totally different development see K. J. Laidler, *Theories of Chemical Reaction Rates* (New York: McGraw–Hill, 1969), pp. 45–53.

[7] Again this is not quite true since the problem is one of quantum, not classical, mechanics. See footnote (4) in this chapter.

between p and $p + dp$, can be estimated if we assume that local equilibrium is established along the reaction coordinate. Naturally the reactant population is being continuously depleted as reaction proceeds so the hypothesis is not strictly true; it is, however, a plausible first approximation. The rate at which systems *approach* the pass can be used to compute the reaction rate. This rate is the reciprocal of the time required for *all systems* in the interval dx to pass a specific point such as x_0 on the reaction profile, Fig. 9.3. Since the systems are distributed in the interval dx, a translation dx is needed in order for all the systems to pass x_0. Each system in the specified momentum range has the same value of p; thus the time required is

$$\mu \, dx/p$$

where μ is the reduced mass for motion along the reaction coordinate, $\mu = m_A m_B/(m_A + m_B)$. The total number of systems approaching the pass per second is

$$C(p, x) \frac{p}{\mu \, dx} \tag{9.4}$$

If the reactants A and B were truly in equilibrium with an A–B complex, the statistical mechanical theory of chemical equilibrium in gases could be applied.[8] The number of such systems N_{AB} would be

$$N_{AB} = N_A N_B (q_{AB}/q_A q_B) \tag{9.5}$$

where q_A, q_B, and q_{AB} are partition functions for A, B, and A–B complex; N_A and N_B are the number of A and B molecules. The energy of the complex is assumed to be separable because we have differentiated motion along the reaction coordinate from all other modes. Thus, far from the pass, where the system is in the flat portion of the reaction profile Fig. 9.3, the energy in the reaction mode is kinetic energy only, $p^2/2\mu$. The energy residing in all modes *except* the reaction coordinate is ϵ_n and the energy of the composite is $\epsilon_n + p^2/2\mu$. The zero of energy for the composite is the same as that of the separated reactants. Motion along the reaction coordinate requires traversing a pass with a classical barrier to reaction $\epsilon^\ddagger$. The total energy in all the other modes is measured from an energy zero set at the top of the pass. From equilibrium statistical thermodynamics the number of composites found with reaction coordinate between x and $x + dx$,

[8] T. L. Hill, *Introduction to Statistical Thermodynamics* (Reading, Mass.: Addison–Wesley, 1960), Chapters 4, 8, and 9. A summary of some important results of statistical thermodynamics is included in Appendix A.

reaction momentum between p and $p + dp$, and ϵ_n is[9]

$$C_n(p, x)\, dp\, dx = \frac{N_{AB}}{q_{AB}} \exp\left(-\frac{\epsilon_n}{kT}\right) \exp\left(-\frac{p^2}{2\mu kT}\right) \frac{dp\, dx}{h} \tag{9.6}$$

Combining (9.6) with (9.5) and summing over all degrees of freedom *except* the reaction coordinate and momentum we find

$$C(p, x)\, dp\, dx = \frac{N_A N_B}{q_A q_B} q_{AB}^{\ddagger} \exp\left(-\frac{p^2}{2\mu kT}\right) \frac{dp\, dx}{h} \tag{9.7}$$

where $q_{AB}^{\ddagger}$ is a partition function for energy disposition in all modes of the complex *except the reaction mode*

$$q_{AB}^{\ddagger} = \sum_n \exp\left(-\frac{\epsilon_n}{kT}\right) \tag{9.8}$$

To compute the population of the complex requires integrating (9.7) over p and x; this is ambiguous since the precise range of values of x that can be considered to correspond to composite systems is unknown. Fortunately the quantity of interest is the rate at which systems approach the pass. This can be determined from the behavior in the reactant valley as long as collisions with a third particle do not perturb the approach to the pass.[10] From (9.4) and (9.7) the rate of approach is

$$N_A N_B \frac{q_{AB}^{\ddagger}}{q_A q_B} \frac{p}{\mu h} \exp\left(-\frac{p^2}{2\mu kT}\right) dp \tag{9.9}$$

which is *independent* of the reaction coordinate. Only systems with $p > 0$ are moving toward the pass and possible reaction. Only a fraction of these descend into the product valley. Introducing a transmission coefficient, $\varkappa(p)$, to account for partial transmission, the reaction rate is found by multiplying (9.9) by $\varkappa(p)$ and integrating over all $p > 0$

$$-\frac{dN_A}{dt} = N_A N_B \frac{q_{AB}^{\ddagger}}{q_A q_B} \int_0^{\infty} \frac{p}{\mu h} \varkappa(p) \exp\left(-\frac{p^2}{2\mu kT}\right) dp$$

[9] See Appendix A for a justification of this expression. The factor of h^{-1} occurs because the formulation is semiclassical. It is against the rules of quantum mechanics to treat reaction coordinate and momentum separately.

[10] This is an excellent approximation in dilute gases. At 1 atm and 500 K the mean free path is ~50 nm, much larger than the range of values of the reaction coordinate that can be considered to correspond to the composite. At very low pressure the approximations used to derive (9.7) are no longer reasonable; under such conditions the concentration of composites is sensitive to the rate at which they traverse the pass. In addition, depletion is significant and (9.5) is no longer a good approximation.

and the second-order rate constant can be identified as

$$k(T) = \frac{q_{\mathrm{AB}}^{\ddagger}}{q_{\mathrm{A}} q_{\mathrm{B}}} \int_0^\infty \frac{p}{\mu h} \varkappa(p) \exp\left(-\frac{p^2}{2\mu kT}\right) dp \tag{9.10}$$

If chemical reaction were a problem in classical mechanics the momentum dependence of the transmission coefficient would only reflect whether there was sufficient energy to surmount the pass. For a pass energy $\epsilon^{\ddagger}$ we have

$$\varkappa(p) = \begin{cases} 0, & p < p^{\ddagger} = (2\mu\epsilon^{\ddagger})^{1/2} \\ 1, & p > p^{\ddagger} \end{cases} \tag{9.11}$$

The problem of computing $\varkappa(p)$ is actually a quantum mechanical one. The momentum dependence is determined by the shape of the reaction profile. Theoretical models show that $\varkappa(p)$ is small but not zero for $p < p^{\ddagger}$ and that for $p > p^{\ddagger}$ it is oscillatory, but generally increasing, until the momentum is large enough so that fragmentation is possible. The qualitative behavior of $\varkappa(p)$ for momenta small enough so that fragmentation is not a problem is shown in Fig. 9.6.[11]

While the quantum and classical forms for $\varkappa(p)$ are strikingly different, the Boltzmann factor in the integrand of (9.10) ensures that the oscillatory structure of the quantum form of $\varkappa(p)$ only slightly influences the rate constant. In Fig. 9.7 the quantities $p\varkappa(p)\exp(-p^2/2\mu kT)$ are plotted for the two forms of $\varkappa(p)$. In both cases roughly the same range of momenta contribute significantly to the integral. Thus (9.10) is approximately

$$\begin{aligned} k_2(T) &\approx \frac{q_{\mathrm{AB}}^{\ddagger}}{q_{\mathrm{A}} q_{\mathrm{B}}} \langle\varkappa\rangle \int_{(2\mu\epsilon^{\ddagger})^{1/2}}^\infty \frac{p}{\mu h} \exp\left(-\frac{p^2}{2\mu kT}\right) dp \\ &= \frac{kT}{h} \frac{q_{\mathrm{AB}}^{\ddagger}}{q_{\mathrm{A}} q_{\mathrm{B}}} \exp\left(-\frac{\epsilon^{\ddagger}}{kT}\right) \langle\varkappa\rangle \end{aligned} \tag{9.12}$$

where $\langle\varkappa\rangle$ is a mean transmission coefficient. Its value, determined by comparing the computed and experimental A-factors, is generally between 0.1 and 0.9; in practice it is often assumed to be $\sim$0.5. As formulated, $k_2(T)$ has units of molecules^{-1} sec^{-1}. By introducing appropriate factors of V and N_0 it can be converted into concentration units, cm^3 mol^{-1} sec^{-1}. In terms of the molar partition function per unit volume,

$$Q = q/(VN_0) \tag{9.13}$$

[11] This problem is treated by H. Eyring, J. Walter, and G. E. Kimball, *Quantum Chemistry* (New York: John Wiley, 1944), pp. 311–326.

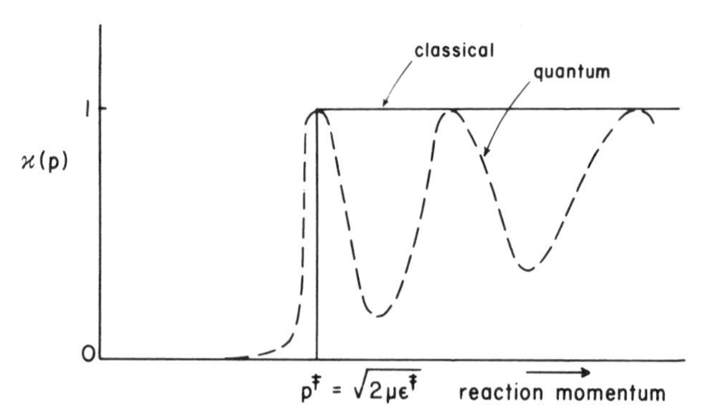

Fig. 9.6. Behavior of transmission coefficient $\varkappa(p)$ as a function of reaction momentum for classical and quantum mechanical models of reaction; a rectangular barrier is assumed.

the rate constant is

$$k_2(T) = \frac{kT}{h} \frac{Q_{AB}^{\ddagger}}{Q_A Q_B} \exp\left(-\frac{\epsilon^{\ddagger}}{kT}\right)\langle\varkappa\rangle \tag{9.14}$$

which is the famous Eyring equation. The rate constant can be calculated from the properties of the potential-energy surface by identifying the reaction coordinate. The pass height determines the Eyring activation energy $\epsilon^{\ddagger}$, here defined as the classical barrier to reaction; as such it does not include the zero point vibrational energy of the activated complex. The shape of the reaction profile determines $\varkappa(p)$ and thus $\langle\varkappa\rangle$; the remainder of the surface near the pass determines $Q_{AB}^{\ddagger}$.

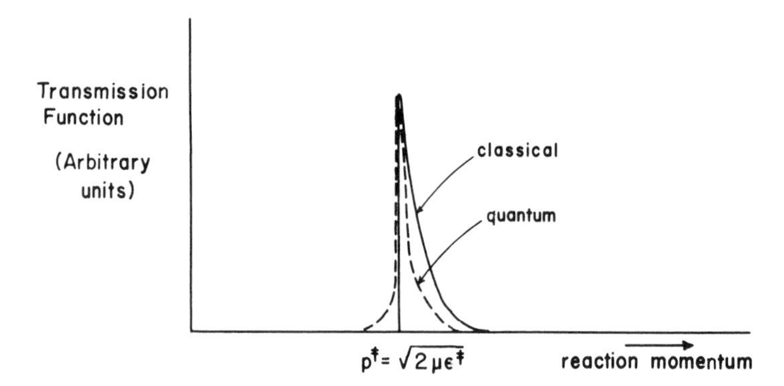

Fig. 9.7. Transmission function, $p\varkappa(p)\exp(-p^2/2\mu kT)$, as a function of reaction momentum for classical and quantum models of $\varkappa(p)$ plotted in Fig. 9.6. Note that oscillations in $\varkappa(p)$ are damped by the Boltzmann factor; $\epsilon^{\ddagger}/kT$ is much greater than 1 corresponding to a significant activation barrier.

As was the case in the hard-sphere model for reaction rates (Section 8.4), $\epsilon^\ddagger$ is not easily computed and thus a complete calculation of $k_2(T)$ is not generally possible. To eliminate $\epsilon^\ddagger$ and determine an expression for the A-factor we write (9.14) as

$$k_2(T) = B(T)\exp(-\epsilon^\ddagger/kT)$$

Using (8.7) the Arrhenius activation energy is

$$E_a = \epsilon^\ddagger N_0 + RT\theta, \qquad \theta = T\frac{d\ln B}{dT}$$

and the A-factor is

$$A = B(T)e^\theta \tag{9.15}$$

Again both A and E_a are temperature dependent but the variation is usually too small to be detected experimentally.

The theory is obviously quite approximate even if $\langle\varkappa\rangle$, $\epsilon^\ddagger$, and $Q^\ddagger_{\text{AB}}$ could all be computed. We began by assuming an equilibrium population of A–B composite systems at points along the reaction profile far from the pass; however, the problem is intrinsically nonequilibrium in nature. We assumed that motion along the reaction coordinate was not coupled to other degrees of freedom of the A—B composite, which is highly unlikely, if only because the moments of inertia of the aggregate vary as the reaction coordinate changes. We assumed the one-dimensional reaction coordinate could be uniquely assigned even though molecular beam experiments indicate that reaction dynamics may be much more complex. Even with these limitations the theory is still a considerable improvement over the hard-sphere model of Section 8.4. Its great achievement is that, through consideration of the partition function $Q^\ddagger_{\text{AB}}$, plausible assumptions as to the nature of the A–B interaction can be introduced and used to calculate, in an admittedly approximate manner, either $k_2(T)$ or, more commonly, the A-factor.

9.4. Model Calculations of $k_2(T)$

The activated complex theory provides a link which connects the potential-energy surface with the chemical rate constant. To apply the theory requires knowing the structure of the potential-energy surface near the saddle point. As discussed in Section 9.2 accurate determination of the surface is a difficult problem. Instead of computing the properties of the surface, the properties of the activated complex are usually estimated. These estimates then provide the data used to determine $k_2(T)$.

Here we treat two specific examples. First we consider the hard-sphere model for chemical reaction and show that activated complex theory leads to a $k_2(T)$ identical with that calculated using kinetic theory arguments in Section 8.4. We then consider the molecular example

$$H + I_2 \rightharpoonup HI + I$$

and, by making reasonable assumptions about the structure of the activated complex, show that the theory provides a means for discriminating between two possible geometries for the activated complex.

In addition to constructing a model for the potential-energy surface or postulating a structure for the activated complex, the reaction coordinate must be identified. Then all other degrees of freedom in the A–B complex can be described in terms of the translational, rotational, and vibrational modes of the complex. If these are assumed to be independent, the computation of $k_2(T)$ requires calculating $Q_{AB}^{\ddagger}$ which, from statistical thermodynamics, is

$$Q_{AB}^{\ddagger} = V^{-1}N_0^{-1}(q_t q_r q_v g_e) \tag{9.16}$$

The q's are partition functions for the various degrees of freedom and g_e is the electronic degeneracy factor for the complex.[12] According to our assumptions only motion along the reaction coordinate leads to chemical change. Other degrees of freedom are presumed unaffected by reaction. Thus the geometry and vibrational frequencies determined at the saddle point can be used to characterize energy disposition in the *nonreactive* modes of the complex. It is this feature of the saddle point which endows the activated complex with special importance.[13] The partition functions for the various degrees of freedom may then be computed using statistical thermodynamics.[14]

[12] We assume that only one electronic state of the complex is important in *thermal* reactions.

[13] It might seem that this assumption of separability implies that knowledge of the saddle point structure can be used to characterize the reactants and products. If each degree of freedom were strictly independent then precise description of the saddle point structure would specify the molecular properties of both reactants and products. Since separability is only approximate, no such extrapolation is possible. In fact, the reaction coordinate evolves from a molecular vibration in a manner unspecified by knowledge of the saddle point region.

[14] T. L. Hill, *Introduction to Statistical Thermodynamics* (Reading, Mass.: Addison–Wesley, 1960), Chapters 4, 8, and 9; or see Appendix A.

The hard-sphere approximation of Section 8.4 is the simplest model for reaction. The molecules A and B are treated as spheres which must approach within a distance σ to react. Only the translational motion of A and B is considered, and from (9.13) and (9.A6) of Appendix A

$$Q_A = q_A/(N_0 V) = N_0^{-1}(2\pi m_A kT/h^2)^{3/2} \tag{9.17}$$

with a similar expression for Q_B. The complex has only translational and rotational degrees of freedom. Relative motion, which would correspond to vibration if AB formed a bound molecule, is the reaction mode. Using (9.16) and (9.A6) we find

$$Q_{AB}^{\ddagger} = \frac{q_t q_r}{N_0 V} = \left(\frac{2\pi MkT}{h^2}\right)^{3/2} \frac{8\pi^2 IkT}{h^2} \frac{1}{N_0} \tag{9.18}$$

where $M = m_A + m_B$. The moment of inertia of the activated complex, $I = \mu\sigma^2$, is determined by σ, the distance of closest approach; μ is the reduced mass $m_A m_B/M$. Combining (9.14), (9.17), and (9.18) we find

$$k_2(T) = N_0\left(\frac{8kT}{\pi\mu}\right)^{1/2} \pi\sigma^2 \exp(-\beta\epsilon^{\ddagger}) \tag{9.19}$$

which, but for the factor of N_0, is identical with (8.14); $\langle\varkappa\rangle$ is exactly one in this model since collision theory is a *classical* mechanical model for chemical reaction.

As a specific molecular example consider the reaction

$$H + I_2 \rightharpoonup HI + I \tag{9.20}$$

In the temperature range 300–700 K the Arrhenius parameters, given in Table 8.1, are $A \sim 2\times10^{14}$ cm^3 mol^{-1} sec^{-1} and $E_a \sim 2$ kJ mol^{-1}. The A-factor, which is about 20% of that estimated using collision theory, may be estimated using (9.14)–(9.16) if reasonable assumptions about the structure of the activated complex of HI_2 are made. Both geometry and vibrational frequencies must be estimated.

At the pass both H—I and I—I distances are somewhat larger than in the corresponding stable molecules. Actual calculations of potential-energy surfaces[15] suggest that this expansion amounts to 20–30%; it affects not only the moments of inertia but also the vibrational frequencies of the HI_2 complex. The parameters required to describe molecular properties are: g, the electronic degeneracy; I, the moment of inertia (for a

[15] See references in footnote (5) in this chapter.

linear molecule); ν, the vibrational frequencies; and σ, the symmetry number. If HI_2 is assumed to be linear with interatomic distances 30% greater than those in HI or I_2, the following set of molecular parameters can be obtained[16]:

H: $g = 2$

I_2: $g = 1$, $\sigma = 2$, $I = 7.5 \times 10^{-45}$ kg m^2, $\nu = 6.45 \times 10^{12}$ sec^{-1}

HI_2: $g = 2$, $\sigma = 1$, $I = 13.0 \times 10^{-45}$ kg m^2, $\nu = 5.3 \times 10^{13}$, 1.45×10^{13}, 1.45×10^{13} sec^{-1}

Both H and HI_2 have electronic degeneracies of 2 since each has a single unpaired electron; I_2, being a symmetric diatomic molecule, has a symmetry number of 2. The diatomic molecule I_2 has two rotational and one vibrational degree of freedom. The linear complex HI_2 has two rotational and four vibrational modes; because one possible vibrational mode corresponds to the reaction coordinate, three are true vibrational degrees of freedom. From (9.14) the preexponential factor $B(T)$ is

$$B(T) = \langle \varkappa \rangle \frac{kT}{h} \frac{Q^{\ddagger}_{HI_2}}{Q_H Q_{I_2}}$$

which may be evaluated using (9.16) and the statistical mechanical expressions for the various partition functions,[17]

$$B(T) = \langle \varkappa \rangle \frac{RT}{h} \left(\frac{g^{\ddagger}}{g_H g_{I_2}} \right) \frac{(2\pi m^{\ddagger} kT/h^2)^{3/2}}{(2\pi m_H kT/h^2)^{3/2} (2\pi m_{I_2} kT/h^2)^{3/2}}$$
$$\times \frac{(8\pi^2 I^{\ddagger} kT/h^2 \sigma^{\ddagger})}{(8\pi^2 I_{I_2} kT/h^2 \sigma)} \frac{\prod_{i=1}^{3} [1 - \exp(-h\nu_i^{\ddagger}/kT)]^{-1} \exp(-h\nu_i^{\ddagger}/2kT)}{[1 - \exp(-h\nu_{I_2}/kT)]^{-1} \exp(-h\nu_{I_2}/2kT)}$$

[16] With a 30% bond expansion, the bond lengths in HI_2 are l(H–I) = 0.21 nm and l(I–I) = 0.35 nm. Calculation of the vibrational frequencies requires estimating the force constants in the activated complex. In the series HF, HCl, HBr, and HI, the quantity [(frequency) × (bond length)] is essentially constant. Assuming this to be generally applicable, the vibrational frequency of HI (6.94×10^{13} sec^{-1}) leads to the estimate given for the stretching frequency of HI_2. The bending frequency in HI_2 cannot be estimated from the properties of HI. However, assuming that HI and HTe bonds are similar, a bending force constant, k_δ, can be determined from the spectroscopic properties of H_2Te (bending frequency 2.58×10^{13} sec^{-1}, bond length 0.165 nm). Using this value for k_δ, the bending frequency for both linear and bent HI_2 may be calculated. Formulas relating force constants to frequencies are given by G. Herzberg, *Infrared and Raman Spectra* (Princeton, N.J.: Van Nostrand, 1945), pp. 168–175.

[17] T. L. Hill, *Introduction to Statistical Thermodynamics* (Reading, Mass.: Addison-Wesley, 1960), Chapters 4, 8, and 9; or see Appendix A.

Table 9.1. Arrhenius A-Factor for the Reaction $H + I_2 \rightarrow HI + I$ Calculated Using Activated Complex Theory[a]

HII angle, degrees	$A \times 10^{-14}$ (cm^3 mol^{-1} sec^{-1})		
	250 K	500 K	750 K
180	0.055 (0.034)	0.055 (0.028)	0.073 (0.042)
90	3.04 (2.70)	2.73 (2.13)	2.95 (2.13)

[a] Two possible geometries for the activated complex are considered. At each geometry the first value is based upon a 30% bond expansion; the values in parentheses are based upon no bond expansion. The experimental value for A in the temperature range 300–700 K is 2×10^{14} cm^3 mol^{-1} sec^{-1}.

Introducing the molecular parameters into this expression yields

$$B(T) = \langle \varkappa \rangle \frac{2.05 \times 10^{14}}{T^{1/2}} \frac{(1 - e^{-310/T})}{(1 - e^{-2560/T})} \frac{(e^{-1820/T})}{(1 - e^{-695/T})^2} \tag{9.21}$$

from which A can be calculated using (9.15). A similar calculation can be carried out assuming that the activated complex has a bent geometry. If an HII angle of 90° is assumed, the following set of estimated molecular parameters is obtained for the HI_2 complex:

$$g = 2,\ \sigma = 1,\ ABC = 3.63 \times 10^{-135}\ \text{kg}^3\ \text{m}^6,\ \nu = 5.3 \times 10^{13},\ 1.45 \times 10^{13}\ \text{sec}^{-1}$$

Here ABC is the product of the three principal moments of inertia.[18] The difference is that one vibrational degree of freedom has been altered to a rotational degree of freedom. By using statistical thermodynamics a result analogous to (9.21) is found. The results of calculations assuming linear and bent configurations for the activated complex are given in Table 9.1 for three temperatures assuming both 30% and no bond expansion. A $\langle \varkappa \rangle$ of 1 was assumed so that the calculation should overestimate the A-factor if the properties postulated for the HI_2 complex are reasonable.

Some interesting points emerge. The calculation is insensitive to the amount of bond expansion assumed. The temperature dependence of A for the two geometries is quite small, which is a general feature so that any temperature in the experimental range can be used in these calculations.

[18] The procedure for calculating the principal moments of inertia is given by I. N. Levine, *Molecular Spectroscopy* (New York: John Wiley, 1975), pp. 197–203.

Table 9.2. Comparison of A-Factors Computed Using Activated Complex Theory and Collision Theory with Experimental Results[a]

Reaction	$A \times 10^{-12}$ (cm^3 mol^{-1} sec^{-1})		
	Experiment	Collision theory	Activated complex theory
$NO + O_3 \rightarrow NO_2 + O_2$	0.80	47	0.44
$NO_2 + F_2 \rightarrow NO_2F + F$	1.6	59	0.12
$NO_2 + CO \rightarrow NO + CO_2$	12	110	6.0
$F_2 + ClO_2 \rightarrow FClO_2 + F$	0.035	47	0.082
$2NOCl \rightarrow 2NO + Cl_2$	9.4	59	0.44
$2ClO \rightarrow Cl_2 + O_2$	0.058	26	0.010
$COCl + Cl \rightarrow CO + Cl_2$	400	65	1.8

[a] D. R. Herschbach, H. S. Johnston, K. S. Pitzer, and R. E. Powell, *J. Chem. Phys.* **25**, 736 (1956).

The *A*-factors for the two geometries are very different. The bent activated complex leads to a reasonable value; the linear complex leads to a gross underestimate. In the bent geometry the calculated *A*-factor is 1.5 times the experimental value, which is remarkable considering the approximate nature of the calculation. In the linear geometry the calculated value is 30–40 times too small. Thus we can argue that this structure for the activated complex is very unlikely.[19] The calculation discriminates against one geometry. It does *not* prove the other to be correct. Nonetheless given a choice between a linear or nearly linear activated complex and one that is strongly bent, there is reason to opt for the latter.

Similar calculations have been carried out to estimate the *A*-factor for various reactions. The results, for some of the examples of Table 8.1, are given in Table 9.2. The calculations did not account for bond-length increases in the activated complex. Since expansion affects both the moments of inertia and the vibrational frequencies in the same way, recalculation would increase the predictions of the theory by factors of 3–5. Even without this correction the predictions of activated complex theory are generally within a factor of 10 of the experimental value. Only when the experimental

[19] It is interesting that the product distribution in molecular beam studies of the $H + I_2$ system also suggests a bent geometry for the activated complex. See D. R. Herschbach, *Disc. Faraday Soc.* **55**, 233 (1973).

A is very large are there gross discrepancies. However, there are other difficulties. According to the theory as developed in Section 9.3, the mean transmission coefficient is less than 1. Thus activated complex theory should *always* overestimate $k_2(T)$, and therefore the A-factor as well. Even recognizing that the method of calculation used in making the predictions of Table 9.2 leads to underestimates, the differences between calculation and experiment are too large to be attributable to neglect of bond expansion in the complex. Whether the differences are due to weaknesses in the theory leading to (9.14) or to errors in the proposed structure of the activated complex is unclear. In any case activated complex theory provides another tool, albeit one which must be laced with a liberal dose of chemical intuition, to be used in studying a chemical reaction. It must be emphasized that agreement between calculated and observed values of A does not "prove" anything. As with all other kinetic arguments *disagreement* indicates that the postulated mechanism or structure for the activated complex is incorrect. A case in point is the $H_2 + I_2$ reaction. The A-factor calculated assuming the Bodenstein mechanism

$$H_2 + I_2 \rightharpoonup 2HI$$

is 4×10^{13} cm^3 mol^{-1} sec^{-1}, half the experimental value.[20] In spite of the satisfactory agreement, there is other kinetic evidence that the reaction does not proceed via the bimolecular mechanism.

9.5. The Kinetic Isotope Effect

Statistical thermodynamics can be used to explain the effect of isotopic substitution on equilibrium constants. Similarly the statistical theory of reaction rates provides a basis for understanding isotope effects in rate constants. Consider a prototype atom–molecule reaction

$$A + BC \rightharpoonup AB + C$$

where B can exist in a number of isotopic forms. Then, from (9.14), the ratio of the rate constants for reaction involving different isotopes B and B′ is

$$\frac{k'}{k} = \frac{\langle \varkappa' \rangle}{\langle \varkappa \rangle} \exp\left(\frac{(\epsilon^\ddagger - \epsilon'^\ddagger)}{kT}\right) \frac{Q^\ddagger(AB'C)}{Q^\ddagger(ABC)} \frac{Q(BC)}{Q(B'C)} \tag{9.22}$$

[20] S. W. Benson, *The Foundations of Chemical Kinetics* (New York: McGraw–Hill, 1960), pp. 285–286; recalculation based upon the quoted data leads to the value given here.

The partition functions are determined by nuclear motion on the potential-energy surfaces for the BC molecule or the ABC complex, and the activation energy is the potential step required to traverse the lowest energy path on the ABC surface. The phenomenon is a direct manifestation of quantum mechanics. In the classical limit (massive species) the ratio is 1.

Simplification of (9.22) is based on the fact that the potential-energy surface is the ground state for electronic motion in a charge distribution determined by the nuclear configuration. Since isotopes have the same nuclear charge, the *potential-energy surface is unaffected by isotopic substitution.* As a consequence the classical barrier height is isotope independent and (9.22) becomes

$$\frac{k'}{k} = \frac{\langle \varkappa' \rangle}{\langle \varkappa \rangle} \frac{Q^{\ddagger}(\mathrm{AB'C})}{Q^{\ddagger}(\mathrm{ABC})} \frac{Q(\mathrm{BC})}{Q(\mathrm{B'C})} \tag{9.23}$$

Furthermore the shape of the surface in the reactant valley and at the saddle point is not changed. The force constants for nuclear motion in either region are invariants. Knowledge of the moments of inertia and vibrational frequencies of one isotopic species allows computation of the same quantities for any other. For a diatomic molecule like BC the results are particularly simple

$$\frac{I'}{I} = \left(\frac{\nu}{\nu'}\right)^2 = \frac{m_{\mathrm{B'}}(m_{\mathrm{B}} + m_{\mathrm{C}})}{m_{\mathrm{B}}(m_{\mathrm{B'}} + m_{\mathrm{C}})}$$

Similar, but more complicated, expressions can be found relating the properties of any molecular species.

Further simplification of (9.23) can be carried out in different ways. In one, the quantities $Q^{\ddagger}$ and Q are determined by a calculation similar to the one used in Section 9.4 to compute the A-factor for the $\mathrm{H} + \mathrm{I}_2$ reaction, leading to an expression involving masses, moments of inertia, and vibrational frequencies. The other method depends upon the solution of the vibrational secular equation in the reactant valley and at the saddle point.[21] The result, for a linear transition state, is

$$\frac{k'}{k} = \frac{\langle \varkappa' \rangle}{\langle \varkappa \rangle} \frac{\nu'^{\ddagger}}{\nu^{\ddagger}} \frac{\Gamma(\mathrm{BC})}{\Gamma(\mathrm{B'C})} \frac{\Gamma_1(\mathrm{AB'C})}{\Gamma_1(\mathrm{ABC})} \left[\frac{\Gamma_2(\mathrm{AB'C})}{\Gamma_2(\mathrm{ABC})}\right]^2 \tag{9.24}$$

[21] Detailed analyses are given by J. Bigeleisen and M. Wolfsberg, *Adv. Chem. Phys.* **1**, 15 (1968); L. Melander, *Isotope Effects on Reaction Rates* (New York: Ronald Press, 1960), Chapter 2.

where

$$\Gamma_i = u_i/\sinh u_i, \qquad u_i = h\nu_i/2kT$$

The frequencies ν_1 and ν_2 correspond, respectively, to the symmetric stretch and the doubly degenerate bend of the activated complex. Motion along the reaction coordinate replaces the asymmetric stretch. The quantity $\nu^{\ddagger}$, which is a computational device without physical meaning as a frequency, is determined by the curvature of the reaction profile at the peak. For a profile like that of Fig. 9.3 it is

$$\nu^{\ddagger} = \frac{i}{2\pi}\left(\frac{-k}{\mu}\right)^{1/2}$$

where k is the curvature (<0) and μ is the reduced mass for motion along the reaction coordinate. Clearly $\nu^{\ddagger}$, while having the dimensions of frequency, has no relation to the rate at which systems cross the pass; that quantity is determined by (9.9).

After the potential-energy surface near the saddle point has been constructed and the properties of the activated complex determined, all factors in (9.24) except for the transmission ratio $\langle\varkappa'\rangle/\langle\varkappa\rangle$ may be immediately computed. The transmission ratio may be calculated if highly simplified models of the reaction profile are used. However, in most instances little error is introduced by assuming that $\langle\varkappa'\rangle/\langle\varkappa\rangle = 1$.[22] The exceptions are for reactions in which isotopes of hydrogen are intimately involved in the reaction process. While both the transmission ratio and the partition function ratios of (9.24) approach 1 for massive reactants, the transmission ratio tends to 1 more rapidly. Therefore its influence on the isotope effect is relatively less significant, except at low temperature.

A system for which the effect of isotopic substitution has been thoroughly investigated is the reaction

$$\text{Cl} + \text{H}_2 \rightharpoonup \text{HCl} + \text{H}$$

The rate law for the rate of disappearance of reactants is

$$-\frac{d[{}^i\text{H}\,{}^j\text{H}]}{dt} = k_{ij}[\text{Cl}][{}^i\text{H}\,{}^j\text{H}], \qquad i, j = 1, 2, 3$$

Experimental values for the isotope effect k_{11}/k_{ij} are plotted in Fig. 9.8. Values computed from (9.24) are included in the figure. The parameters

[22] R. E. Weston, Jr., *Science* **158**, 332 (1967).

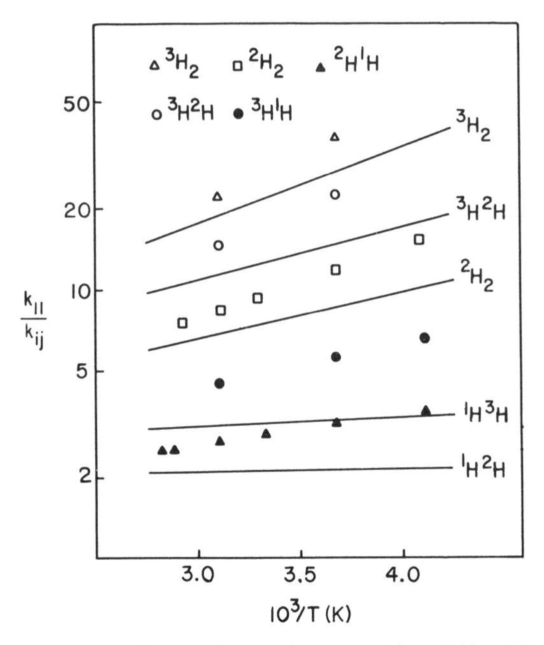

Fig. 9.8. Kinetic isotope effect for the reaction $H_2 + Cl \rightharpoonup HCl + H$. Solid lines calculated using (9.24). [Adapted from R. E. Weston, Jr., *Science* **158**, 332 (1967), where original references are given.]

required to evaluate (9.24) were calculated using a semiempirical potential-energy surface; variation of the transmission ratio was taken into account. Note that the kinetic isotope effect can be quite large; $^3H_2$ at the lower temperature studied (273 K) reacts 35 times slower than $^1H_2$. The calculated values are in fair agreement with the data. The trends are correctly predicted and numerical discrepancies are not large. It appears that deviations become more severe at lower temperatures. Whether these are due to the weaknesses of transition-state theory itself, the model used for the potential-energy surface, or the way in which the transmission ratio has been estimated is unclear.

Isotopic substitution in the $Cl + H_2$ system leads to a "normal" isotope effect, i.e., $[k(\text{light})]/[k(\text{heavy})] > 1$. In other systems this ratio is less than 1—the so-called "inverse" isotope effect. The isotope effect is determined by the difference between the *shape* of the potential-energy surface at the reactant configuration and at the pass. Ignoring vibrational excitation, the ratio (9.24) reflects differences in zero point energies (Problem 9.15). In the example considered the zero point energy of the activated complex is less than that of the reactants—the activated complex is "floppy."

If the activated complex were "constrained," as is the case when the isotopic reactants are atoms, the opposite would be true; an "inverse" isotope effect is expected.[23]

9.6. Unimolecular Reaction—General Features

In Chapter 5 we examined the phenomenology of unimolecular reactions and their pressure-dependent order of reaction. The simple mechanistic model (5.18) predicted the rate law

$$-\frac{d[\mathrm{A}]}{dt} = k^*[\mathrm{A}] \tag{9.25}$$

with k^*, a concentration-dependent "first-order" rate constant given by

$$k^* = k_\infty/\{1 + k_d/k_-[\mathrm{A}]\} \tag{9.26}$$

where k_∞ is the high-pressure unimolecular rate constant and k_d/k_- is the ratio of the rate of reaction of activated molecules to their rate of deactivation. To provide a theoretical model, we must recognize that activated molecules are formed with a distribution of energies and that the rate constant for decomposition to products must be energy dependent, as follows from a rudimentary molecular picture of the decomposition process. As an example consider the reaction

$$C_2H_5CHO \rightarrow C_2H_6 + CO$$

For products to form, the ethyl–carbonyl C–C bond must stretch and the H atom on the carbonyl group must approach the ethyl group. Only if sufficient internal energy is concentrated in a "critical mode" that leads to reaction can decomposition occur. The total internal energy of a molecule is distributed among *all* its internal modes. Only rarely is it concentrated in a single vibrational mode. Thus, if the internal energy ϵ is just greater than the threshold energy ϵ_0, few molecules will be collisionally activated to a state for which sufficient energy is in the critical mode; the decomposition probability is thus small. As ϵ increases, the likelihood of finding an energy ϵ_0 in the critical mode also increases thus enhancing the decomposition probability.

[23] K. J. Laidler, *Chemical Kinetics*, 2nd ed. (New York: McGraw–Hill, 1965), p. 97.

These ideas on energy transfer can be formalized by considering the rate of formation and the rate of destruction of excited molecules with energies between ϵ and $\epsilon + d\epsilon$. Excited molecules are formed via collisions. If n is the concentration of unexcited A molecules, the total collision rate as computed in (2.21) is $\gamma n^2/2$, where γ, assuming the molecules to be hard spheres of diameter d, is[24]

$$\gamma = 4\pi d^2(kT/\pi m)^{1/2} \tag{9.27}$$

If we assume that reaction does *not* significantly affect energy distribution among the low-energy states of the A molecule, the collisions between *unexcited* molecules produce molecules with an energy distribution typical of a gas at equilibrium. The approximation, that reaction does not affect the partitioning among low-energy states, is similar to the one used in estimating the number of composite systems in the development of activated complex theory (Section 9.3). If the equilibrium probability density is $P(\epsilon)$, the total rate of production of excited molecules with energies in the range ϵ to $\epsilon + d\epsilon$ is

$$(\gamma n^2/2)(2)[P(\epsilon)\,d\epsilon] = \gamma n^2 P(\epsilon)\,d\epsilon \tag{9.28}$$

A factor of 2 is introduced because in the collision process either of the A molecules might be excited.

While the population of low-lying states identified with A is not altered by reaction, the population of the states identified with A*, i.e., those with sufficient energy to react, is affected. The total concentration of A* molecules with energy between ϵ and $\epsilon + d\epsilon$ is $n\xi(\epsilon)P(\epsilon)\,d\epsilon$, where $\xi(\epsilon)$ is a depletion factor accounting for reaction. Since A*–A* collisions are highly unlikely, they can be ignored. The total collision rate of A* molecules of the specified energy is $\gamma n[n\xi(\epsilon)P(\epsilon)\,d\epsilon]$. Since the overwhelming majority of A–A* collisions lead to deactivation, the total rate of destruction of A* molecules of energy ϵ is

$$\gamma n^2 \xi(\epsilon)P(\epsilon)\,d\epsilon + k(\epsilon)[n\xi(\epsilon)P(\epsilon)\,d\epsilon] \tag{9.29}$$

[24] We ignore any relation between collision frequency and the internal energy of the products of the collision. Furthermore, the effect of the intermolecular potential is ignored. To lift these restrictions a more general formulation of the problem, based upon the "master equation" must be constructed. The master equation focuses upon the transition probabilities between molecular states. For a discussion and a general review of unimolecular reactions see D. C. Tardy and B. S. Rabinovitch, *Chem. Rev.* **77**, 369 (1977).

The second term accounts for the effect of reaction; $k(\epsilon)$ is the reaction rate constant for molecules of energy ϵ. Once a steady state is established (9.28) and (9.29) are equal and we find

$$\xi(\epsilon) = \frac{\gamma n}{k(\epsilon) + \gamma n} = \frac{1}{1 + k(\epsilon)/\gamma\beta p} \tag{9.30}$$

where the ideal gas law, $n = \beta p$, has been used. At high pressure there is little depletion of excited states due to reaction. At low pressure the effect is important; molecules react too rapidly for the excited-state population to be maintained by collisional activation. At low energy, where $k(\epsilon) = 0$, there is also no depletion.

The total rate of product formation is found by integrating the second term in (9.29) over all energies greater than ϵ_0, the threshold for reaction

$$-\frac{d[\mathrm{A}]}{dt} = \frac{d[\mathrm{Prod}]}{dt} = \int_{\epsilon_0}^{\infty} k(\epsilon)[nP(\epsilon)\xi(\epsilon)\, d\epsilon]$$

so that, using (9.25) and (9.30) the unimolecular rate constant k^* is

$$k^* = \int_{\epsilon_0}^{\infty} \frac{P(\epsilon)k(\epsilon)\, d\epsilon}{1 + k(\epsilon)/\gamma\beta p} \tag{9.31}$$

which has as its high- and low-pressure limits

$$k_\infty = \int_{\epsilon_0}^{\infty} P(\epsilon)k(\epsilon)\, d\epsilon, \qquad k_0 = \gamma\beta p \int_{\epsilon_0}^{\infty} P(\epsilon)\, d\epsilon \tag{9.32}$$

Note that at low pressure the rate constant is determined by the fraction of activated molecules. If $k(\epsilon)$ is assumed to be constant for $\epsilon \geq \epsilon_0$, (9.31) reduces to the Lindemann expression (9.26).

9.7. Unimolecular Reaction—RRKM Theory

To apply (9.31) requires estimating the energy dependence of the excited-state population $P(\epsilon)$ and the decomposition rate constant $k(\epsilon)$. Progress to this end, based upon treating molecular vibration as harmonic, was made by Hinshelwood,[25] Rice and Ramsperger,[26] and Kas-

[25] C. N. Hinshelwood, *Proc. Roy. Soc.* (*London*) **A113**, 230 (1926).
[26] O. K. Rice and H. C. Ramsperger, *J. Am. Chem. Soc.* **49**, 1617 (1927); **50**, 612 (1928).

sel.[27] The modern picture, based, as is activated complex theory,[28] upon identifying the reaction coordinate (the critical coordinate in unimolecular vocabulary), proposes the following three-step process to describe unimolecular decomposition

$$A + A \rightleftharpoons A^* + A$$
$$A^* \rightleftharpoons A^\dagger \rightharpoonup \text{products}$$

In the first step the molecule is collisionally excited. There is then intramolecular energy transfer as A^* undergoes various conformational changes. Among the states of A^* is a subset that describes the critical conformation $A^\dagger$, defined so that there is an *even chance* that molecules in the configuration $A^\dagger$ react.[29] The distinction between the *isoenergetic* A^* and $A^\dagger$ is apparent in considering the reaction cyclopropane ⇀ propene. Unless sufficient energy resides in a C–C vibrational mode a cyclopropane molecule does not isomerize, no matter how energized it is. As energy is redistributed the molecule evolves into the critical configuration from which it may react or revert back to states of the activated molecule.

The RRKM theory proposes a model to describe energy transfer among molecules belonging to the classes A^* and $A^\dagger$. It assumes the following:

(i) In collisional processes that form A^* molecules of energy ϵ, molecules are randomly distributed among all degenerate states at that energy (*strong collision* assumption).

(ii) *Vibrational* energy redistribution within states of A^* is rapid with respect to the rate of unimolecular reaction (*random access* assumption).

Strong collision means that there is no preference for populating a subset of states of A^* in the excitation process. It is the antithesis of laser activation (discussed in Section 6.8) in which particular vibrational–rotational states are populated. *Random access* means that molecules produced in particular states of A^* are rapidly transformed into states of $A^\dagger$; there are no states that are weakly coupled.[30] Since rotational–vibrational (R–V) energy

[27] L. S. Kassel, *J. Phys. Chem.* **32**, 225, 1065 (1928).

[28] R. A. Marcus and O. K. Rice, *J. Phys. Chem.* **55**, 894 (1951); R. A. Marcus, *J. Chem. Phys.* **20**, 359 (1952).

[29] The form of RRKM theory presented here is adapted from W. L. Hase, in *Modern Theoretical Chemistry*, W. H. Miller, ed. (New York: Plenum Press, 1976), Vol. 2, Part B, Chapter 3.

[30] Non-RRKM behavior is discussed by W. L. Hase, in *Modern Theoretical Chemistry*, W. H. Miller, ed. (New York: Plenum Press, 1976), Vol. 2, Part B, Chapter 3; and D. L. Bunker, *Theory of Elementary Gas Reaction Rates* (Oxford: Pergamon Press, 1966), Chapter 3.

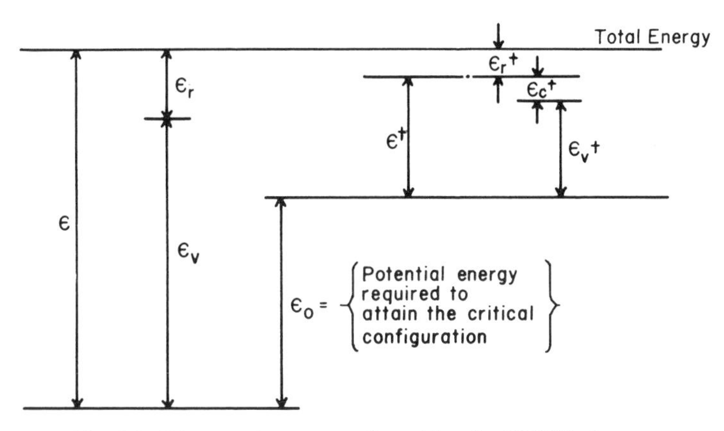

Fig. 9.9. Diagram for energy disposition in RRKM theory.

transfer is slow, only V–V transfer can significantly affect unimolecular reaction.(31)

The energetics of the species A^* and $A^\dagger$ are illustrated in Fig. 9.9. For all states considered the energy, ϵ, is greater than a threshold value, ϵ_0, required to form molecules in the states of $A^\dagger$. The energy of $A^\dagger$ available to drive reaction is only $\epsilon^\dagger = \epsilon - \epsilon_0 - \epsilon_r^\dagger$, since due to conservation of angular momentum, some of the internal energy of the molecule cannot be converted to vibrational energy. Of the total useful energy, a portion, $\epsilon_c^\dagger$, is concentrated in a critical coordinate (the analog to the reaction coordinate of activated complex theory).(32)

The molecular states that correspond to either A^* or $A^\dagger$ are highly excited ones; as a result they are nearly continuous. To compute $P(\epsilon)$ and $k(\epsilon)$ it is convenient to reformulate the idea of degeneracy in terms of a *density of states*, $N(\epsilon)$. The *total number of states* (the degeneracy) in the energy range ϵ to $\epsilon + d\epsilon$ is $N(\epsilon)\,d\epsilon$; $N(\epsilon)$ is the number of states per unit energy. With this definition, the equilibrium probability for a molecule to be in a state of energy ϵ, $P(\epsilon)\,d\epsilon$, is

$$P(\epsilon)\,d\epsilon = [N(\epsilon)e^{-\beta\epsilon}/q_i]\,d\epsilon \tag{9.33}$$

(31) R–V transfer requires a change of angular momentum, which is not possible in an *isolated* molecule unless it undergoes a pure rotational transition. Typical rotational lifetimes at very low pressure are $\sim 10^{-6}$ sec, much longer than the decomposition time.

(32) Note that ϵ_r and $\epsilon_r^\dagger$ are not necessarily equal. Although angular momentum is conserved as A^* evolves to $A^\dagger$, the rotational energy may change since the molecular structure is *not* invariant. The separation of rotational and vibrational degrees of freedom is only approximate in both configurations.

where q_i is the partition function for the *internal* degrees of freedom of the molecule.

To compute $k(\epsilon)$ consider the basic RRKM assumptions. They imply *no preference* for populating states of the same energy. Thus the states of A^* and $A^\dagger$ are populated in proportion to their degeneracies. At an energy ϵ and a specified angular momentum the ratio $[A^\dagger]/[A^*]$ is determined statistically

$$\frac{[A^\dagger]}{[A^*]} = \frac{g_r{}^\dagger N(\epsilon^\dagger)}{g_r N(\epsilon_v)} = \frac{N(\epsilon^\dagger)}{N(e_v)}$$

where $N(\epsilon^\dagger)$ is the density of vibrational states of $A^\dagger$ molecules with vibrational energy $\epsilon^\dagger$ and likewise $N(\epsilon_v)$ for A^* molecules with vibrational energy ϵ_v. The rotational degeneracies $g_r{}^\dagger$ and g_r are equal since angular momentum is conserved; as a result no rotational transition can occur as the molecule evolves from A^* to $A^\dagger$. The density of states $N(\epsilon^\dagger)$ is further decomposed since the critical mode is presumed to be distinguishable from all other vibrational modes of $A^\dagger$,

$$N(\epsilon^\dagger) = \sum_{\epsilon_v^\dagger=0}^{\epsilon^\dagger} g(\epsilon_v{}^\dagger) N_c(\epsilon^\dagger - \epsilon_v{}^\dagger = \epsilon_c{}^\dagger)$$

Here $N_c(\epsilon_c{}^\dagger)$ is the density of states of the critical mode at energy $\epsilon_c{}^\dagger$ and $g(\epsilon_v{}^\dagger)$ is the degeneracy of the other vibrational modes at energy $\epsilon_v{}^\dagger$. This decomposition, treating the critical coordinate as continuous and all other vibrational modes as discrete, is analogous to distinguishing a specific reaction coordinate in activated complex theory. Defining an energy-dependent frequency of passage from $A^\dagger$ to products, $\nu(\epsilon_c{}^\dagger)$, the expression for the reaction rate constant of A^* molecules is

$$k(\epsilon) = \sum_{\epsilon_v^\dagger=0}^{\epsilon^\dagger} \frac{\nu(\epsilon_c{}^\dagger)}{2} \left[\frac{g(\epsilon_v{}^\dagger) N_c(\epsilon_c{}^\dagger)}{N(\epsilon_v)}\right] \tag{9.34}$$

The factor of $\frac{1}{2}$ accounts for the fact that half the $A^\dagger$ molecules revert to A^* molecules. The bracketed expression is the probability that a molecule with total energy ϵ and vibrational energy ϵ_v has an energy $\epsilon_c{}^\dagger$ concentrated in the critical coordinate.

The product $\nu(\epsilon_c{}^\dagger)N_c(\epsilon_c{}^\dagger)$ can be estimated from uncertainty principle arguments. A density of states $N_c(\epsilon_c{}^\dagger)$ implies a separation between levels $\sim 1/N_c(\epsilon_c{}^\dagger)$ and a corresponding energy uncertainty $\lesssim 1/N_c(\epsilon_c{}^\dagger)$. Since

$$(\text{lifetime}) \times (\text{energy uncertainty}) \sim h$$

we find that[33]

$$\text{lifetime} \sim hN_c(\epsilon_c^{\dagger})$$

The lifetime is twice the reciprocal of the passage frequency $\nu(\epsilon_c^{\dagger})$.[34] Then

$$\nu(\epsilon_c^{\dagger})N_c(\epsilon_c^{\dagger}) \sim 2/h$$

and with (9.34)[35]

$$k(\epsilon) = \frac{1}{hN(\epsilon_v)} \sum_{\epsilon_v^{\dagger}=0}^{\epsilon^{\dagger}} g(\epsilon_v^{\dagger}) \tag{9.35}$$

In this form $k(\epsilon)$ depends upon ϵ_r, ϵ_v, and $\epsilon_r^{\dagger}$ since, from Fig. 9.9, $\epsilon^{\dagger} = \epsilon - \epsilon_0 - \epsilon_r^{\dagger} = \epsilon_v - \epsilon_0 + (\epsilon_r - \epsilon_r^{\dagger})$. It is qualitatively correct; as $\epsilon \rightharpoonup \epsilon_0$, $\epsilon^{\dagger}$ decreases and fewer vibrational states of $A^{\dagger}$ are accessible. As a result $k(\epsilon) \rightharpoonup 0$.

Calculations based upon (9.35), (9.33), and (9.31) require detailed models to determine the density of states in the critical configuration. The problems are similar to those which arise in the application of activated complex theory, but possibly more difficult. Useful expressions for k^* are found by assuming rotational and vibrational degrees of freedom are separable. The elaboration of (9.33) is then

$$P(\epsilon_v, \epsilon_r)\, d\epsilon_v\, d\epsilon_r = \frac{N(\epsilon_v)N(\epsilon_r)\exp[-\beta(\epsilon_v + \epsilon_r)]}{q_v q_r}\, d\epsilon_v\, d\epsilon_r \tag{9.36}$$

where $N(\epsilon_v)$ and q_v are the density of states and partition function for vibration, respectively, and likewise $N(\epsilon_r)$ and q_r for rotation. Substituting this and (9.35) into (9.31) yields

$$k^* = \frac{1}{hq_r q_v} \int_0^{\infty} \int_0^{\infty} \frac{\sum_{\epsilon_v^{\dagger}=0}^{\epsilon^{\dagger}} g(\epsilon_v^{\dagger})N(\epsilon_r)\exp[-\beta(\epsilon_r + \epsilon_v)]\, d\epsilon_r\, d\epsilon_v}{1 + k(\epsilon)/\gamma\beta p} \tag{9.37}$$

Since the sum over vibrational states of $A^{\dagger}$ is zero unless $\epsilon_r + \epsilon_v = \epsilon > \epsilon_0$, the range of integration has been formally extended. It is generally more

[33] Careful analysis is required to demonstrate that h, and not $\hbar$, appears here. O. K. Rice, *J. Phys. Chem.* **65**, 1588 (1961).

[34] The factor of 2 reflects the even chance of proceeding from A^{+} to products.

[35] A different argument, based upon treating motion along the critical coordinate as equivalent to translational motion in a one-dimensional box, yields the same result. See W. L. Hase, in *Modern Theoretical Chemistry*, W. H. Miller, ed. (New York: Plenum Press, 1976), Vol. 2, Part B, Chapter 3.

convenient to express k^* in terms of the properties of the critical configuration. Since the rotational degeneracy of both A^* and $A^\dagger$ are the same we find

$$N(\epsilon_r)\, d\epsilon_r\, d\epsilon_v = N(\epsilon_r^\dagger)\, d\epsilon_r^\dagger\, d\epsilon^\dagger$$

If we assume ϵ_0 is the same for all states in the critical configuration then, incorporating the energy constraints of Fig. 9.9, k^* is

$$k^* = \frac{\exp(-\beta\epsilon_0)}{hq_rq_v} \int_0^\infty \int_0^\infty \frac{\sum_{\epsilon_v^\dagger=0}^{\epsilon^\dagger} g(\epsilon_v^\dagger) N(\epsilon_r^\dagger) \exp[-\beta(\epsilon_r^\dagger + \epsilon^\dagger)]\, d\epsilon_r^\dagger\, d\epsilon^\dagger}{1 + k(\epsilon)/\gamma\beta p} \tag{9.38}$$

The high-pressure limit determined from (9.38) is

$$k_\infty = \frac{\exp(-\beta\epsilon_0)}{hq_rq_v} \left(\int_0^\infty N(\epsilon_r^\dagger) \exp(-\beta\epsilon_r^\dagger)\, d\epsilon_r^\dagger\right) \left(\int_0^\infty \sum_{\epsilon_v^\dagger=0}^{\epsilon^\dagger} g(\epsilon_v^\dagger) \exp(-\beta\epsilon^\dagger)\, d\epsilon^\dagger\right)$$

The integrals are evaluated in Appendix B and the result is

$$k_\infty = \frac{kT}{h} \exp(-\beta\epsilon_0) \frac{q_r^\dagger q_v^\dagger}{q_rq_v} \tag{9.39}$$

where $q_v^\dagger$ is the vibrational partition function for all *noncritical* vibrational modes of $A^\dagger$. The result is identical to that found by applying activated complex theory and assuming a quasi-equilibrium population of critical molecules $A^\dagger$. At high pressure, where depletion is not significant, the two approaches must be equivalent. At low pressure the correspondence cannot hold.

The low-pressure limit is most easily deduced directly from (9.31) and (9.33); it is

$$k_0 = \frac{\gamma\beta p}{q_i} \left[\int_{\epsilon_0}^\infty N(\epsilon) e^{-\beta\epsilon}\, d\epsilon\right]$$

which is precisely the result of collision theory since the bracketed expression is the fraction of molecules with $\epsilon \geq \epsilon_0$. To relate k_0 to the properties of the critical configuration is more difficult and can only be done approximately. Various prescriptions have been suggested.[36]

[36] See W. L. Hase, in *Modern Theoretical Chemistry*, W. H. Miller, ed. (New York: Plenum Press, 1976), Vol. 2, Part B, Chapter 3.

9.8. Critique of RRKM Theory—Isomerization of Methyl Isocyanide

The RRKM theory was designed to account for the pressure dependence of the unimolecular rate constant. The theory has been successfully applied to many reactions, e.g., cyclopropane $\rightharpoonup$ propene,[37] $C_2H_5 \rightharpoonup C_2H_4 + H$,[38] and $CH_3NC \rightharpoonup CH_3CN$.[39] Of these the latter is a particularly interesting example. The theory accounts for the experimental results over a range of temperature yet there is evidence that energy transfer within the CH_3NC molecule is not described by the RRKM hypotheses. The falloff curves calculated using (9.38) provide a test of RRKM theory. The falloff ratios in isomerization reactions are fairly insensitive to the choice of critical mode or to moderate variation of the structure and the vibrational frequencies in the critical configuration as long as the molecular model reproduces the high-pressure Arrhenius A-factor.[40] Thus the specifics of the dynamical model of the isomerization process are not too important, which limits the rigor of the test.

Making reasonable estimates of the molecular parameters and adjusting d to make theory and experiment coincide when $k^*/k_\infty = 0.1$ yields the results of Fig. 9.10. The corresponding values of d are given in Table 9.3. The agreement between theory and experiment is gratifying. An indication of the theory's consistency is the behavior of d. Its temperature variation is small and its value is very close to the hard-sphere diameter determined from viscosity measurements, 0.45 nm.

Both observations suggest that the observed agreement is not fortuitous since d, in RRKM theory, establishes the cross section for collisions in which there is substantial energy transfer. In any reasonable theory this cross section must be close to that which determines the transport coefficients and must not be very temperature sensitive.

The structure of the critical configuration was tested by means of isomerization studies on a series of isocyanides: CD_3NC, CH_2DNC, C_2H_5NC, and C_2D_5NC.[41] With the same critical configuration as was

[37] M. C. Lin and K. J. Laidler, *Trans. Faraday Soc.* **64**, 927 (1968).
[38] J. V. Michael and G. N. Suess, *J. Chem. Phys.* **58**, 2807 (1973).
[39] F. W. Schneider and B. S. Rabinovitch, *J. Am. Chem. Soc.* **84**, 4215 (1962).
[40] G. M. Weider and R. A. Marcus, *J. Chem. Phys.* **37**, 1835 (1962).
[41] F. W. Schneider and B. S. Rabinovitch, *J. Am. Chem. Soc.* **85**, 2365 (1963); B. S. Rabinovitch, P. W. Gilderson, and F. W. Schneider, *J. Am. Chem. Soc.* **87**, 158 (1965); K. M. Maloney and B. S. Rabinovitch, *J. Phys. Chem.* **73**, 1652 (1969); K. M. Maloney, S. P. Pavlou, and B. S. Rabinovitch, *J. Phys. Chem.* **73**, 2756 (1969).

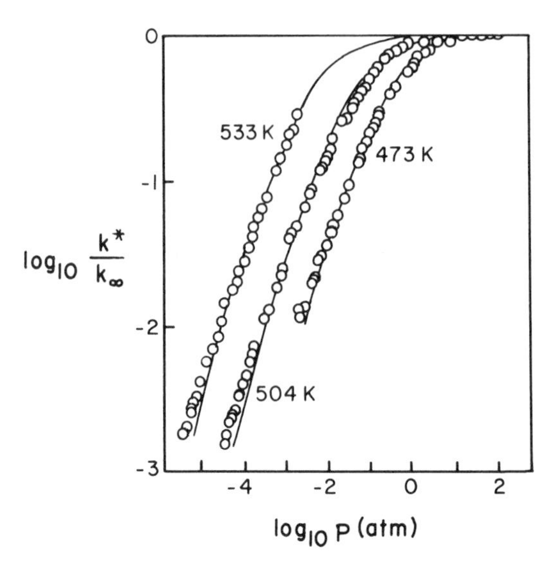

Fig. 9.10. Pressure dependence of the unimolecular rate constant for CH_3NC isomerization. For clarity the 533 K curve is displaced one log p unit to the left and the 473 K curve is displaced one log p unit to the right. Solid lines are calculated values based upon RRKM theory, adjusted to coincide with experiment when $k^*/k_\infty = 0.1$. [Adapted from F. W. Schneider and B. S. Rabinovitch, *J. Am. Chem. Soc.* **84**, 4215 (1962).]

Table 9.3. Temperature Dependence of Collision Diameter as Required by RRKM Theory[a]

T (K)	d (nm)
473	0.43
504	0.47
533	0.44

[a] The experimental value is 0.45 nm.

used to correlate the CH_3NC data, good agreement between theoretical and experimental falloff ratios was found in all cases.

There remain unresolved questions. Trajectory calculations, based upon classical treatment of intramolecular energy transfer, suggest that the *random access* hypothesis does not apply to CH_3NC.[42] It appears that

[42] D. L. Bunker, and W. L. Hase, *J. Chem. Phys.* **59**, 4621 (1973); H. L. Harris and D. L. Bunker, *Chem. Phys. Lett.* **11**, 433 (1971).

there may not be facile energy transfer between states of the critical configuration and some of the states of the activated molecule. How this is to be interpreted in light of the agreement depicted in Fig. 9.10 is, at present, unclear. Possibly a quantum treatment of energy flow would establish CH_3NC as obeying the RRKM assumptions. Alternatively, it may be that the RRKM approach is not sufficiently sensitive to its own assumptions, at least for the CH_3NC isomerization reaction. The fact that the falloff ratio, for isomerization reaction, is not too dependent on the detailed structure of the critical configuration[43] gives this idea some credibility.

If the RRKM assumption of random access is invalid, different models must be formulated in order to obtain an expression for $k(\epsilon)$, if (9.31) is to be used. Slater[44] investigated the other extreme model and assumed no vibrational energy transfer between normal modes of a molecule.[45]. This approach is too restrictive. Recent work attempts to steer a middle (and more difficult) course.[46] The dynamics of vibrational energy transfer are incorporated in the theory. Effects attributable to vibrational excitation of modes both strongly and weakly coupled to the critical coordinate can then be exhibited.

Experimental tests of the RRKM hypotheses have also been developed. By means of chemical activation molecules are formed with sufficient energy to undergo unimolecular processes directly. Typical activation processes are H-atom addition to olefins, radical recombination, and CH_2 addition or insertion. Molecules formed in this way have non-Boltzmann energy distributions; their rate of decomposition can be used to test the *random access* hypothesis introduced previously. A rather different formulation for the rate constant is needed.[47] The decomposition of C_2H_5, formed by chemical activation, can be accounted for using the same parameters for the critical configuration that were used to fit the thermal decomposition

[43] G. M. Weider and R. A. Marcus, *J. Chem. Phys.* **37**, 1835 (1962).

[44] N. B. Slater, *Theory of Unimolecular Reactions* (Ithaca, N. Y.: Cornell University Press, 1959).

[45] Such a model assumes, in effect, that molecular vibration is adequately treated as harmonic. Theories of crystals, which might be viewed as enormously large molecules, have shown conclusively that anharmonicity is needed to properly account for nonequilibrium behavior; see I. Prigogine, *Nonequilibrium Statistical Mechanics* (New York: Interscience, 1962), Chapter 2.

[46] W. M. Gelbart, S. A. Rice, and K. F. Freed, *J. Chem. Phys.* **52**, 5718 (1970); **57**, 4699 (1972).

[47] W. L. Hase, in *Modern Theoretical Chemistry*, W. H. Miller, ed. (New York: Plenum Press, 1976), Vol. 2, Part B, Chapter 3.

data.[48] More often, thermal and chemical activation experiments cannot be carried out on the same molecule. Instead, experiments are performed on a homologous series for which a single critical configuration is suitable throughout. The most extensive kinetic studies, involving chemiactivated radicals formed by H-atom addition to olefins, can be rationalized using the RRKM approach.[49]

9.9. Thermodynamic Analogy

In addition to theoretical estimates of the rate constant based on postulated structures for the activated complex, the statistical theory of rate constants can be used to correlate kinetic trends in homologous series of reactions. For this purpose we construct a "thermodynamic" interpretation of the rate constant.

The reaction rate, in terms of activated complex theory, is the number of A–B composite systems that traverse the pass per second. We computed it by treating separately two aspects of the problem. First local equilibrium was assumed and the number of A–B complexes calculated. Then progress along the reaction coordinate was considered and the rate of barrier crossing determined. From (9.10)–(9.12) this frequency ($\sec^{-1}$) can be identified as $kT\langle\varkappa\rangle/h$. Assuming $\langle\varkappa\rangle = 1$, the reaction rate is

$$-\frac{dN_A}{dt} = \frac{kT}{h} N_{AB}^{\ddagger} \tag{9.40}$$

where $N_{AB}^{\ddagger}$ is to be interpreted as the number of activated complexes with energy greater than the pass energy; these complexes are presumed to differ from excited reactants or products by having molecular configurations which closely resemble the structure of the activated complex. Naturally this identification is empirical. Transforming to concentration units (9.40) becomes

$$-\frac{dn_A}{dt} = \frac{kT}{h} n_{AB}^{\ddagger} = k_2(T) n_A n_B$$

from which the rate constant may be identified as

$$k_2(T) = \frac{kT}{h} \frac{n_{AB}^{\ddagger}}{n_A n_B} \tag{9.41}$$

[48] J. V. Michael and G. N. Suess, *J. Chem. Phys.* **58**, 2807 (1973).
[49] C. W. Larson and B. S. Rabinovitch, *J. Chem. Phys.* **52**, 5181 (1970).

The ratio $n_{AB}^{\ddagger}/n_A n_B$ has the form of an equilibrium constant for formation of activated complex. For systems in which the activity of the various species equals their concentration (dilute gases, dilute solutions of non-electrolytes) (9.41) *defines* a totally empirical function, the activation equilibrium constant $K^{\ddagger}(T)$,

$$k_2(T) \equiv \frac{kT}{h} K^{\ddagger}(T) \tag{9.42}$$

Continuing in this vein, the corresponding enthalpy, entropy, and Gibbs free energy of activation are

$$-RT \ln K^{\ddagger}(T) = \Delta G_0^{\ddagger} = \Delta H_0^{\ddagger} - T\,\Delta S_0^{\ddagger}$$

and the rate constant is[50]

$$k_2(T) = \frac{kT}{h} \exp(-\Delta G_0^{\ddagger}/RT) = \frac{kT}{h} \exp(\Delta S_0^{\ddagger}/R) \exp(-\Delta H_0^{\ddagger}/RT) \tag{9.43}$$

These quasi-thermodynamic parameters can be related to the Arrhenius activation energy by means of (8.7); using (9.42) the result is

$$E_a = RT + RT^2 \frac{d \ln K^{\ddagger}}{dT}$$

As the equilibrium constant $K^{\ddagger}$ is based on concentrations

$$\frac{d \ln K^{\ddagger}}{dT} = \frac{\Delta U_0^{\ddagger}}{RT^2}$$

and

$$\begin{aligned} E_a &= RT + \Delta U_0^{\ddagger} = \Delta H_0^{\ddagger} + RT - \Delta(PV)^{\ddagger} \\ &= \Delta H_0^{\ddagger} + RT(1 - \Delta n_g^{\ddagger}) \end{aligned} \tag{9.44}$$

where $\Delta n_g^{\ddagger}$ is the change in the number of moles of gas in the activation process. For gas-phase binary reactions $\Delta n_g^{\ddagger} = -1$.

[50] Units must be treated carefully in (9.43). The left-hand side has dimensions of a second-order rate constant while the right-hand side appears to have dimension t^{-1}. A choice of standard state is implicit in the definition of the free energy of activation, $\Delta G_0^{\ddagger}$. For details see A. C. Norris, *J. Chem. Ed.* **48**, 797 (1971).

9.10. Gas-Phase Reactions

The thermodynamic analogy is not to be used as a method for establishing absolute values for the various activation parameters; these are of limited significance since they derive from an equilibrium thermodynamic interpretation of intrinsically nonequilibrium properties. Furthermore, to accept such numbers uncritically ascribes a measure of definiteness to the activated complex which is unwarranted. On the other hand, trends and similarities may be useful in helping to characterize reaction mechanism. In Table 9.4 the values of $\Delta S_0^\ddagger$ and $\Delta H_0^\ddagger$ calculated from (9.43) and (9.44) are given for a number of gas-phase reactions. For the bimolecular reactions the value of $\Delta S_0^\ddagger$ depends upon the choice of standard state; for rate constants in units of $cm^3\ mol^{-1}\ sec^{-1}$ the natural standard state is a concentration of 1 mol cm^{-3}.

Table 9.4. Activation Parameters for Some Gas-Phase Reactions

Reaction	$\Delta H_0^\ddagger$ (kJ)	$\Delta S_0^\ddagger$ (JK^{-1})[a]	Reference
$C_2H_5Cl \rightarrow C_2H_4 + HCl$	232	+11.7	b
$C_2H_4Cl_2 \rightarrow C_2H_3Cl + HCl$	220	+16.5	b
Cyclopropane → Propene	270	+50.0	b
Cyclobutene → *trans*-Butadiene	133	+15.6	b
$CH_3 + C_2H_4 \rightarrow C_3H_7$	24	−20.4	c
$H + C_2H_4 \rightarrow C_2H_5$	3.3	+26.5	c
$CH_3 + CH_3 \rightarrow C_2H_6$	−8.3	+19.4	d
$CH_3 + CO \rightarrow CH_3CO$	7.9	−80.1	d
$CH_3 + C_2H_6 \rightarrow CH_4 + C_2H_5$	35.1	−49.7	e
$CH_3 + C_2H_4 \rightarrow CH_4 + C_2H_3$	33.5	−50.1	e
Isobutene + HCl → t-C_4H_9Cl	92.5	−70.9	e

[a] Standard state for $\Delta S_0^{\neq}$ for binary reactions is a concentration of 1 mol cm^{-3}.
[b] S. W. Benson and H. E. O'Neal, *Kinetic Data on Gas Phase Unimolecular Reactions* (Washington: National Bureau of Standards, 1970).
[c] J. A. Kerr and M. J. Parsonage, *Evaluated Kinetic Data on Gas Phase Addition Reactions* (London: Butterworths, 1972).
[d] V. N. Kondratiev, *Rate Constants of Gas Phase Reactions*, translation by L. J. Holtschlag (Washington: National Bureau of Standards, 1972).
[e] S. W. Benson, *The Foundations of Chemical Kinetics* (New York: McGraw-Hill, 1960).

The activation entropies for the four unimolecular reactions are positive and can be interpreted if we assume that the molecule, in its activated state, is less ordered than in the ground state. Disordering occurs because a bond must be weakened or broken for reaction to be possible. The cyclopropane isomerization is a particularly good example. As a C–C bond breaks the molecule passes through an intermediate, possibly a diradical,

The intermediate has greater rotational and vibrational freedom than either reactant or product. Thus it is not surprising that $\Delta S_0^\ddagger > \Delta S_{rxn}$; the values are 50 and 28 J K^{-1}, respectively.

The numerical values of $\Delta S_0^\ddagger$ for binary reactions have no particular significance since they depend critically on the choice of reference state. It is no longer meaningful to suggest that $\Delta S_0^\ddagger > 0$ indicates an activated complex that is less ordered than the reactants. On the other hand, trends are important. The more negative the value of $\Delta S_0^\ddagger$, the more severe are the orientational requirements for reaction to occur, i.e., the pass on the potential-energy surface is narrow. The more positive the value of $\Delta S_0^\ddagger$, the fewer are the orientational constraints corresponding to a much broader pass on the potential-energy surface. One possible explanation for positive $\Delta S_0^\ddagger$ is that, in the activated complex, the reactants are very loosely bound leading to a large moment of inertia; this is assumed to be the case for CH_3 dimerization.

The trends in the other bimolecular examples are reasonably understandable. There are fewer orientational constraints in H-atom addition to ethylene than in methyl radical addition; the values of $\Delta S_0^\ddagger$ are in accord with this interpretation. The hydrogen-transfer reactions of methyl with ethane and ethylene have the same entropy of activation, indicative of similar orientational requirements. The addition of CH_3 to CO has a large negative $\Delta S_0^\ddagger$, reflecting the severe geometric constraints necessary for this reaction to be possible.

It is the Arrhenius *A*-factor which is related to $\Delta S_0^\ddagger$ and thus to mechanistic models for the structure of the activated complex. The value of $\Delta H_0^\ddagger$ is determined by the barrier to reaction and it cannot be simply related to a model for the reaction intermediate.

Alternatively, the thermodynamic analogy can be used to provide estimates of rate constants. In equilibrium thermodynamics one uses tabulations of bond enthalpies and entropies to estimate ΔH_0 and ΔS_0 for compounds for which thermodynamic measurements are unavailable. Similar, but less reliable, methods have been devised for estimating $\Delta H_0^{\ddagger}$ and $\Delta S_0^{\ddagger}$. While they require numerous assumptions about the structure of the activated complex, these are really not much different in spirit from guessing the mechanical properties of the activated complex.[51]

9.11. Solution Reactions

The activation entropy for solution reactions can be interpreted in similar fashion. The solvent exchange reactions

$$Cr(H_2O)_5X^{2+} + H_2O^* \rightharpoonup Cr(H_2O)_4(H_2O^*)X^{2+} + H_2O$$

have been studied with $X = Cl^-$, Br^-, and NCS^-. They are interesting because, with each ligand, two types of exchange kinetics are found.[52] Four of the water molecules exchange slowly; $\Delta S_0^{\ddagger}$ is between 0 and 12 J K^{-1}. The fifth exchanges much more rapidly; $\Delta S_0^{\ddagger}$ is between 57 and 90 J K^{-1}. These observations may be related to a possible exchange mechanism.

The complex has the octahedral structure shown in Fig. 9.11a. Exchange of water may be imagined as occurring through a variety of intermediate structures; the most extreme examples are shown in Figs. 9.11b and c. In the first case an extra water molecule is bound before another molecule leaves; the activated complex is seven coordinate. In the second case a water molecule is lost to solvent in the activation process; the complex is five coordinate. The difference between the two intermediates is striking. Two water molecules bound to the seven-coordinate complex are free to mingle with the solvent if the complex is five coordinate. The seven-coordinate complex is thus the more ordered and correspondingly should have a lower value of $\Delta S_0^{\ddagger}$. As four of the water molecules exchange with $\Delta S_0^{\ddagger} < 12$ J K^{-1} we can propose that, for *cis* exchange, a seven-coordinate intermediate forms. For *trans* exchange, on the other hand, a water molecule leaves before another one is attached; the intermediate is five coordinate.

[51] For a discussion of this approach see S. W. Benson, *Thermochemical Kinetics* (New York: Wiley-Interscience, 1976).

[52] T. W. Swaddle, *Coord. Chem. Rev.* **14**, 217 (1974); D. E. Bracken and H. W. Baldwin, *Inorg. Chem.* **13**, 1325 (1974).

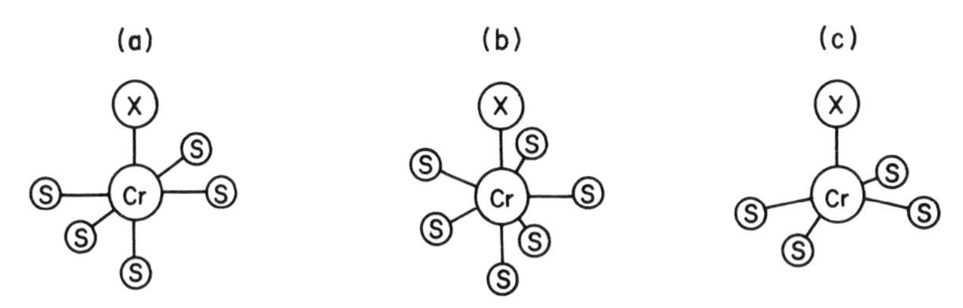

Fig. 9.11. Structures of the (a) octahedral CrS_5X complex ion; (b) seven-coordinate intermediate; (c) five-coordinate intermediate. The solvent S is water.

9.12. Primary Salt Effect

In the process of identifying the quasi-thermodynamic activation parameters in Section 9.9 we assumed that activity could be equated with concentration. While reasonable for the examples discussed in Sections 9.10 and 9.11 it is obviously inappropriate if we wish to consider ionic reactions. From (9.41) the rate constant is related to the ratio $n_{AB}^{\ddagger}/n_A n_B$ which can no longer be set equal to an equilibrium constant. It is the activities, not the concentrations, which are related to the equilibrium constant

$$\frac{a_{AB}^{\ddagger}}{a_A a_B} = K^{\ddagger}(T) \tag{9.45}$$

Since $a = \gamma n$, where γ is the activity coefficient, (9.45) can be rewritten

$$\frac{n_{AB}^{\ddagger}}{n_A n_B} = K^{\ddagger}(T) \frac{\gamma_A \gamma_B}{\gamma_{AB}^{\ddagger}}$$

which can be substituted in (9.41) with the result

$$k_2(T) = \frac{kT}{h} K^{\ddagger}(T) \frac{\gamma_A \gamma_B}{\gamma_{AB}^{\ddagger}} = k_0(T) \frac{\gamma_A \gamma_B}{\gamma_{AB}^{\ddagger}} \tag{9.46}$$

The quantity $k_0(T)$ is the rate constant in very dilute solution where all the activity coefficients are unity. The formula (9.46) is obviously applicable to any system but the activity corrections are most dramatic for ionic reactions.

In ionic solution the ion activity coefficients are strongly charge dependent; from Debye–Hückel theory we have

$$\log \gamma_A = -\tfrac{1}{2} B z_A{}^2 I^{1/2} \tag{9.47}$$

where z_A is the ionic charge and I is the ionic strength of the solution $I = \sum z_i^2 C_i$. In water at 25°C the constant B is 1.0. Since the activated complex in the reaction

$$A^{z_A} + B^{z_B} \rightleftharpoons (AB)^{z_A+z_B} \rightharpoonup \text{products}$$

has a total charge $z_A + z_B$ we may now use (9.46) to determine the effect of ionic strength on the rate constant. From (9.46) and (9.47) we have

$$\begin{aligned} \log k_2(T) &= \log k_0(T) - \tfrac{1}{2}BI^{1/2}[z_A^2 + z_B^2 - (z_A + z_B)^2] \\ &= \log k_0(T) + z_A z_B B I^{1/2} \end{aligned} \tag{9.48}$$

Thus increasing the ionic strength speeds up reactions between similarly charged ions and slows down those between ions of opposite charge, a point already mentioned in our discussion of diffusion-controlled reactions

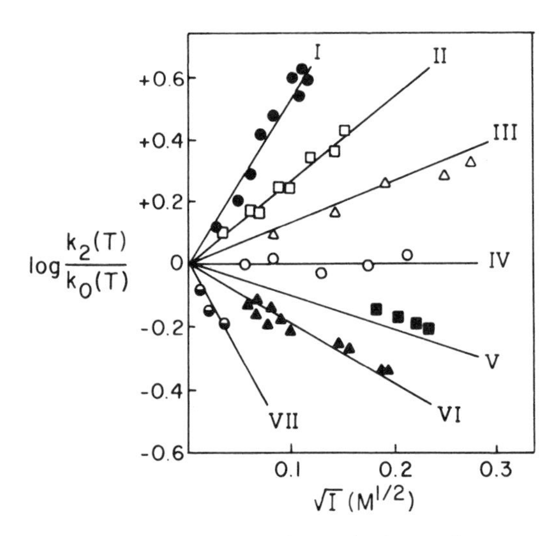

Fig. 9.12. Effect of ionic strength on rates of some ionic reactions:

(I) $Co(NH_3)_5Br^{2+} + Hg^{2+} + H_2O \rightharpoonup [Co(NH_3)_5H_2O]^{3+} + HgBr^+$;
(II) $S_2O_8^{2-} + I^- \rightharpoonup I_3^- + 2SO_4^{2-}$;
(III) $[O_2N\text{-}N\text{-}CO_2Et]^- + OH^- \rightharpoonup N_2O + CO_3^{2-} + EtOH$;
(IV) Sucrose $+ OH^- \rightharpoonup$ inversion;
(V) $H_2O_2 + H^+ + Br^- \rightharpoonup H_2O + \frac{1}{2} Br_2$;
(VI) $Co(NH_3)_5Br^{2+} + OH^- \rightharpoonup Co(NH_3)_5OH^{2+} + Br^-$
(VII) $Fe^{2+} + Co(C_2O_4)_3^{3-} \rightharpoonup Fe^{3+} + Co(C_2O_4)_3^{4-}$.

Some reactions have not been balanced. [Adapted from S. W. Benson, *The Foundations of Chemical Kinetics* (New York: McGraw–Hill, 1960), p. 525, where original references are given.]

in Section 5.7. Some data illustrating the effect of ionic strength are plotted in Fig. 9.12. The expression (9.48) accounts for the observations at ionic strengths where Debye–Hückel theory is reliable.[53]

The effect is readily understood qualitatively. Increasing ionic strength *always* reduces the magnitude of the electrostatic effects since the ions become more shielded by their ion atmospheres. Thus repulsion between ions of like sign is reduced allowing them to approach one another more easily; the result is an increase in $k_2(T)$. Attraction between ions of opposite sign is also reduced; they approach one another less readily and $k_2(T)$ falls.

9.13. *Diffusion-Limited Kinetics—Debye Theory*

Activated complex theory and the RRKM approach to unimolecular decomposition emphasize the specific conformational rearrangements that are required for reaction to take place. In gas-phase reactions such considerations must be dominant. However, in solutions, reactants that have diffused together cannot simply collide and separate. They remain constrained by the surrounding solvent molecules which must be displaced if the particles are to move apart. The reactants remain trapped, typically for 10^{-11} sec, and while so constrained undergo $\sim$100 collisions with one another. At the same concentrations species in both gas and solution phase have the same collision frequency. The difference is that in a gas all collisions are independent while in a solution, two particles, having encountered one another, make multiple collisions before separating. If the chemical rearrangement is facile it is the *encounter frequency*, governed by diffusion, which sets an upper bound to the bimolecular reaction rate.

To compute the maximum rate of a bimolecular chemical reaction in solution assume that, after the molecules have diffused together, reaction takes place with each encounter.[54] In order to simplify the mathematics we center our attention, and the coordinate system, on a representative molecule of type B. Molecules of type A diffuse through the solution and, occasionally, encounter the B molecule. Details of molecular structure are suppressed. Both A and B are spherical; the distance of closest approach

[53] Ionic strength can affect reactions of neutral species. An example is the reaction of CO(aq) with hemoglobin [L. H. Parkhurst and Q. H. Gibson, *J. Biol. Chem.* **242**, 5762 (1967)]. Here changing ionic strength greatly alters the activity of the dissolved CO thus changing the rate constant.

[54] The treatment in this section is modeled after that given by I. Amdur and G. G. Hammes, *Chemical Kinetics* (New York: McGraw–Hill, 1966), pp. 59–64.

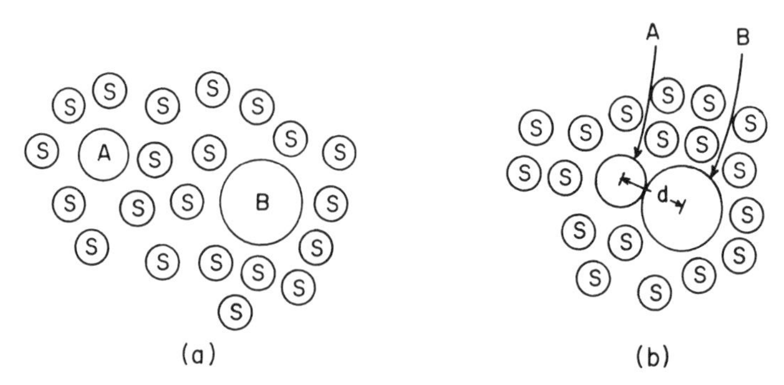

Fig. 9.13. Model for AB relative diffusion through solvent. (a) A and B separated by solvent sheath. (b) A and B have diffused into contact when reaction occurs; the contact distance is d.

is d. Since reaction takes place at each encounter, the concentration of A molecules at the surface of a B molecule is zero; at large distances from the B molecule the molecular concentration of A is that of the bulk solution, n_A^0. In terms of r, the AB distance, the boundary conditions are

$$r = \begin{cases} d, & n_A = 0 \\ \infty, & n_A = n_A^0 \end{cases} \tag{9.49}$$

Since reaction occurs whenever $r = d$, the rate of reaction of A with a representative B molecule is the total flux of A at the surface of a sphere of radius d, as indicated in Fig. 9.13.

To continue, generalize the ionic flux equation (3.15) to consider effects due to any potential

$$-\mathbf{J}_A = D(\nabla n_A + \beta n_A \nabla U_A)$$

Here U_A is the potential energy of interaction between A and B molecules. The net diffusion coefficient, D, is $D_A + D_B$, which accounts for centering the coordinate system on a B molecule. As the molecules and the intermolecular potential are chosen spherically symmetric, $\mathbf{J}_A$ is orientation independent and directed along $\mathbf{r}$, the A–B radius vector. Thus we have

$$J_A = -D\left(\frac{dn_A}{dr} + \beta n_A \frac{dU_A}{dr}\right) \tag{9.50}$$

The total flux across a surface of radius r is

$$I_A = -4\pi r^2 J_A \tag{9.51}$$

Since the system is in a steady state, I_A is constant; it is the total flux of A molecules directed *radially* toward the representative B molecule. Combining (9.50) and (9.51) we find that

$$I_A = 4\pi r^2 D \exp(-\beta U_A) \frac{d}{dr} [n_A \exp(\beta U_A)] = \text{const} \tag{9.52}$$

If D is assumed concentration independent, (9.52) can be manipulated into a form that provides a useful expression for I_A. Multiply both sides of the equation by $\exp(-\beta U_A)/r^2$ and integrate over r; the result is

$$\begin{aligned} I_A \int_d^\infty \frac{\exp(-\beta U_A)}{r^2}\, dr &= 4\pi D \int_d^\infty \left(\frac{d}{dr} [n_A \exp(\beta U_A)]\right) dr \\ &= 4\pi D \int_0^{n_A{}^0} d[n_A \exp(\beta U_A)] = 4\pi D n_A{}^0 \end{aligned} \tag{9.53}$$

where we have used the boundary conditions (9.49) and the fact that the A–B interaction $U_A(r)$ must be zero at large A–B separation. Since I_A represents the rate at which A molecules reach the surface of the representative B molecule, it is the rate of reaction of A per B molecule. Multiplying by $n_B{}^0$ to account for the population of B we find the rate law,

$$-\frac{dn_A}{dt} = 4\pi d(D_A + D_B) f n_A{}^0 n_B{}^0$$

where $D_A + D_B$ has been substituted for D, and

$$f = \left[d \int_d^\infty \frac{\exp(\beta U_A)}{r^2}\, dr\right]^{-1} \tag{9.54}$$

from which the rate constant can be found. Expressing it in the conventional units M^{-1} sec^{-1} yields the rate constant for formation of an encounter complex,

$$k_{\text{diff}} = 4\pi d(D_A + D_B) f \times 10^3 N_0 \tag{9.55}$$

with d and the D's measured in SI units.

This treatment, an elaboration by Debye of an approach introduced by Smoluchowski,[55] is based upon a succession of rather gross assumptions. The equations which describe bulk diffusion were assumed to remain accurate even at distances where A–B interaction is significant. We then im-

[55] M. V. Smoluchowski, *Z. Physik. Chem.* **92**, 129 (1917); P. Debye, *Trans. Electrochem. Soc.* **82**, 265 (1942).

posed a steady state and assumed the D's to be concentration independent. Using a more realistic approach only slightly alters the results, except for reactions with very short half-times.

The same approach may be used to calculate the maximum rate constant for dissociation of the encounter complex. The only difference is the choice of boundary conditions. To replace (9.49) we have

$$r = \begin{cases} \infty, & n_{\mathrm{AB}} = 0 \\ d, & n_{\mathrm{AB}} = 1/\Delta V, \ \Delta V \equiv 4\pi d^3/3 \end{cases} \tag{9.56}$$

since at infinite A–B separation the complex does not exist and at contact there is one encounter complex in a volume $\sim 4\pi d^3/3$. The arguments which led to (9.53) can be modified to yield

$$I_{\mathrm{AB}}/df = 4\pi(D_{\mathrm{A}} + D_{\mathrm{B}})\{-\exp[\beta U_{\mathrm{A}}(d)]/\Delta V\}$$

from which the maximum dissociation rate constant is

$$k_{\mathrm{diss}} = \frac{4\pi(D_{\mathrm{A}} + D_{\mathrm{B}})\, df \exp[\beta U_{\mathrm{A}}(d)]}{\Delta V} \tag{9.57}$$

since I_{AB} is the flux per encounter complex, $d \ln[\mathrm{AB}]/dt$.[56] Combining (9.55) and (9.57) yields an equilibrium constant for the formation of the complex

$$K = \frac{k_{\mathrm{diff}}}{k_{\mathrm{diss}}} = \frac{4\pi d^3}{3} 10^3 N_0 \exp[-\beta U_{\mathrm{A}}(d)] \tag{9.58}$$

in units of M^{-1}. This expression, which is valid for any nonchemically bound complex, was first derived to treat ion-pair formation in liquids.[57]

The limitation to spherical molecules is a more severe restriction; it may be relaxed by introducing a steric factor which reduces the factor 4π appearing in (9.52), (9.53), and (9.55). Rough quantitative evaluation of steric factors have been made.[58] It might seem that such factors are inconsistent with the idea of an encounter followed by multiple collisions. However, nonspherical molecules must rotate while they are trapped in the encounter complex if they are to assume a reactive configuration (assuming that in the initial moments of the encounter the configuration was non-

[56] The first derivation of (9.57) was given by M. Eigen, *Z. Phys. Chem.* **NF1**, 176 (1954).
[57] R. M. Fuoss, *J. Am. Chem. Soc.* **80**, 5059 (1958).
[58] R. M. Noyes, *Prog. React. Kinetics* **1**, 129 (1961).

reactive). Since molecules remain together for $\sim 10^{-11}$ sec there is time for ~ 10 rotations, which would seem to allow a sampling of many orientations. However, if the reactants are fairly strongly associated with the surrounding solvent, any such rotation would be significantly hindered, and the reaction rate correspondingly less than the limiting value for spherical molecules.

The most common application of (9.55) and (9.57) is to the problem of ionic reactions in solution. If the solution is very dilute the AB potential energy is

$$U_A(r) = z_B z_A e^2/\epsilon r \tag{9.59}$$

where ϵ is the dielectric constant of the bulk solvent. For this simple potential (9.54) can be integrated with the result

$$f = \beta U_A(d)/\{\exp[\beta U_A(d)] - 1\} \tag{9.60}$$

The value of k_{diff} can then be estimated using (9.55). From the Nernst–Einstein equation (3.13) and the data of Table 3.1 we find that ion diffusion coefficients at infinite dilution are ~ 0.6–2×10^{-9} m^2 sec^{-1} at 25°C (except for H^+ and OH^- which are 9.3×10^{-9} and 5.3×10^{-9}, respectively). Typical encounter distances are ~ 0.5 nm and in water at 25°C $\epsilon \sim 80$. Values of k_{diff} computed on the basis of these assumptions are plotted in Fig. 9.14

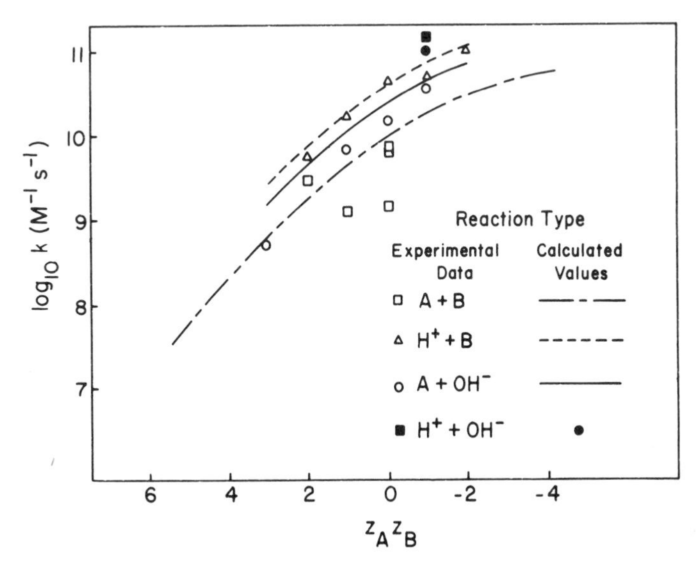

Fig. 9.14. Comparison of experimental and calculated values of k_{diff} for diffusion-limited reactions. Values calculated using (9.55).

as a function of $z_A z_B$; for comparison, the experimental rate constants quoted in Table 5.3 for ions with the same charge product are also indicated. The agreement between theory and experiment is excellent, which suggests the model represents accurately the major features which account for the magnitude of the rate constants. The same approach allows estimation of k_{diss} from (9.57). For uncharged species in water at 25°C the result is $k_{\text{diss}} \sim 3.1 \times 10^{10}$ sec^{-1}. At concentrations $\gtrsim 10^{-4}$ M where the ion-atmosphere affects conductance and equilibrium properties, it also affects k_{diff} and k_{diss}. The potential energy (9.59) must be modified to account for the shielding in which case (9.54) can only be integrated numerically. The net result is to somewhat reduce the effects attributable to ionic charge.[59]

The calculation of k_{diff} and k_{diss} was based on two assumptions:

(i) chemical reaction of the encounter complex is instantaneous;
(ii) the system is in a steady state.

Neither assumption is universally valid. As a result, a measured rate constant, even if of the order of magnitude expected for k_{diff}, cannot be directly identified as k_{diff}.

If the rate constant for reaction of the encounter complex is k_2, a steady-state treatment, presuming the mechanism

$$\mathrm{A} + \mathrm{B} \underset{k_{\text{diss}}}{\overset{k_{\text{diff}}}{\rightleftharpoons}} (\mathrm{A} \cdots \mathrm{B}) \xrightarrow{k_2} \text{products}$$

yields the rate law

$$-\frac{d[\mathrm{A}]}{dt} = \frac{k_2 k_{\text{diff}}}{k_2 + k_{\text{diss}}} [\mathrm{A}][\mathrm{B}] = k_{\text{app}}[\mathrm{A}][\mathrm{B}]$$

so that unless $k_2 \gg k_{\text{diss}}$ the measured value of the apparent rate constant does not depend on k_{diff} alone. While k_2 is hard to estimate, the related quantity $k' = k_2 k_{\text{diff}}/k_{\text{diss}}$ can be given a physical interpretation. It is the *intrinsic* rate constant for the A + B reaction, i.e., the rate constant which would be observed if the reactants were placed in contact.[60] As such k'

[59] It is worth noting that the method of calculation must be significantly altered for more concentrated solutions to insure that the values of k_{diff} found by letting A diffuse toward a stationary B or vice versa be the same. See L. Bass and W. J. Greenhalgh, *Trans. Faraday Soc.* **62**, 715 (1966); A. M. Watts, *Trans. Faraday Soc.* **62**, 2219, 3189 (1966).

[60] R. M. Noyes, *Prog. React. Kinetics* **1**, 129 (1961); T. R. Waite, *J. Chem. Phys.* **28**, 103 (1958); **32**, 21 (1960).

is not necessarily limited by the encounter frequency (in solution) or the collision frequency (in a gas). In terms of k', the apparent rate constant is

$$k_{app} = k'k_{diff}/(k' + k_{diff}) \tag{9.61}$$

In general, given the uncertainty in the estimate of k_{diff}, it is hard to disentangle chemical and diffusive contributions to k_{app}.[61]

A steady state is not established instantaneously. After A and B form an encounter complex and react, the mean concentrations of *unreacted* A and B in the vicinity are below their steady state values. Diffusion of A and B is required to reestablish the steady state. As a result there is an induction period before steady-state behavior is observed. At shorter times the binary rate constant, k_{app}, is time dependent, and greater than its steady-state value,[62]

$$k_{app}(t) = \frac{k'k_{diff}}{k' + k_{diff}}\left\{1 + \frac{d}{1 + k_{diff}/k'}\,\frac{1}{[\pi(D_A + D_B)t]^{1/2}}\right\} \tag{9.62}$$

Choosing the same values for d and D that were used in estimating k_{diff} we find that, as long as the reaction is diffusion limited ($k' > k_{diff}$), $\sim 10^{-7}$ sec is required for k_{app} to differ by less than 1% from its steady-state value. After $\sim 10^{-9}$ sec the difference is $\sim 10\%$.

The effect is significant for reactions with short half-times. In an ordinary diffusion-limited reaction under conditions where the species B is in great excess, the half-time is $1/(k_{diff}[B])$. If $k_{diff} \sim 10^{10}\ M^{-1}\ sec^{-1}$, a steady-state treatment of the kinetics will suffice as long as $[B] \lesssim 0.1\ M$.[63] If [B] is larger, the effect of transient behavior cannot be ignored. Another instance where transients are important is in the kinetics of fluorescence quenching. Here, the intrinsic lifetimes of the photoexcited species are typically $\sim 10^{-9}$ sec. The photostationary state is not established and effects attributable to the time dependence of k_{app} can be observed.

(61) For the practicing kineticist this is usually not an issue since the major significance of the diffusion limit is that derived binary rate constants cannot exceed k_{diff}. The individual contributions to k_{app} have been distinguished in an analysis of the kinetics of iodine-atom recombination in CCl_4 [R. M. Noyes, *J. Am. Chem. Soc.* **86**, 4529 (1964)].

(62) R. M. Noyes, *Prog. React. Kinetics* **1**, 129 (1961); T. R. Waite, *J. Chem. Phys.* **28**, 103 (1958); **32**, 21 (1960).

(63) The measured rate constants will be $\sim 10\%$ too large, but efforts to obtain greater accuracy are usually unwarranted.

Appendix A. Statistical Thermodynamics

This section is meant only as a compilation of some of the important results of statistical thermodynamics. No proof will be given.[64] The fundamental quantity which relates the mechanical properties of molecules to the thermodynamic properties of a dilute gas is the molecular partition function $q(T)$

$$q(T) = \sum_n \exp\left(-\frac{\epsilon_n}{kT}\right) \tag{9.A1}$$

where ϵ_n are the energy levels of the molecule of interest. The thermodynamic consequence is that $q(T)$ determines the molar chemical potential via the relation

$$\mu = -RT \ln q(T)/N \tag{9.A2}$$

Thus, if a chemical equilibrium is established in the reaction

$$A + B \rightleftharpoons AB$$

we know that

$$\mu_A + \mu_B = \mu_{AB}$$

or using (9.A2) and taking antilogarithms

$$\frac{N_{AB}}{N_A N_B} = \frac{q_{AB}}{q_A q_B} \tag{9.A3}$$

so that knowledge of the partition functions determines the equilibrium constant.

Another important result is that statistical thermodynamics allows calculation of the fraction of molecules, f_n, in a particular energy level. The expression is

$$f_n = \frac{\exp(-\epsilon_n/kT)}{q} \tag{9.A4}$$

If this is summed over all energy levels it reduces to

$$\sum_n f_n = 1$$

which is of course required.

[64] A fuller discussion can be found in texts on this subject, e.g., T. L. Hill, *Introduction to Statistical Thermodynamics* (Reading, Mass.: Addison–Wesley, 1960), Chapters 4, 8, and 9.

Application of (9.A1) and (9.A4) requires knowledge of the molecular energy levels. A good approximation at low energy is to assume that the various degrees of freedom are independent and that the energy of a molecule can be decomposed into electronic, vibrational, rotational, and translational contributions,

$$\epsilon = \epsilon_e + \epsilon_v + \epsilon_r + \epsilon_t$$

The consequence is that the partition function (9.A1) may be factored into a product of contributions from each degree of freedom

$$q = q_e q_v q_r q_t \tag{9.A5}$$

where each q is now calculated by an expression like (9.A1); the energy levels now refer to those corresponding to a particular degree of freedom. When expressions for the energy levels, deduced from quantum mechanics, are introduced into (9.A1) the following results are obtained:

$$q_t = \left(\frac{2\pi mkT}{h^2}\right)^{3/2} V$$

$$\left.\begin{aligned} q_r &= \frac{8\pi^2 IkT}{\sigma h^2} && \text{(linear molecules)} \\ &= \frac{8\pi^2 (ABC)^{1/2}(2\pi kT)^{3/2}}{\sigma h^3} && \text{(nonlinear molecules)} \end{aligned}\right\} \text{rigid rotors}$$

$$q_v = \prod_i \frac{\exp(-h\nu_i/2kT)}{[1 - \exp(-h\nu_i/kT)]} = \prod_i \left[2 \sinh\left(\frac{h\nu_i}{2kT}\right)\right]^{-1} \quad \text{harmonic oscillators}$$

$$q_e = g \tag{9.A6}$$

In these expressions V is the volume of the system; m is the molecular mass; I is the moment of inertia (linear molecules); ABC is the product of the principal moments of inertia (nonlinear molecules); ν_i are the vibrational frequencies; g is the electronic degeneracy; and h is Planck's constant. The quantity σ is the symmetry number which takes into account the number of ways a molecule can be rotated to exchange identical nuclei; σ equals 2 and 3 for I_2 and NH_3, respectively.

The expression for q_r is not accurate when used to determine the properties of isotopes of H_2 at temperatures below 1000 K. However, the errors are less than 10% if $T > 250$ K. The approximate form for q_r is thus adequate at moderate temperatures when used to estimate $q^\ddagger$ since the uncertainties of activated complex theory surely exceed 10%.

In some instances it is convenient to use classical rather than quantum mechanics. The analog to (9.A1) is

$$q = \int \frac{dp\,dx}{h} e^{-\epsilon/kT} \tag{9.A7}$$

for one classical degree of freedom. An integration over position (x) and momentum (p) replaces the sum over energy levels. The energy ϵ is given by the classical expression

$$\epsilon = \frac{p^2}{2m} + V(x)$$

the sum of kinetic and potential energies. The factor h is introduced so that classical and quantum mechanics correspond in the limit where they are known to lead to the same predictions; if it were not included the classical partition function (9.A7) would also not be dimensionless.

Appendix B. Evaluation of k_∞ in RRKM Theory

To determine k_∞ requires evaluating the integrals

$$\int_0^\infty N(\epsilon_r^\dagger) \exp(-\beta\epsilon_r^\dagger)\, d\epsilon_r^\dagger \tag{9.B1}$$

$$\int_0^\infty \sum_{\epsilon_v^\dagger=0}^{\epsilon^\dagger} g(\epsilon_v^\dagger) \exp(-\beta\epsilon^\dagger)\, d\epsilon^\dagger \tag{9.B2}$$

Since $N(\epsilon_r^\dagger)\, d\epsilon_r^\dagger$ is just the rotational degeneracy (9.B1) is equivalent to

$$\sum_{\epsilon_r^\dagger} g(\epsilon_r^\dagger) \exp(-\beta\epsilon_r^\dagger) = q_r^\dagger \tag{9.B3}$$

the rotational partition function of the critical configuration.

To evaluate (9.B2) note that $\sum_{\epsilon_v^\dagger=0}^{\epsilon^\dagger} g(\epsilon_v^\dagger)$ is a discontinuous function of $\epsilon^\dagger$,

$$\sum_{\epsilon_v^\dagger=0}^{\epsilon^\dagger} g(\epsilon_v^\dagger) = \begin{cases} g_0^\dagger, & 0 \le \epsilon_v^\dagger < \epsilon_1^\dagger \\ g_0^\dagger + g_1^\dagger, & \epsilon_1^\dagger \le \epsilon_v^\dagger < \epsilon_2^\dagger, \quad \text{etc.} \end{cases}$$

where $g_i^\dagger$ and $\epsilon_i^\dagger$ are the degeneracy and energy of the discrete vibrational states of $A^\dagger$. Hence (9.B2) is

$$\int_0^{\epsilon_1^\dagger} \exp(-\beta\epsilon^\dagger) g_0^\dagger \, d\epsilon^\dagger + \int_{\epsilon_1^\dagger}^{\epsilon_2^\dagger} \exp(-\beta\epsilon^\dagger)(g_0^\dagger + g_1^\dagger) \, d\epsilon^\dagger + \cdots$$

$$= \sum_{\epsilon_i^\dagger} g_i^\dagger \int_{\epsilon_i^\dagger}^{\infty} \exp(-\beta\epsilon^\dagger) \, d\epsilon^\dagger$$

$$= \frac{1}{\beta} \sum_{\epsilon_i^\dagger} g_i^\dagger \exp(-\beta\epsilon_i^\dagger) = kTq_v^\dagger \tag{9.B4}$$

i.e., kT times the partition function for noncritical vibrational modes of the critical configuration.

Problems

9.1. Show that the moment of inertia of a linear triatomic molecule, XYZ, is

$$I = [m_x(m_y + m_z)R_1^2 + 2m_x m_z R_1 R_2 + m_z(m_x + m_y)R_2^2]/M$$

where the XY distance is R_1, the YZ distance is R_2, and M is the mass of the molecule.

9.2. In the temperature range ~400 K, the *A*-factor for the reaction

$$Cl + H_2 \rightarrow HCl + H$$

is 1.2×10^{13} mole^{-1} sec^{-1}. Calculate the *A*-factor for this reaction assuming a *linear* transition state with bonds 30% longer than in H_2 or HCl. Assume that the vibrational frequencies in the activated complex are sufficiently large so that none are excited. The bond lengths in H_2 and HCl are 0.0742 nm and 0.1275 nm, respectively. The electronic degeneracies are $g_{Cl} = 6$, $g_{H_2} = 1$, $g^\ddagger = 2$. (Estimation of the vibrational frequencies of HHCl is *not* required.)

9.3. For the reaction of Problem 9.2 compute the *A*-factor assuming the ClHH angle to be 90°. For an XYZ molecule with such geometry the moment of inertia product is

$$ABC = (I_{11} + I_{22})(I_{11}I_{22} - I_{21}^2)$$

where

$$I_{11} = m_x(m_y + m_z)R_1^2/M$$

$$I_{22} = m_z(m_x + m_y)R_2^2/M$$

$$I_{21} = m_x m_z R_1 R_2/M$$

Does the Eyring theory allow discrimination between the two extreme geometries for the $Cl + H_2$ reaction?

9.4. Recalculate the A-factors in Problems 9.2 and 9.3 assuming no bond expansion in the activated complex. Does this alter your choice for the more probable geometry?

9.5. A calculation of the A-factor for the $H_2 + I_2$ reaction using Bodenstein's mechanism leads to agreement with the data. Show that either of Sullivan's postulated mechanisms (Problem 4.1) leads to the identical result if the geometry of the activated complex is the same in each case. (*No* numerical calculation is needed.)

9.6. For the reaction

$$NO + NO_2Cl \rightharpoonup NO_2 + NOCl$$

the Arrhenius parameters are $E_a = 28.9\ \text{kJ mol}^{-1}$ and $A = 0.83 \times 10^{12}$ $\text{cm}^3\ \text{mol}^{-1}\ \text{sec}^{-1}$.

(a) Compute $\Delta H^\ddagger$, $\Delta S^\ddagger$, and $K^\ddagger$ at 350 K.
(b) Assuming $p_{NO} = p_{NO_2Cl} = 0.1$ atm, compute $p^\ddagger$ and $n^\ddagger$ at 350 K.

9.7. (a) Using the data of Table 9.4 determine $n^\ddagger$ in the isobutene + HCl reaction at 400 K; the pressure of each of the reactants is 1 atm. To what molecular concentration does your value of $n^\ddagger$ correspond?
(b) How do you define the activated complex? Interpret the result of part (a) within this context.

9.8. $\Delta S_0^\ddagger$ is much larger for the reaction cyclopropane $\rightharpoonup$ propene than for the reaction cyclobutene $\rightharpoonup$ *trans*-butadiene. Why?

9.9. Look up S^0 for HCl(g), C_2H_4(g), and C_2H_5Cl(g) at 298 K. Use these values and the data of Table 9.4 to compute $\Delta S_0^\ddagger$ for the reaction

$$HCl(g) + C_2H_4(g) \rightharpoonup C_2H_5Cl(g)$$

Compare your value with that for the addition of HCl to isobutene. Comment on the difference.

9.10. (a) Extend the thermodynamic analogy of Section 9.9 to define an activation volume $\Delta V_0^\ddagger$ in terms of the pressure dependence of the rate constant.
(b) The pressure dependence of the rate constant for the solvent exchange reaction

$$Ir(NH_3)_5{}^{18}OH_2{}^{3+} + {}^{16}OH_2 \rightharpoonup Ir(NH_3)_5{}^{16}OH_2{}^{3+} + {}^{18}OH_2$$

was measured in 0.01 M $HClO_4$ at 70.5°C [S. B. Tong and T. W. Swaddle,

Inorg. Chem. **13**, 1538 (1974)]:

k (sec^{-1})$\times 10^5$:	3.82	4.01	4.56	4.95	5.79, 5.76	6.00
p (kbar):	0.001	0.059	1.027	2.02	3.31	4.00

What is $\Delta V_0^\ddagger$?

(c) Would you expect any correlation between $\Delta V_0^\ddagger$ and $\Delta S_0^\ddagger$? Explain.

9.11. Stokes' law permits an estimate to be made of ionic radii. Combine this with the Nernst–Einstein relation (3.13) and the expression for the diffusion limited rate constant to show that

$$k_{\text{diff}} = \frac{2}{3}\,\frac{RT}{\eta}\left(\frac{|z_A|}{\lambda_A} + \frac{|z_B|}{\lambda_B}\right)\left(\frac{\lambda_A}{|z_A|} + \frac{\lambda_B}{|z_B|}\right)\times 10^3 f$$

and that

$$d = \frac{eF}{6\pi\eta}\left(\frac{|z_A|}{\lambda_A} + \frac{|z_B|}{\lambda_B}\right)$$

Use these results with the data of Table 3.1 to compute k_{diff} and d for the reaction, at 25°C,

$$H^+ + OH^- \rightarrow H_2O$$

At this temperature $\eta = 0.89\times 10^{-3}$ kg m^{-1} sec^{-1} and $\epsilon = 80$. The experimental value of k_{diff} is 1.4×10^{11} M^{-1} sec^{-1}.

9.12. Carry out the same calculation described in Problem 9.11 to estimate d and k_{diff} for the reactions

$$NH_4^+ + OH^- \rightarrow NH_4OH$$
$$OH^- + HCO_3^- \rightarrow CO_3^{2-} + H_2O$$
$$H^+ + HCO_3^- \rightarrow CO_2 + H_2O$$

Compare your results with the experimental data quoted in Table 5.3.

9.13. The general theory of diffusion-limited rate constants may be combined with the theory of the primary salt effect to estimate the effect of ionic strength on k_{diff}. Assume mean diffusion coefficients 1.3×10^{-9} m^2 sec^{-1} and encounter distances of 0.5 nm for ions in water at 25°C.

(a) Use (9.55), (9.59), and (9.60) to show that, when $I \rightarrow 0$,

$$k_0(T) = 10^{10}\gamma/(e^\gamma - 1), \qquad \gamma \equiv 1.4 z_A z_B$$

(b) Combine this with (9.48) and estimate $k_2(T)$ for the reaction

$$A^{z_A} + B^{z_B} \rightarrow \text{product}$$

at ionic strengths 0, 10^{-4} and 10^{-2} for values of $z_A = 1, 2, 3$, $z_B = \pm 1, \pm 2, \pm 3$.

(c) From the formula of (b), determine the value of I at which k_{diff} is the same if the ionic charge product $z_A z_B$ is $+q$ or $-q$.
(d) What does the result of (c) suggest about the range of validity of the general formula deduced in (b)?

9.14. The RRKM model may be treated phenomenologically as

$$A + A \underset{k_-'}{\overset{k_+'}{\rightleftharpoons}} A^* + A$$

$$A^* \underset{k_-''}{\overset{k_+''}{\rightleftharpoons}} A^\ddagger \xrightarrow{k_3} \text{products}$$

where the rate determining step is the decomposition of $A^\ddagger$.
(a) What is the decomposition rate $-d[A]/dt$, in a steady state?
(b) What are the high- and low-pressure limits?
(c) Is this phenomenological approach any improvement over Lindemann's treatment? Why?

9.15. Show that, if vibrational excitation is ignored, (9.24) reduces to

$$\frac{k'}{k} = \frac{\langle \varkappa' \rangle}{\langle \varkappa \rangle} \frac{\nu'^\ddagger}{\nu^\ddagger} \exp\left[\frac{h}{2kT}(\Delta\nu' - \Delta\nu)\right]$$

where $\Delta\nu = \nu_1^\ddagger(ABC) + 2\nu_2^\ddagger(ABC) - \nu(BC)$; thus the isotope effect, under these limiting conditions, reflects differences in zero point energy.

9.16. At 273 K the isotope effect, k_1/k_2, is 35 for the reactions:

(1) $Cl + H_2 \rightarrow HCl + H$
(2) $Cl + T_2 \rightarrow TCl + T$

Determine the isotope effect for the reverse reactions. The vibrational frequencies are 1.32×10^{14} sec^{-1} for H_2 and 8.97×10^{13} sec^{-1} for HCl. Those for T_2 and TCl can be estimated since

$$\nu_{AB} = \frac{1}{2\pi}\left(\frac{k(m_A + m_B)}{m_A m_B}\right)^{1/2}$$

and the force constant is determined by the shape of the potential surface.

9.17. Demonstrate the equation for k_{diss}, (9.57).

9.18. Estimate k_{diss} as a function of $z_A z_B$ for ionic complexes in water at 25°C. Assume $D_A + D_B = 2.6 \times 10^{-9}$ m^2 sec^{-1} and $d = 0.5$ nm.

9.19. What is K_{eq} for ion-pair formation in the reaction

$$Na^+ + Ac^- = (Na^+ \cdots Ac^-)$$

Assume $d = 0.4$ nm, 0.5 nm, and 0.6 nm. At what separation is K_{eq} (a) maximum, (b) minimum? Can you ascribe physical significance to these results?

9.20. Singlet–singlet excitation transfer, as discussed in Section 6.5.2, takes place at distances of 6–7 nm. Use data from Problem 9.13 and estimate the diffusion limit for this process in water. (The largest values observed are $\sim 3 \times 10^{11}\ M^{-1}\ \text{sec}^{-1}$.) Is the process diffusion-limited?

9.21. (a) Use data from Problems 9.13 and 9.20 to estimate the time required to establish a steady state when studying singlet–singlet excitation transfer. (b) Would transient behavior complicate studies made using pulsed-laser flash photolysis?

9.22. As the intrinsic rate constant varies, the time required to establish a steady state does also. For reactions for which the measured rate constant is less than $\frac{1}{4}$ the diffusion limit, under what conditions, if any, does non-steady-state behavior complicate the interpretation of an experiment?

General References

I. Amdur and G. G. Hammes, *Chemical Kinetics* (New York: McGraw–Hill, 1966), Chapters 2 and 4.

H. S. Johnston, *Gas Phase Reaction Rate Theory* (New York: Ronald Press, 1966).

K. J. Laidler, *Theories of Chemical Reaction Rates* (New York: McGraw–Hill, 1969), Chapters 2, 3, and 6.

Potential-Energy Surfaces

D. G. Truhlar and R. E. Wyatt, *Adv. Chem. Phys.* **36**, 141 (1977).

Activated Complex Theory

K. J. Laidler and A. Tweedale, *Adv. Chem. Phys.* **21**, 113 (1971).

W. H. Miller, *Acc. Chem. Res.* **9**, 306 (1976).

Models for Calculation of Rate Constants

D. L. Bunker, *Acc. Chem. Res.* **7**, 195 (1974).

10

A Dynamical Model for Chemical Reaction

10.1. Introduction

If the reaction cross section, $Q(\epsilon, i, j)$, is known, the rate constant for the corresponding chemical reaction can be calculated from (8.6) and (8.11). To interpret the details of molecular beam experiments even more information is needed. The hard-sphere model was obviously far too naive to give reliable estimates of the reaction cross section. Improvement, by considering the dynamics of reaction across a realistic potential-energy surface such as that described in Section 9.2, is a formidable quantum mechanical problem. As already mentioned it has not been solved, except for low-energy $H + H_2$ chemical reaction.

In this chapter we shall try to clarify this chemically significant problem by studying the dynamics of the collision process A + BC in extremely simplified terms.[1] We shall not attempt to calculate $Q(\epsilon, i, j)$. Instead we shall see how changing various molecular parameters influences reaction probability and affects energy transfer. The effect of varying the collision energy, the reactant mass, and the shape of the reaction profile will be investigated.

The results of our analysis will show that slight variations in the nature of the ABC potential-energy surface, in the collision energy of the reactants, or in the initial vibrational energy of the BC molecule can have dramatic effects on the collision dynamics. In our model we shall emphasize collinear

[1] The treatment is adapted from B. H. Mahan, *J. Chem. Ed.* **51**, 308, 377 (1974).

collisions and ignore effects due to rotational motion. The model, while highly restrictive, nonetheless illustrates many aspects of reaction dynamics.

The fact that so many features of chemical reaction can be reproduced by a simple model emphasizes the point that chemical reaction is not a mysterious problem of rearrangement and recombination. It is, on the other hand, directly determined by the type of collision taking place and the nature of the interparticle forces.

10.2. Kinetic Energy of a Triatomic System

In order to describe the dynamics of the triatomic system A + BC a convenient coordinate system must be constructed. Six degrees of freedom are required to completely describe the *relative* motion of the three particles. These may be chosen in many ways. The most familiar set of coordinates is that which we used in describing the ABC potential-energy surface in Section 9.2: the AB and BC distances, the A–B–C angle, and three angular coordinates to specify the orientation of the ABC system in space. Such a coordinate set is shown in Fig. 10.1, where the *vectors* $\mathbf{r}_{AB}$ and $\mathbf{r}_{BC}$ provide all the information required to specify the relative motion and the rotational motion of the three particles. While a natural set of coordinates, it is not convenient for calculation. The discussion is greatly simplified if we describe the system in terms of BC relative motion and the relative motion of A with respect to the center of mass of BC; these coordinates are also shown in Fig. 10.1.

Complete analysis of the dynamics of collision in the A + BC system is extremely difficult even if A, B, and C are atoms. To avoid the complications introduced by molecular rotation we limit consideration to collinear configurations such as those shown in Fig. 10.2; then only A–B

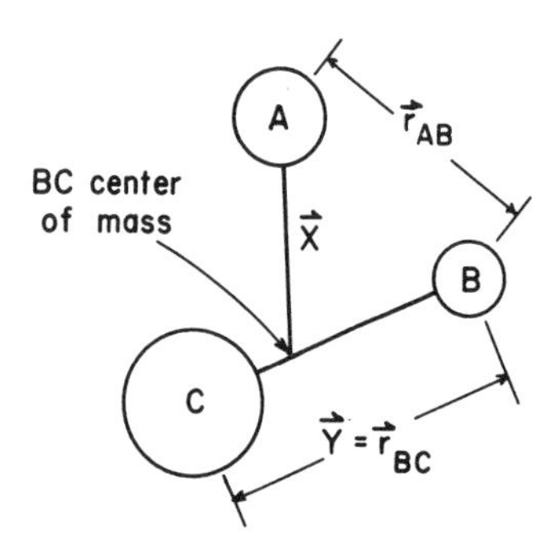

Fig. 10.1. Coordinates for atom–diatom system.

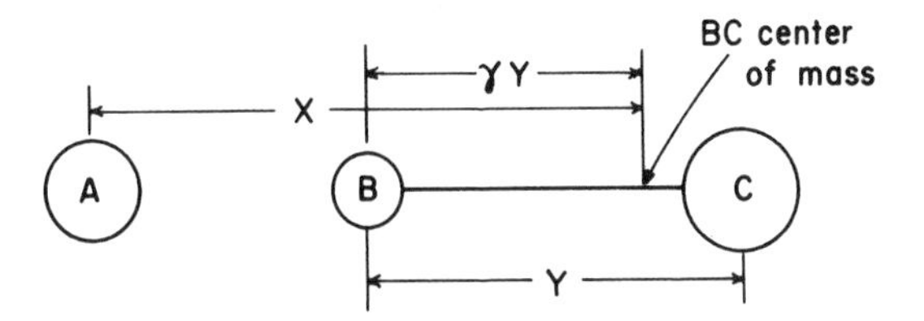

Fig. 10.2. Coordinates for collinear A–BC system.

collisions are possible. Many features of reaction dynamics can be illustrated even if the model is simplified further and A, B, and C are assumed to obey the laws of classical mechanics.

For the collinear system the total kinetic energy of relative motion is the sum of the vibrational kinetic energy of the BC diatom and the kinetic energy of A relative to the center of mass of BC. Letting m_A, m_B, and m_C denote the atomic masses yields

$$T = \frac{1}{2}\left(\frac{m_B m_C}{m_B + m_C}\right)\dot{Y}^2 + \frac{1}{2}\left(\frac{m_A(m_B + m_C)}{m_A + m_B + m_C}\right)\dot{X}^2 \qquad (10.1)$$

Here $m_B m_C/(m_B + m_C)$ is the reduced mass for BC vibration and $m_A(m_B + m_C)/(m_A + m_B + m_C)$ the reduced mass for A–BC relative motion; $\dot{Y}$ and $\dot{X}$ are the corresponding velocities. The kinetic energy is then particularly simple. The chemically more intuitive coordinates r_{AB} and r_{BC} may be expressed in terms of X and Y; using Fig. 10.2 we find

$$r_{BC} = Y, \qquad r_{AB} = X - \gamma Y \qquad (10.2)$$

with $\gamma \equiv m_C/(m_B + m_C)$. A further coordinate transformation yields an even simpler expression than (10.1). Choosing

$$x = X \quad \text{and} \quad y = Y/\eta \qquad (10.3)$$

with $\eta = [m_A(m_B + m_C)^2/m_B m_C M]^{1/2}$, where $M = m_A + m_B + m_C$, the kinetic energy becomes

$$T = \frac{1}{2}\,\frac{m_A(m_B + m_C)}{M}\,(\dot{x}^2 + \dot{y}^2) \qquad (10.4)$$

In addition to its obvious interpretation as the relative kinetic energy of the three collinear particles, (10.4) may also be interpreted as the kinetic energy of a *single* particle of mass $m_A(m_B + m_C)/M$ moving on a planar surface.

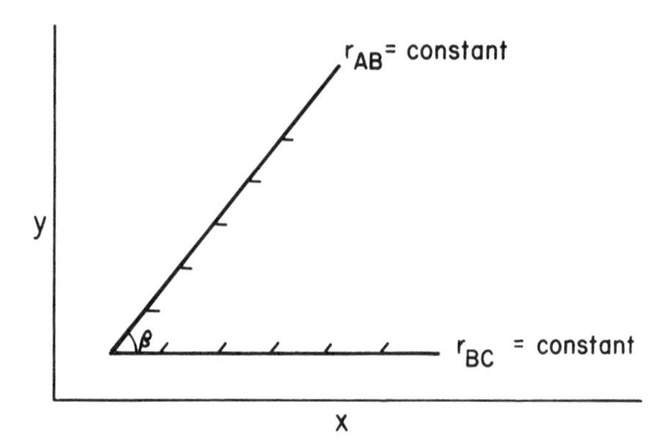

Fig. 10.3. Mapping of molecular coordinate system into Cartesian (x, y) system. Lines of constant r_{AB} and r_{BC} intersect at an angle $\beta = \tan^{-1}(m_B M/m_A m_C)^{1/2}$.

The relation between x and y and the chemically interesting coordinates r_{AB} and r_{BC} may be found from (10.2) and (10.3),

$$y = r_{BC}/\eta = (x - r_{AB})/\gamma\eta \tag{10.5}$$

Lines of constant r_{BC} run parallel to the x axis and lines of constant r_{AB} have a slope of $(\gamma\eta)^{-1}$ in the x–y coordinate system; the angle between these lines is determined by the atomic masses

$$\tan\beta = (\gamma\eta)^{-1} = (m_B M/m_A m_C)^{1/2} \tag{10.6}$$

The mapping is shown in Fig. 10.3.

10.3. *Model for Inelastic Collision*

To utilize the representation we have just presented it is necessary to consider the interaction potential (the potential-energy surface) of the linear ABC system and determine the corresponding potential energy for motion of the equivalent *single* particle in the x–y plane. The simplest problem of chemical interest is a model for energy transfer in inelastic collisions, which was the mechanism we proposed to understand unimolecular decomposition and bimolecular radical recombination reactions. Assuming the square-well, hard-sphere potentials of Fig. 10.4, the corresponding potential-energy surface in the x–y plane is given by Fig. 10.5. The regions where $V = +\infty$ are configurations in which r_{AB} and/or r_{BC} are less than their

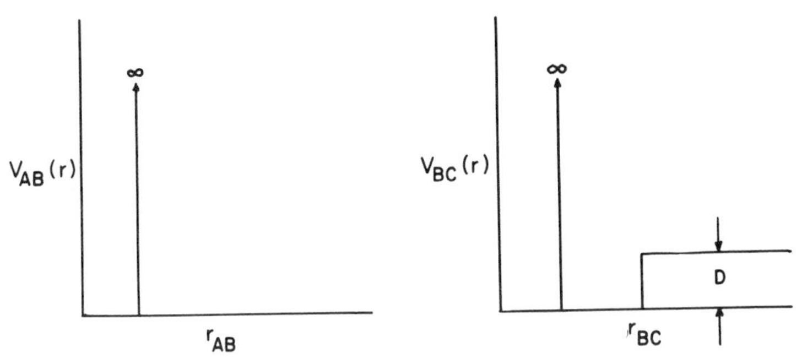

Fig. 10.4. Interaction potentials for hard-sphere square-well oscillator model.

minimum allowed values. The trough where $V = 0$ corresponds to the well in the BC potential, and when $V = D$ the diatom is dissociated.

The simplest example is the somewhat nonphysical trajectory shown in Fig. 10.5. Here the initial motion is parallel to the x axis; y is constant and from (10.5) r_{BC} is constant. In dynamical terms the A atom is approaching a BC diatom which has no vibrational energy. The representative particle collides elastically with wall 1 (an AB collision) and is reflected; the new trajectory makes an angle $\pi - 2\beta$ with the initial one. Reflection then occurs at wall 2 (a BC collision). At this step the representative particle is either reflected or transmitted depending upon whether the BC diatom has acquired sufficient vibrational energy to overcome the dissociation barrier. If transmitted, the y component of kinetic energy decreases in order to conserve energy in crossing the step. The y component of velocity is reduced, accounting for the deflection which is analogous to the phe-

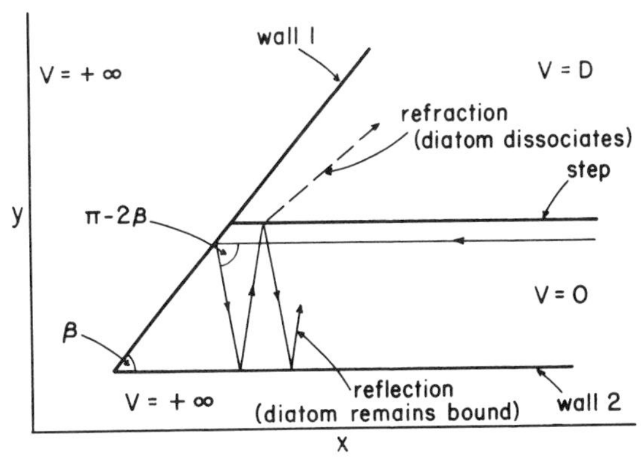

Fig. 10.5. Potential-energy surface and trajectory of representative particle for collinear hard-sphere A interacting with square-well oscillator BC.

nomenon of refraction of light. Thus we shall designate the deflection of the particle as it enters a region of different potential as refraction.

Since, from Figs. 10.1 and 10.2, X represents the distance between A and the center of mass of BC, the relative velocity of A and BC is, from (10.3), $\dot{x} = v_x$, which, in terms of a trajectory, is just the projection of the velocity vector along the x axis. The final relative velocity of A and BC is

$$v_x = v\cos(\pi - 2\beta) = -v\cos 2\beta$$

where v is the *magnitude* of the initial velocity vector. Using (10.3) and (10.1), the kinetic energy of relative A–BC motion after collision is

$$E' = \frac{m_A(m_B + m_C)}{2M} v^2 \cos^2 2\beta = E_0 \cos^2 2\beta$$

where E_0 is the initial kinetic energy of the ABC system, which, in this example, is the initial relative A–BC kinetic energy. If BC remains bound the rest of the energy resides in the vibrational mode of BC, E_v':

$$E_v' = \Delta E = E_0 - E' = E_0 \sin^2 2\beta$$

Dissociation occurs if $E_v' > D$. The fractional energy transfer is a function of particle mass only; using (10.6) we find

$$\frac{\Delta E}{E_0} = \sin^2 2\beta = \left(\frac{2\tan\beta}{1+\tan^2\beta}\right)^2 = \frac{4m_A m_B m_C M}{(m_A + m_B)^2(m_B + m_C)^2} \tag{10.7}$$

If A is lighter than B the relative energy transfer is small

$$\frac{\Delta E}{E_0} \approx \frac{4m_A m_C}{m_B(m_B + m_C)}, \qquad m_A \ll m_B$$

while if A is heavy the energy transfer is independent of A

$$\frac{\Delta E}{E_0} \approx \frac{4m_B m_C}{(m_B + m_C)^2}, \qquad m_A \gg m_B, m_C$$

The mass effect may be roughly correlated with the activation (deactivation) efficiency of buffer gases in unimolecular decompositions (bimolecular recombinations).

As an example, catalysis efficiency in the isomerization of cyclopropane to propene can be interpreted, in part, in terms of the mass of the buffer-gas molecules. To activate C_3H_6, energy is most likely transferred first to a C–H bond by collision of a buffer-gas molecule with the H atom; the extra

Table 10.1. Energy Transfer Efficiency to an Isolated HC Bond by Buffer Gases of Various Mass Assuming the Model of Section 10.3

Buffer mass (molecular analog)	Efficiency computed from (10.7)	Relative efficiency in cyclo-C_3H_6 isomerization from Table 5.1
4 (He)	0.77	0.060
40 (Ar)	0.35	0.053
2 (H_2)	0.95	0.24
28 (N_2)	0.39	0.060

energy can then be partitioned among other vibrational modes and eventually weaken a C–C bond. From (10.7) the relative energy transfer may be estimated by considering the head-on collision of a buffer gas A with the CH moiety so that B = 1 and C = 12. The ratio $\Delta E/E_0$, computed for buffer gases of mass 4, 40, 2, and 28 (He, Ar, H_2, N_2), is tabulated in Table 10.1; the results are in qualitative agreement with the experimental activation efficiencies of Table 5.1. A buffer of mass 4 is more efficient than one of mass 40; one of mass 2 is better than one of mass 28. Considering the approximate nature of the model being used—effects due to multiple collisions, molecular structure, molecular rotation, and intermolecular forces having all been ignored—no better agreement could be expected.

The dynamics of the hard-core, square-well problem is particularly simple for the initial conditions just considered. Sometimes multiple collisions occur as shown in Fig. 10.6a. The representative particle collides with wall 1 more than once, i.e., A hits B, then B hits C, and finally B and A collide again before atom and diatom separate. If, as in Fig. 10.6a, BC is initially vibrationless, multiple collisions can only occur when $\beta < 60°$; as m_B decreases these become more common and are predominant if $\beta < 50°$. Such repeated collisions significantly increase the energy transfer; instead of (10.7) the inelasticity is

$$\frac{\Delta E}{E_0} = \sin^2 2n\beta \tag{10.8}$$

where n is the number of AB collisions that occurred.

If the BC diatom initially has vibrational energy (as it must) the incoming path of the representative particle is no longer parallel to the x

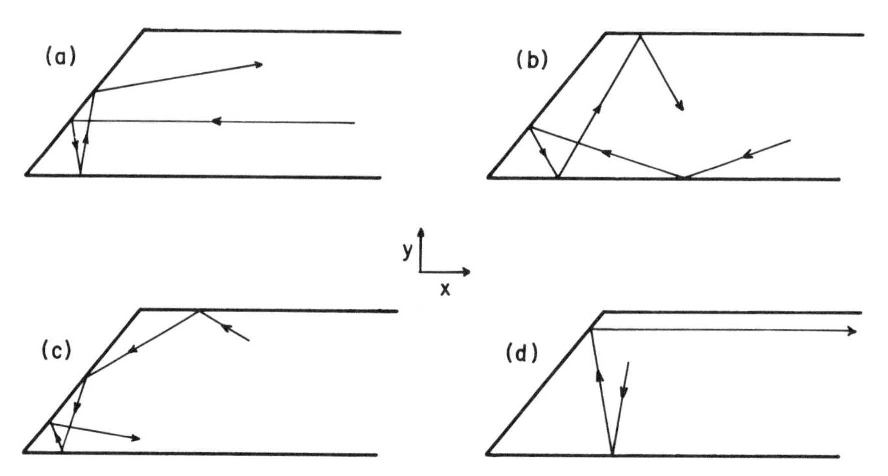

Fig. 10.6. Possible trajectories with potential-energy surface of Fig. 10.5. (a) Multiple AB encounter. (b) BC has initial vibrational energy. (c) BC has initial vibrational energy and multiple AB collisions occur. (d) BC has zero *final* vibrational energy.

axis. Two such trajectories are shown in Fig. 10.6b and c. The energy-transfer calculation is more complicated but, on the average, less energy is transferred to the vibrational mode. The possibility of multiple collisions is much greater. For some trajectories, such as the one shown in Fig. 10.6d, collision leads to vibrational deactivation, which affects bimolecular recombination kinetics.

In bimolecular recombinations the vibrational energy is much larger than the relative translational energy. Under such circumstances (10.8) also describes the fractional energy transfer from vibration to translation (see Problem 10.2). Let us consider the reaction

$$\mathrm{Br} + \mathrm{Br} + \mathrm{M} \rightarrow \mathrm{Br_2} + \mathrm{M}$$

In highly vibrationally excited molecules the quantum mechanical aspects of particle motion are less significant and the details of the interparticle potential also less important. Just as with high-temperature gas-phase transport, discussed in Chapter 2, the repulsive part of the potential has the most significant effect on reaction dynamics. A study of the recombination kinetics of Br_2 with the buffer gases He, Ne, Ar, and Kr shows the steady trend predicted by (10.8)—the more massive the rare gas atom, the more effective it is at stabilizing Br_2.[2]

[2] J. K. K. Ip and G. Burns, *J. Chem. Phys.* **51**, 3414 (1969).

Mass effects do not always predominate. In the vibrational deexcitation of CO($v=1 \rightarrow v=0$), He is far more effective than Ne in direct contradiction of (10.8).[3] Here the model potential of Fig. 10.5 is far from appropriate; the details of the interaction are significant and vibrational quantization is very important.

10.4. *Effects Due to More Complex Potential-Energy Surfaces*

The model may be altered to describe the effect of atom–diatom interactions other than hard-sphere repulsions. Assume A is attracted equally by both B and C. This may be incorporated by adding a well to the potential trough and assuming the force acts between A and the center of mass of BC. The modified potential-energy surface is shown in Fig. 10.7a. The wall is at constant x, which, from (10.3) and Fig. 10.2, is equivalent to a constant distance between A and the center of mass of BC. For this trajectory, the representative particle accelerates when it enters the well, the total kinetic energy becoming $E - \Delta$, where $\Delta < 0$. After it strikes wall 1 its y component of kinetic energy can be shown to be $(E - \Delta) \sin^2 2\beta$ by analogy with the previous calculation. As it leaves the well only the x *component* of kinetic energy is affected. Thus the net inelasticity is increased ($\Delta < 0$)

$$\frac{\Delta E}{E} = \left(1 - \frac{\Delta}{E}\right) \sin^2 2\beta \tag{10.9}$$

If the interaction were repulsive ($\Delta > 0$), the inelasticity would decrease; if E_t, the relative A–BC translational energy, is less than Δ, the particle cannot overcome the barrier and is reflected at the discontinuity.

If A is preferentially attracted to B, assume the attraction is to B alone and modify the potential surface by introducing a step at constant r_{AB} as shown in Fig. 10.7b. For the initial conditions shown the particle is refracted upon entering the well, reflected at wall 1, and refracted again upon leaving the well. Geometric considerations show that the well has *not* altered the energy transfer. The inelasticity is still given by (10.7) and (10.8).

The other limiting case occurs when A is attracted to C, which can be modeled by incorporating a potential well with a discontinuity at a constant separation r_{AC} as shown in Fig. 10.7c. The inelasticity calculations

[3] R. C. Millikan, *J. Chem. Phys.* **38**, 2855 (1963).

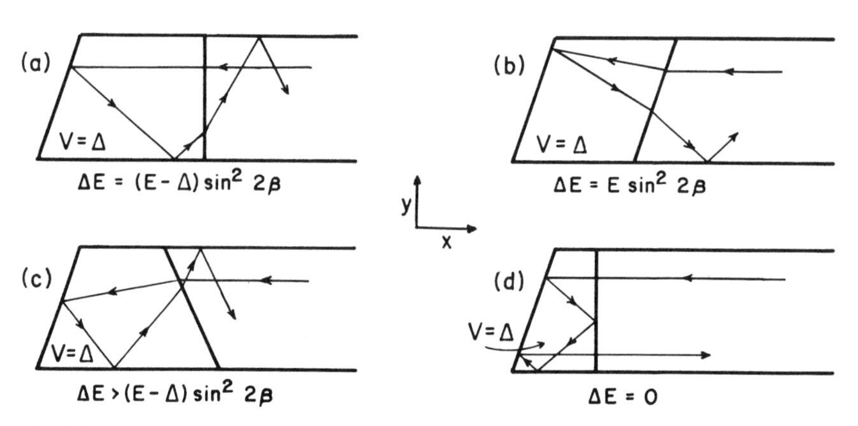

Fig. 10.7. Potential-energy surface features which complicate trajectory analysis; $\varDelta$ is the depth of the potential well. (a) Inelasticity altered due to potential step perpendicular to x axis. (b) If A only attracted to B, $\varDelta E_v$ is not affected. (c) A is attracted to C. (d) If potential well is deep enough temporary trapping occurs.

for the trajectory shown are no longer simple. There is, however, an increase in energy transfer when the interaction is attractive.

The atom–diatom attraction not only affects energy transfer but it also introduces the possibility of "sticky" collisions. A possible trajectory is shown in Fig. 10.7d. After reflection at wall 1 the representative particle attempts to recross the step but its x component of kinetic energy is not sufficient. Here the kinetic energy of the particle is $E - \varDelta$ while it is in the well; the condition for crossing the step on the *first* try is

$$(E - \varDelta) \cos^2 2\beta \geq |\varDelta| \tag{10.10}$$

If (10.10) is not satisfied the particle is reflected and thus temporarily trapped. As diagrammed in Fig. 10.7d the ABC complex undergoes a complete AB vibration; the second AB collision converts enough energy into relative kinetic energy of A and BC so that dissociation occurs. Given different initial conditions the complex may survive many AB vibrations before dissociating.

Qualitatively, trapping takes place because of the attractive force between atom and diatom and because there is an internal degree of freedom (the BC vibrational mode). Some of the relative kinetic energy resides temporarily in the internal degree of freedom. By increasing the well depth, increasing the number of vibrational degrees of freedom, or decreasing the relative kinetic energy, the chances of a sticky collision are favored and longer-lived complexes will be formed. This correlates with another feature of bimolecular recombination kinetics as presented in Section 5.5. In atomic recombination long-lived complexes cannot form and a third particle is required *immediately* to take away the excess kinetic energy. Methyl radicals form a long-lived complex; only at low pressures is the time between collisions long enough to allow dissociation instead of collisional deactivation. The phenyl radical complex has such a long lifetime that experiments have never been run at a pressure so low that dissociation is competitive with collisional deactivation.

10.5. Model for Thermoneutral Chemical Reaction

The model presented to study inelastic energy transfer for a collinear atom–diatom system can be extended to treat the exchange reaction

$$A + BC \rightharpoonup AB + C$$

The simplest potential-energy surface, shown in Fig. 10.8, is one where AB and BC are assumed to be bound by identical square-well potentials and where the potential energy remains the same whether two or three particles are interacting. It is a primitive model for a thermoneutral exchange reaction. The potential-energy surface for linear H_3 (the portion corresponding to the regions in which all three particles in H_3 are interacting is shown in Fig. 10.9) has a significant barrier in the trough when all interatomic distances are small. However, if the relative kinetic energy is large, the details of the trough are less significant than its general shape. Thus the model surface is an aid to understanding qualitative features of *high-energy* atom–molecule collision processes (the restriction to collinear trajectories is still a significant limitation).

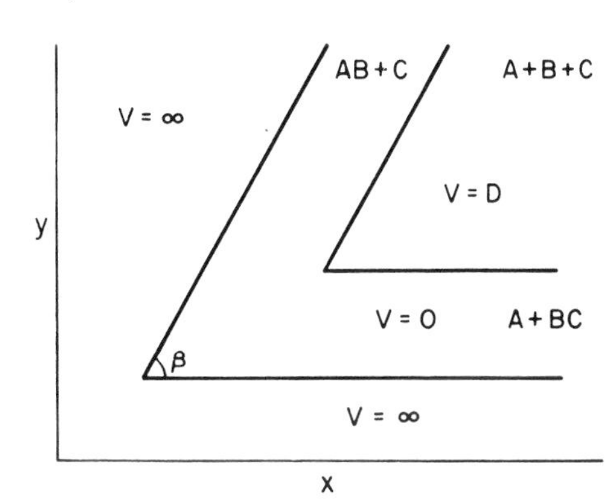

Fig. 10.8. A square-trough potential-energy surface for the reaction $A + BC \rightharpoonup AB + C$.

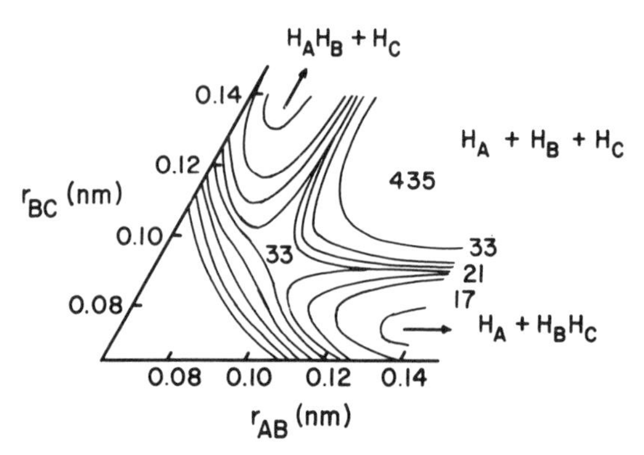

Fig. 10.9. A portion of the potential-energy surface for the exchange reaction $H + H_2 \rightarrow H_2 + H$ assuming collinear geometry. The energies are in units of kJ mol^{-1}. [Adapted from R. E. Weston, Jr., *J. Chem. Phys.* **31**, 892 (1959).]

As an example consider $\beta = 60°$, which corresponds to an unlimited number of mass ratios, among them A:B:C = 1:1:1, 1:2:6, 2:3:5, etc. Different combinations can be deduced from (10.6). From Fig. 10.10a we see that, since both reactant and product channels have the same width, all horizontal trajectories (zero vibrational energy) are reactive; the products formed have zero vibrational energy. The result is peculiar to the choice $\beta = 60°$. For other β this trajectory is not always reactive; in the example of Fig. 10.10b, large values of y lead to inelastic collisions.

If the reactants have vibrational energy, the initial trajectory is no longer horizontal. An angle $|\theta|$ determines the partitioning of energy between vibration and translation; it is defined as

$$\tan\theta = \dot{y}/\dot{x}$$

where velocities are measured at the standard reference line perpendicular to the x axis shown in Fig. 10.10c. We then find that

$$E_t = E\cos^2\theta, \qquad E_v = E\sin^2\theta \tag{10.11}$$

where E is the total kinetic energy of the system; E_t and E_v are the initial relative A–BC kinetic energy and BC vibrational energy, respectively. As long as $\beta \geq 60°$ there are only three types of reactive trajectories, those shown in Figs. 10.10c,d, and e. Since both $\dot{y}$ and $\dot{x}$ are negative at the reference line in Fig. 10.10c, θ is positive. In Fig. 10.10d θ is negative, since $\dot{y} > 0$ and $\dot{x} < 0$ at the reference line. In Fig. 10.10e θ is positive for the two trajectories shown.

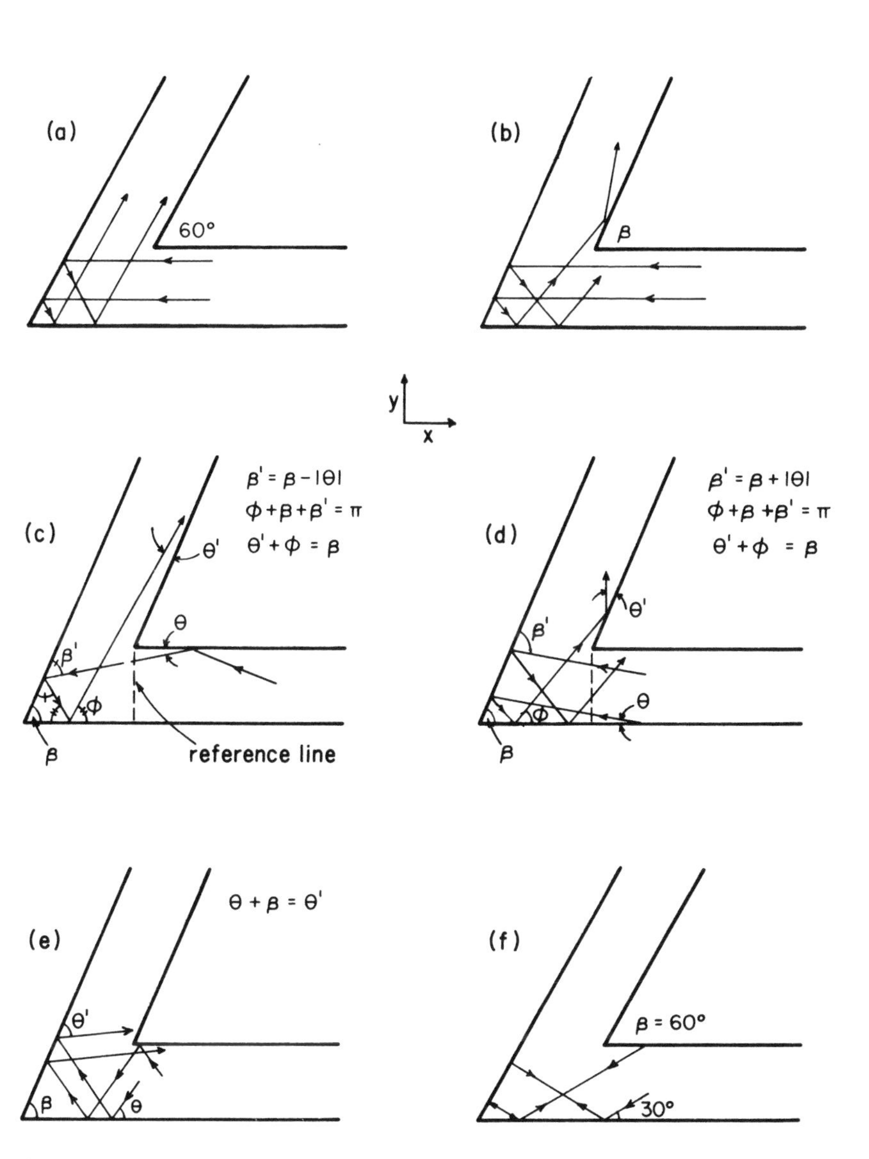

Fig. 10.10. Effects of initial vibrational energy and phase on product formation and energy distribution. The diagrams at the top show trajectories with no initial vibrational energy for (a) $\beta = 60°$, (b) $\beta > 60°$ surface. (c) Construction for deducing θ' for collision sequence where A hits B, then B hits C. (d) Effect of phase on reactivity. (e) Construction for finding θ when collision sequence is B hits C, then A hits B. (f) Trajectories with $\theta = 30°$ on $\beta = 60°$ surface.

Table 10.2. Summary of Reaction Possibilities on the Potential-Energy Surface of Figure 10.8, with $\beta = 60°$, as a Function of Vibrational Energy of Reactants, as Characterized by $|\theta|$ [Equation (10.11)]

$|\theta|$, degrees	Extent of reaction	$|\theta'|$, degrees	E_v'/E_v
0	Total	$|\theta|$	1
0–30	Partial	$|\theta|$	1
30	None		
30–60	Partial	$120 - |\theta|$	$\frac{1}{4}(3^{1/2}\cot|\theta| + 1)^2$
60–90	Total	$120 - |\theta|$	$\frac{1}{4}(3^{1/2}\cot|\theta| + 1)^2$

The $\beta = 60°$ case is particularly simple. As θ increases from 0° to 30° an increasing fraction of trajectories lead to inelastic collisions. Paths with $\theta < 0$ and large values of y at the reference line are the first to be lost from the product channel (Fig. 10.10d); for $-30° < \theta < -16.1°$ there are *no* reactive trajectories. For $\theta > 0$ reaction is not impeded until $\theta = 16.1°$; the number of reactive paths then decreases until, when $\theta = 30°$, all paths are blocked (Fig. 10.10f). As $|\theta|$ increases reaction is again possible; for $|\theta| > 60°$ all paths are reactive. Thus reactivity depends not only on the ratio E_v/E_t (as determined by $|\theta|$) but also upon the vibrational *phase* of the reactants (as determined by the sign of θ and the value of y at the reference line). The various possibilities are summarized in Table 10.2.

The angle $|\theta'|$ characterizes energy disposition in the product channel, $E_v' = E\sin^2\theta'$. For each of the trajectories in Figs. 10.10c,d, and e, θ' is a different function of θ. Noting that the sum of the angles of a triangle is π radians, we find that when $0 < \theta < \beta/2$ (Fig. 10.10c)

$$\beta' = \beta - |\theta|, \qquad \varphi + \beta + \beta' = \pi, \qquad \varphi + \theta' = \beta \tag{10.12}$$

and hence

$$\theta' = 3\beta - |\theta| - \pi \qquad (0 < \theta < \beta/2) \tag{10.13}$$

The case $-\pi/2 + \beta < \theta < 0$ (Fig. 10.10d) differs only in that $\beta' = \beta + |\theta|$ and thus

$$\theta' = 3\beta + |\theta| - \pi \qquad (-\pi/2 + \beta < \theta < 0) \tag{10.14}$$

For all other values of θ (Fig. 10.10e) we obtain

$$\theta' = \pi - \beta - |\theta| \qquad (\beta/2 < \theta < \pi/2, -\pi/2 < \theta < -\pi/2 + \beta) \tag{10.15}$$

The case $\beta = 60°$ is again the simplest. Energy disposition in the products depends on the relative energy but not on the phase of the reactants. From the value of $|\theta'|$ the ratio E_v'/E_v may be calculated; these are included in Table 10.2. In the range $30° < |\theta| < 60°$ the products have more vibrational energy than the reactants and the AB formed dissociates if $E_v' > D$.

Changing initial conditions affect both reactivity and energy disposition. For fixed E_v, increasing E_t corresponds to decreasing $|\theta|$. For $\beta = 60°$ we find from (10.11) and Table 10.2 that when $E_t < E_v/3$ all trajectories are reactive. For $E_t > E_v/3$ the reaction probability drops until, when $E_t = 3E_v$, no reaction occurs. With a further increase of E_t, the reaction probability rises again. For other choices of β product energy disposition as well as reactivity depend upon both phase and $|\theta|$.

10.6. Exoergic Reactions

In order to discuss anything other than thermoneutral reactions, the potential-energy surface must be modified. The easiest way to represent an exoergic reaction is to introduce a single step in the potential energy trough. The location and orientation of the step can be varied to model different situations. The simplest case arises if the step is located in the product channel as shown in Fig. 10.11a; as drawn it corresponds to a repulsion between C and the center of mass of AB. Trajectories which were reactive for the potential surface of Fig. 10.8 remain that way. When the representative particle crosses the step it is refracted. Only the component of velocity perpendicular to the step is affected. The excess energy appears as relative translational energy of the products

$$E_t' = E \cos^2 \theta' - \Delta, \qquad E_v' = E \sin^2 \theta' \tag{10.16}$$

where θ' is the angle characterizing energy disposition in the product channel for the simple surface of Fig. 10.8. When the exoergicity is due to BC repulsion the step is parallel to the x axis (Fig. 10.11b) and the excess energy affects both translation and vibration.

The effect of a step in the reactant channel is quite different. Figures 10.11c and d illustrate dynamics on potential surfaces where the exoergicity is due to attraction between A and the center of mass of BC. In Fig. 10.11c the relationship of initial phase to reactivity is shown by considering two trajectories for which the reactants have zero initial vibrational energy. For the path with large y the representative particle is reflected back toward

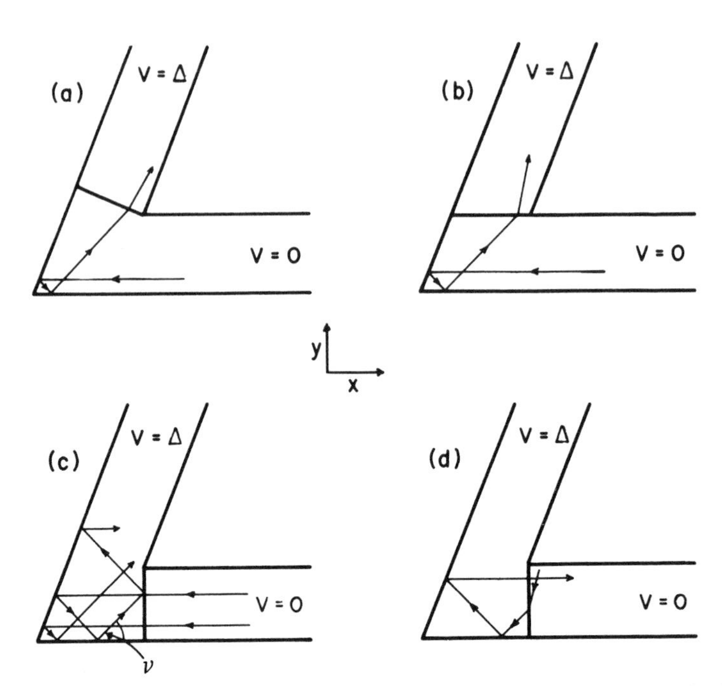

Fig. 10.11. Effects of exoergicity ($\Delta < 0$) on product formation and energy distribution. (a) Step in exit channel corresponding to repulsion between C and center of mass of AB. Final translational energy altered by step. (b) Step in exit channel due to BC repulsion. Both translational and vibrational energy of products are affected. (c) Step in entrance channel due to attraction between A and center of mass of BC. Large y trajectory which would have led to inelastic collision becomes reactive if E is small. (d) Step in entrance channel. Collision between A and vibrationally excited BC leads to inelastic energy transfer instead of reaction because of the exoergicity.

the reagent channel; if the reaction were thermoneutral ($\Delta = 0$) the trajectory would be nonreactive. From Fig. 10.5 we see that $\nu = \pi - 2\beta$. Since the total kinetic energy in the region to the left of the step is $E - \Delta$, where $\Delta < 0$, the x component of the kinetic energy is

$$(E - \Delta)\cos^2 \nu = (E - \Delta)\cos^2 2\beta$$

Thus if

$$(E - \Delta)\cos^2 2\beta < -\Delta \quad \text{or} \quad E < -\Delta \tan^2 2\beta \tag{10.17}$$

the representative particle is reflected when it returns to the step. In this example the representative particle enters the product channel after reflection at the step. Note that if the initial kinetic energy is too *large* reaction does *not* occur. The exoergicity permits *low-energy* trajectories which would otherwise be *nonreactive* to enter the product channel. In addition, energy disposition in the products is affected. For the small y path, θ', from (10.13), is $3\beta - \pi$; similarly for large y, when reaction occurs, θ' is β. Since the total kinetic energy is $E - \Delta$ the final vibrational energy is

$$E_v' = \begin{cases} (E - \Delta)\sin^2\beta, & \text{large } y \\ (E - \Delta)\sin^2 3\beta, & \text{small } y \end{cases} \tag{10.18}$$

Of the total exoergicity $-\Delta$, only a fraction appears as product vibrational energy, the amount depending upon y as well as β. The large $|\theta|$ trajectory of Fig. 10.11d corresponds to an initial state for which $E_v > E_t$. Were $\Delta = 0$, this trajectory would lead to the product channel; however, the new reaction dynamics requires that it return to the potential step and attempt to cross. Since the x component of kinetic energy is greater after the AB collision than before, the representative particle recrosses the step; the result is inelastic collision. Of the two effects the enhanced low-energy reactivity is particularly significant chemically. It correlates with the observation that exoergic reactions have large low-energy cross sections. The fact that reaction probability *drops* with an increase in E_t may account for negative Arrhenius activation energies in some exothermic reactions.

The disposition of the excess energy differs markedly depending upon the placement of the step in the potential-energy surface. When the exoergicity is due to a step in the product channel (termed "late" or "repulsive" energy release) as shown in Figs. 10.11a and b, refraction at the step leads to a major portion of the exoergicity appearing as relative *translational* energy. The situation that occurs when the step is in the reactant channel (termed "early" or "attractive" energy release) is quite different. From Fig. 10.11c it is clear that much of the exoergicity has appeared as *vibrational* energy in the product BC. The results, distinguishing attractive and repulsive potential-energy surfaces, are similar to those obtained using far more sophisticated models.[4] In addition to the shape of the surface, the mass ratios of the interacting particles also affect energy disposal, as reflected by the way in which E_v', from (10.16) or (10.18), depends upon β. Naturally the distinction—attractive or repulsive—is an oversimplification. The step need not occur primarily in one channel or the other. It may appear closer to the midpoint of the path in which case the energy release is termed "mixed," indicating that neither translational nor vibrational energy transfer predominates.

[4] A good account of more realistic treatments of reaction dynamics is given by J. C. Polanyi, *Acc. Chem. Res.* **5**, 161 (1972).

10.7. Endoergic Reactions

Endoergic reactions can be treated by introducing potential steps in which $\Delta > 0$. Analysis is simplest when the step is placed in the reactant channel (Fig. 10.12a). If $E_t = E\cos^2\theta < \Delta$, the A atom is reflected at the step and reaction is forbidden. In this example, representing repulsion between A and the center of mass of BC, only the relative *translational* energy is useful for surmounting the barrier. If the particle has sufficient energy to cross the barrier it is refracted. After passing the barrier the total kinetic energy is $E_1 = E - \Delta$; the vibrational energy is unaltered, thus

$$E_{1v} = E_1 \sin^2\theta_1 = E\sin^2\theta = E_v$$

and

$$\sin\theta_1 = [E/(E-\Delta)]^{1/2}\sin\theta \tag{10.19}$$

The further course of reaction is determined by the angle θ_1. Any trajectory

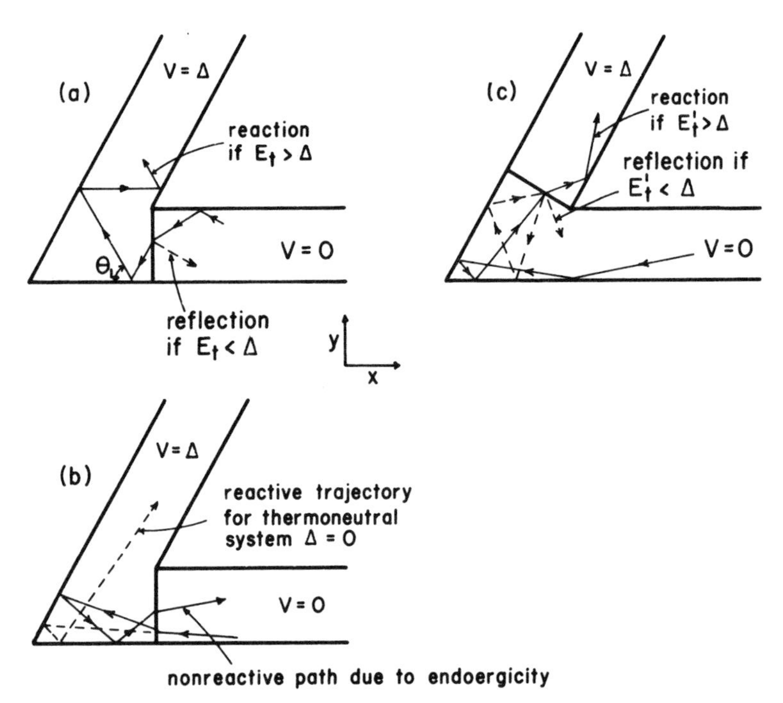

Fig. 10.12. Effect of endoergicity ($\Delta > 0$) on product formation and energy distribution. (a) Step in entrance channel. Reaction only occurs if $E_t = E\cos^2\theta > \Delta$. (b) Step in entrance channel. Effect of endoergicity in modifying trajectories to decrease reactivity. (c) Step in product channel. Reaction only occurs if $E_t' = E\cos^2\theta' > \Delta$.

which leads back to the step is necessarily nonreactive. Since $|\theta_1| > |\theta|$, initial trajectories corresponding to small vibrational energy (small $|\theta|$) may, after refraction, be redirected along a nonreactive path as can be seen in Fig. 10.12b. Thus not only does the barrier prevent reaction if $E_t < \Delta$, it also reduces the reaction probability (when E_v is small) for many trajectories if $E_t > \Delta$.

A barrier in the product channel (Fig. 10.12c) can only be surmounted if $E_t' = E\cos^2\theta' > \Delta$. In this example the endoergicity is due to attraction between C and the center of mass of AB. Here a small amount of vibrational energy is far more effective in promoting reaction than is increasing the translational energy. As long as $\beta - \pi/2 < \theta < 2\beta - \pi/2$, (10.13) or (10.14) are valid for all β and

$$\theta' = 3\beta - \theta - \pi$$

so that

$$E_t' = E\cos^2(3\beta - \theta) = E(\cos 3\beta \cos\theta + \sin 3\beta \sin\theta)^2 \qquad (10.20)$$

As usual E_t' depends on the sign of θ. If $\beta > \pi/3$, E_t' is larger for θ positive. Using (10.11) and choosing the phase which maximizes E_t', (10.20) becomes

$$E_t' = (E_t^{1/2}|\cos 3\beta| + E_v^{1/2}|\sin 3\beta|)^2 \qquad (10.21)$$

If the initial total energy is E it is completely available for surmounting the barrier only if

$$\frac{E_v}{E_t} = \tan^2 3\beta$$

Thus, unless $\beta = 60°$, a small amount of vibrational energy is *required* in order to most efficiently utilize the available translational energy. The effect is clearly demonstrated in the reaction[5]

$$He + H_2^+ \rightarrow HeH^+ + H, \qquad \Delta E = 77 \text{ kJ mol}^{-1}$$

where, near threshold, vibrational energy in H_2^+ is many times more effective in promoting reaction than is relative translational energy. The relative reaction probability as a function of vibrational state and *total* energy is tabulated in Table 10.3. At the lowest total energy, which is only slightly above threshold, the effectiveness of vibrational excitation is dramatic. One quantum of vibration is 27.5 kJ mol^{-1}, an excitation equivalent to

[5] W. A. Chupka and M. E. Russell, *J. Chem. Phys.* **49**, 5426 (1968).

Table 10.3. Relative Probability for the Reaction $H_2^+ + He \rightarrow HeH^+ + H$ as a Function of E, the Total Energy (Translational and Vibrational), and of the Vibrational Quantum Number, v, of the H_2^+[a]

v	E (kJ mol^{-1})			
	96	192	288	385
0	0.06	0.10	0.13	0.17
1	0.49	0.35	0.31	0.25
2	1.95	0.93	0.55	0.34
3	—	1.70	0.99	0.56
4	—	2.35	1.22	0.68
5	—	2.49	1.70	0.89

[a] W. A. Chupka, in *Ion Molecule Reactions*, J. L. Franklin, ed. (New York: Plenum Press, 1972), Vol. 1, p. 73.

increasing the E_v/E_t ratio from 0 to 0.4; it leads to an eightfold increase in reaction probability. A second quantum means E_v/E_t increases to 1.3; the reaction probability increases by another factor of 4. Also interesting is the effect of increasing energy at constant vibrational excitation. In the ground vibrational state, reaction probability increases slightly as the total energy increases. In all other vibrational states the effect of increasing the total energy is counterproductive. Accurate quantum mechanical calculations have corroborated the model for the potential-energy surface; the 77 kJ mol^{-1} barrier is almost entirely in the product channel.[6]

A similar situation, in which vibrational excitation enhanced reactivity, has been mentioned in a different context. Laser activation in the decomposition of D_3BPF_3 seemed to lead to much more efficient excitation (Section 6.8). Of course for such a complex molecule no potential surface has been calculated. No correlation with a product channel barrier is presently possible.

Combinations of wells and barriers may be used to model more complex potential-energy surfaces. In the examples we considered both reactivity and energy disposition were significantly affected by the position and nature of a single well or barrier. Further structure in the potential surface, such as placing steps in both reactant and product channels,

[6] P. J. Brown and E. F. Hayes, *J. Chem. Phys.* **55**, 922 (1971).

greatly complicates the dynamics.[7] Since even for the simpler examples slight changes in initial conditions may alter the outcome of interaction, it is clear how difficult it is to deduce a potential-energy surface from the results of molecular beam experiments.

For the much simpler problem of an intermolecular potential between argon atoms, discussed in Section 2.8, we saw that only by combining data from different types of experiments could a reliable potential be constructed. In that case the potential was a function of one variable only, the argon–argon distance. For the atom–molecule system the potential-energy surface is a function of the three independently variable atomic distances. To vary the parameters of such a potential in order to reproduce molecular beam scattering data is a herculean task; some efforts toward this goal are discussed in Section 10.9. To carry out a similar program for reactions involving more complicated molecules is presently beyond consideration. At present reactive scattering data are used to establish the general features of a potential-energy surface, not the precise details.

10.8. Noncollinear Geometries

The model, based as it is on the assumption of collinear collision geometry, is a gross oversimplification. Even if the BC diatom had no rotational angular momentum, real A–BC trajectories occur with nonzero impact parameter. Two such possibilities are illustrated in Fig. 10.13. After the first AB collision takes place, some of the collision energy must be converted to rotational energy, either of the BC diatom or, if a single collision leads to reaction, of the AB diatom. The consequences are particularly important if multiple collisions are predicted for the collinear geometry.

An example is the isotope effect in the reaction of Kr^+ with HD at high collision energy.[8] The facts are as follows:

(i) At very low collision energies the cross section ratio $Q(KrH^+)/Q(KrD^+)$ is slightly greater than 1; the ratio peaks at $E_{coll} \sim 100$ kJ mol^{-1} and then drops steadily. For $E_{coll} > 180$ kJ mol^{-1} the ratio is less than 1;

[7] A model, classical trajectory calculation for collinear $H + H_2 \rightarrow H_2 + H$, using a realistic potential function, was set up as an experiment in the application of programming skill and of Runge–Kutta numerical integration methods [J. M. White, *Physical Chemistry Laboratory Experiments* (Englewood Cliffs, N.J.: Prentice–Hall, 1975), pp. 301–326].

[8] S. Chivalak and P. M. Hierl, *J. Chem. Phys.* **67**, 4654 (1977).

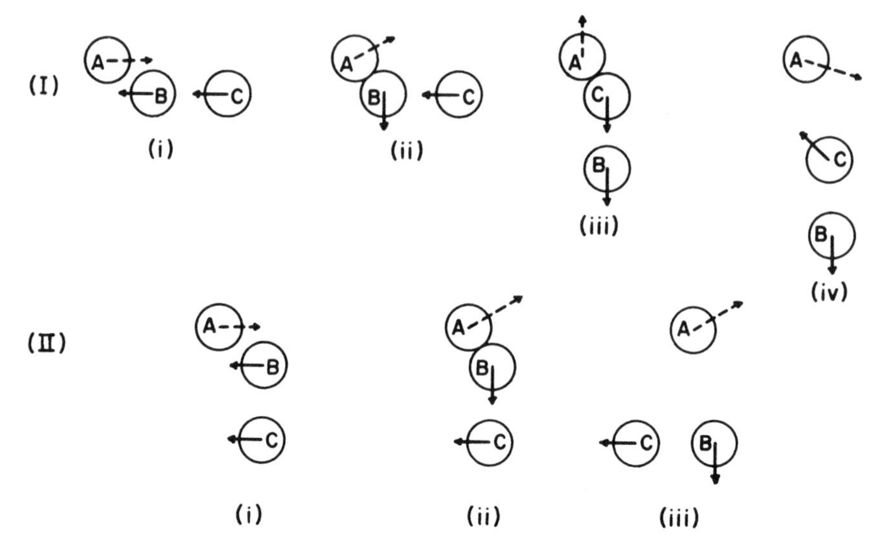

Fig. 10.13. Effect of noncollinearity on CM reaction dynamics for a thermoneutral reaction for particles A, B, and C of mass 86, 1, and 2, respectively—a model for $Kr^+ + HD$ dynamics. Length of vectors indicates velocity. Particle A is so heavy that it is nearly stationary; direction of motion is indicated. Case I: (i) initial trajectory; (ii) instant after AB collision; (iii) instant after AC collision; (iv) instant after AC distance reaches its maximum. Note that an AC diatom has formed even though initial collision was with B. The AC diatom has vibrational and rotational energy. Case II: (i) initial trajectory; (ii) instant after AB collision; (iii) BC remains bound—no reaction occurs. The BC diatom has vibrational and rotational energy.

KrD^+ is then the predominant product. The effect is illustrated in Fig. 10.14.

(ii) KrH^+ is *forward* scattered at all energies studied; the cross section *decreases* by a factor of 20 over the energy range considered (8 kJ mol^{-1} $\lesssim E_{coll} \lesssim$ 300 kJ mol^{-1}).

(iii) At low collision energies KrD^+ is forward scattered; as the energy increases it is back scattered. The cross section drops by ~4 as the collision energy increases from 8 to 300 kJ mol^{-1}.

Consider high collision energies. Here the details of the potential surface are less significant. Our model may be applicable since experiment suggests that E_{coll} is substantially larger than any barrier to reaction.[9] Furthermore, the exoergicity (~30 kJ mol^{-1}) is small. The high collision energy data seem to suggest that KrD^+, as it is back scattered, is formed

[9] K. R. Ryan and I. G. Graham, *J. Chem. Phys.* **59**, 4260 (1973); P. F. Fennelly, J. D. Payzant, R. S. Hemsworth, and D. K. Bohme, *J. Chem. Phys.* **60**, 5115 (1974).

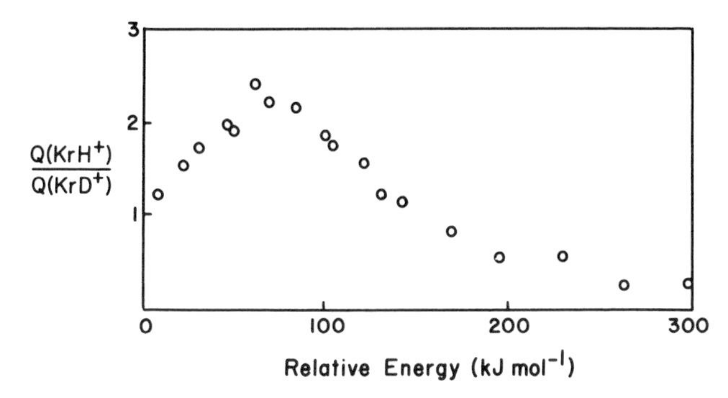

Fig. 10.14. Kinetic isotope effect in the $Kr^+ + HD$ system [S. Chivalak and P. M. Hierl, *J. Chem.* **67**, 4654 (1977)].

mainly in a recoil (small impact parameter) process. At the same energies KrH^+ is forward scattered, which is suggestive of a large impact parameter process.

Now consider small impact parameter processes and assume that reactive trajectories are collinear, or nearly so. Ignore exoergicity and model the reaction with an atom–diatom potential such as that of Fig. 10.10. Two cases arise, corresponding to the geometries Kr^+HD and Kr^+DH with $\beta = 34.8°$ and $55.2°$, respectively. If the diatom is presumed to have no initial vibrational energy, trajectory analysis indicates that for the two values of β all collisions in both geometries are reactive (Problem 10.10) thus suggesting that equal amounts of KrD^+ and KrH^+ are formed. The difference between the two paths is that Kr^+ collides with the H end of the DH molecule twice before KrH^+ can separate. Only a single collision between Kr^+ and the D end of the DH molecule is needed to form KrD^+.

At nonzero impact parameter rotation greatly changes the predictions of the model. Because KrD^+ can form in a single collision, collisions with the D end of the molecule still lead to KrD^+, which is formed with some rotational energy. The consequences of induced rotation upon collisions of Kr^+ with the H end of the molecule are strikingly different. Because the H atom is *not* transferred on the first collision, the HD molecule starts to rotate as shown in Fig. 10.13. Depending upon the initial conditions the Kr^+ might separate or it might take part in another collision with either the H or the D end of the molecule. As collision energy or impact parameter increase, more angular momentum is imparted so the Kr^+ ion's second collision is more likely to be with the D end of the molecule.[10]

[10] If the impact parameter is larger than the hard-core diameter the model is no longer remotely applicable. The arguments presented are useful at smaller impact parameters.

As a consequence the amount of KrD^+ formed in recoil processes is enhanced; the amount of KrH^+ is decreased.

The qualitative features of the high-energy cross-section ratio and scattering patterns can now be understood. Smaller impact parameters favor formation of KrD^+ because of rotation and the requirement of multiple collision for formation of KrH^+. The major recoil product is KrD^+, as reflected in its scattering pattern. Recoil processes produce little KrH^+; only large impact parameter processes, which lead to forward scatter, are effective. The data reflect these considerations. The enhanced cross section for KrD^+ is also understandable. As collision energy increases large impact parameters are less likely to lead to reaction (Sections 8.5–8.9). Thus KrH^+ formation becomes increasingly improbable. Smaller impact parameters, which are now more significant, favor KrD^+ formation. The result is an enhancement of $Q(KrD^+)$, as observed.

Problem 10.3 illustrates that for collinear trajectories in which the diatom has no initial vibrational energy ($\theta = 0$), multiple collision is mandatory if $\beta < \pi/4$. According to (10.6) such values of β are possible only if $m_B < m_C$. At nonzero values of the impact parameter b, since reaction and separation cannot occur, the BC diatom starts to rotate after collision of A with the B end of the molecule. The larger the collision energy or the larger the value of b,[11] the greater the angular velocity. The second collision is more likely to involve the heavy end of the molecule. Sufficient energy transfer is now possible; separation may occur *without* further collisions. In general, therefore, the collinear approximation overestimates effects due to multiple collisions. At nonzero impact parameters more than two collisions are quite unlikely unless the potential surface favors complex formation.

10.9. Realistic Scattering Calculations

The procedure we have outlined, a classical mechanical trajectory analysis based upon a chosen potential-energy surface, has been applied in a far more sophisticated fashion to a number of triatomic systems. The restriction to collinear collision geometries has been lifted and more realistic potential-energy surfaces have been used. We consider two examples, the reactions $H + D_2 \rightarrow HD + D$ and $Ar^+ + D_2 \rightarrow ArD^+ + D$, to demon-

[11] There are limits on the range of b for which these considerations apply. See footnote (10) in this chapter.

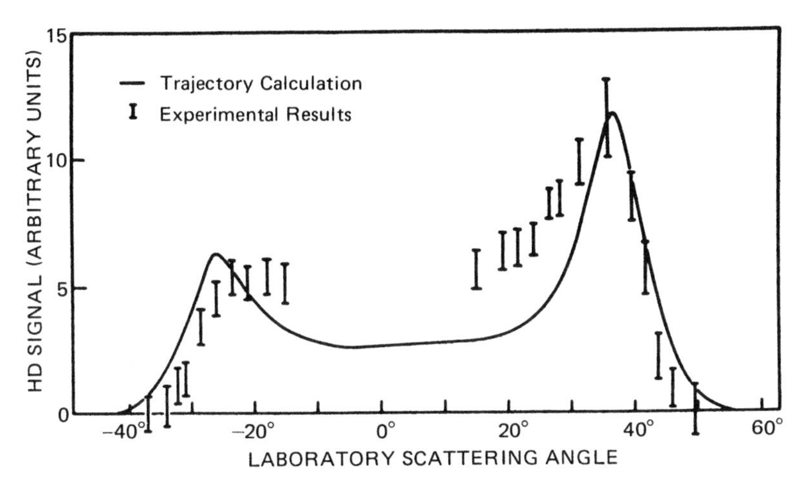

Fig. 10.15. In-plane angular distribution of HD formed in the reaction $H + D_2 \rightharpoonup HD + D$. The trajectory calculation is by P. Brumer and M. Karplus, *J. Chem. Phys.* **54**, 4955 (1971); the experimental data are from J. Geddes, H. F. Krause, and W. L. Fite, *J. Chem. Phys.* **52**, 3296 (1970). The calculated curve is normalized to agree with experiment at $+ 36°$. [From P. Brumer and M. Karplus, *J. Chem. Phys.* **54**, 4955 (1971), reprinted by permission, American Institute of Physics.]

strate the accuracy with which scattering patterns can be calculated.[12] In particular, we shall find that good results can sometimes be obtained using most approximate model potentials.

10.9.1. $D + H_2 \rightharpoonup HD + H$

Classical trajectory calculations for the hydrogen exchange reaction have been used to compute the angular distribution of the HD produced.[13] The potential-energy surface was a realistic one.[14] Theory and experiment are compared in Fig. 10.15 for low-energy collisions. Agreement is quite good; the calculation reproduces the general shape of the experimental scattering curve. The results presumably provide a test of the potential-

[12] Another interesting system for which scattering patterns have been calculated and compared with experiment is the reaction of H^+ with D_2. Depending upon energy, various channels are open. The possibilities are: $D^+ + HD$ ($E > 4$ kJ mol^{-1}); $H + D_2^+$, $D + HD^+$ ($E > 180$ kJ mol^{-1}); $H + D^+ + D$, $H^+ + D + D$ ($E > 440$ kJ mol^{-1}) [J. R. Krenos, R. K. Preston, R. Wolfgang, and J. C. Tully, *J. Chem. Phys.* **60**, 1634 (1974)].

[13] P. Brumer and M. Karplus, *J. Chem. Phys.* **54**, 4955 (1971).

[14] R. N. Porter and M. Karplus, *J. Chem. Phys.* **40**, 1105 (1964).

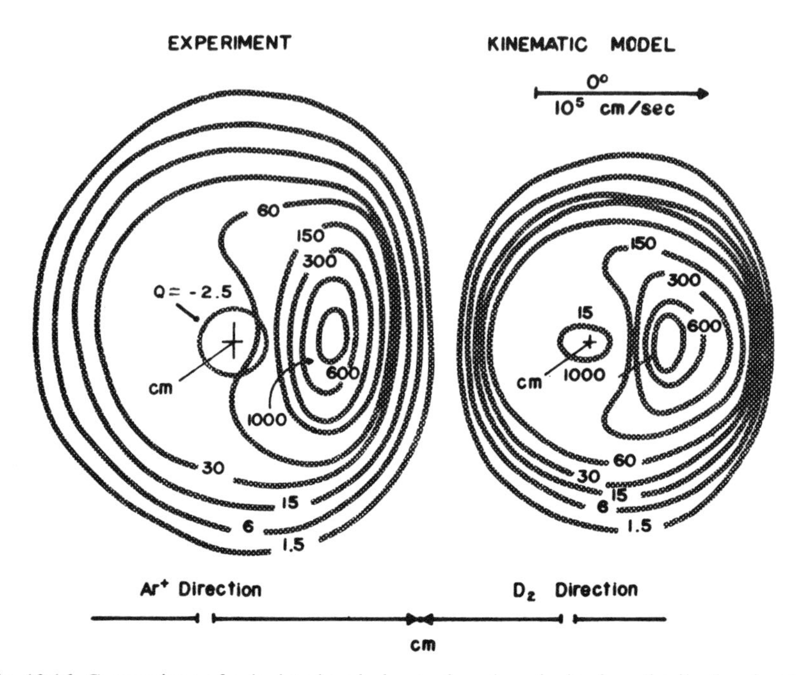

Fig. 10.16. Comparison of calculated and observed angle–velocity flux distribution for the reaction $Ar^+ + D_2 \rightharpoonup ArD^+ + D$ at a collision energy of 260 kJ mol^{-1}. Calculated values from T. F. George and R. J. Suplinskas, *J. Chem. Phys.* **54**, 1037 (1971); experimental data from M. Chiang, E. A. Gislason, B. H. Mahan, C. W. Tsao, and A. S. Werner, *J. Chem. Phys.* **52**, 2698 (1970). [From Z. Herman and R. Wolfgang, in *Ion–Molecule Reactions*, J. L. Franklin, ed. (New York: Plenum Press, 1972), Vol. 2, p. 580, reprinted by permission.]

energy surface since the collision energies are only somewhat above threshold.

10.9.2. $Ar^+ + D_2 \rightharpoonup ArD^+ + D$

Contour maps for the angle–velocity flux of ArD^+ produced in this reaction have been computed on the basis of a highly approximate potential-energy surface.[15] The model treats Ar^+ as a hard sphere and D_2 as a pair of hard spheres nearly in contact. The relative motion prior to Ar^+–D_2 collision is governed by the ion-induced-dipole potential between Ar^+ and D_2. During collision the ion and each atom are treated as hard spheres. An ArD^+ ion is presumed to have formed if the relative translational energy of the nuclei is less than the bond dissociation energy. As the products

[15] T. F. George and R. J. Suplinskas, *J. Chem. Phys.* **54**, 1037 (1971).

separate motion is again controlled by an ion-induced-dipole potential. The model potential is essentially the same one we have treated in this chapter, and is extended to incorporate the long-range ion-induced-dipole interaction.

A comparison of calculation and experiment, for a collision energy of 260 kJ mol^{-1}, is presented in Fig. 10.16. The agreement is remarkable, considering the crudity of the model. Presumably this may be partly rationalized since the collision energy is high, a condition under which the detailed structure of the potential-energy surface is less important. In any case, it is clear from this example that agreement of the sort illustrated provides little basis for assessing the accuracy of a potential-energy surface.[16] The problem is similar to that of establishing an intermolecular potential from limited data, first discussed in Chapter 2.

Problems

Problems 10.1–10.4 are designed to exhibit some general features of nonreactive collisions assuming the potential-energy surface of Fig. 10.5.

10.1. Show that for $\beta = \pi/2n$ all collinear collisions are *elastic* (no interconversion of translational and vibrational energy).

10.2. Consider a collinear collision for which the initial partitioning of the energy is

$$E_t = E_0 \cos^2 \theta, \qquad E_v = E_0 \sin^2 \theta$$

$\theta > 0$ if $\dot{y}/\dot{x} > 0$ when the first AB collision occurs.

(a) Show that, if the trajectory leads to m AB collisions, the vibrational energy after interaction is

$$E_v = E_0 \sin^2 \theta', \qquad \theta' = \pi + \theta - 2m\beta$$

(b) Show that $\Delta E_v/E_v = 1 - \sin^2 \theta'/\sin^2 \theta$ and that, if $E_v \gg E_t$,

$$\Delta E_v/E_v \rightharpoonup \Delta E_v/E_0 \rightharpoonup \sin^2 2m\beta$$

(c) Compute $\Delta E_v/E_0$ in Br_2 recombination when the buffer gas is He, Ne, Ar, Kr. (The ratio of the recombination rate constants is 1 : 1.4 : 2.2 : 2.9 at ~300 K.)

[16] A detailed critique of the model has been given by M. Henchman, in *Ion–Molecule Reactions*, J. L. Franklin, ed. (New York: Plenum Press, 1972), Vol. 1, pp. 206–208.

10.3. (a) Show that the only possible values of θ are $-\pi/2 < \theta < \beta$.
(b) For $\beta > \pi/4$ show that only *one* AB collision is possible if $\theta < 3\beta - \pi$.
(c) For $\beta > \pi/4$ show that there are *two* AB collisions if $2\beta - \pi/2 < \theta < \beta$.
(d) Energy disposition is determined by $|\theta|$, not θ. For $|\theta| < \beta$ show that the fraction of collisions with $\theta = -|\theta|$ is $\frac{1}{2}[1 + \tan|\theta| \cot \beta]$.
(e) For $3\beta - \pi < \theta < 2\beta - \pi/2$ there can be 1 or 2 AB collisions. Show that, of the trajectories with θ in this range, a fraction $2/[1 - \cot \beta \tan(\beta - \theta)]$ leads to only a single collision.

10.4. Consider systems for which $\beta = 50°$ and $70°$.
(a) In each system what are the possible values of $|\theta'|$ if $|\theta| = 0°, 15°, 30°, 45°, 60°, 75°$?
(b) What fraction of the trajectories with the $|\theta|$'s of (a) lead to a particular $|\theta'|$?
(c) From your results deduce the qualitative features that encourage vibrational excitation. What is the effect of the mass of the buffer gas? What is the effect of increasing the vibrational energy in the molecule?

10.5. Consider a potential surface such as that of Fig. 10.7d, where there is a *potential well.*
(a) Show that, for a path where there is no initial vibrational energy, the collision is *elastic* if the path only involves *one* reflection at the step.
(b) Now show that, even if there is initial vibrational energy, a collision involving only *one* reflection at the step is elastic.
(c) What is the condition under which there can be more than one reflection at the step, i.e., formation of a long-lived collision complex?

In Problems 10.6–10.8 do not consider multiple collisions.

10.6. (a) For the potential surface of Fig. 10.11b calculate $E_{t'}$ for the trajectory shown as a function of E_0, Δ, and β.
(b) Construct a potential-energy surface for which the exothermicity is due to A–B attraction and determine $E_{t'}$ for a trajectory where $\theta = 0$ ($E_v = 0$).

10.7. Assume $|\Delta| \gg E_0$ and determine $E_{t'}/|\Delta|$ and $E_{v'}/|\Delta|$ for the two "repulsive" surfaces of Figs. 10.11a and b; for the two "attractive" surfaces of Fig. 10.11c and the one constructed in Problem 10.6b. Note the qualitative differences.

10.8. (a) Construct a potential surface for "mixed" energy release by placing the step so that it connects the *corners* of the surface of Fig. 10.10.
(b) Determine $E_{t'}$ for an initially vibrationless system.
(c) When $|\Delta| \gg E_0$ compare your results for $E_{v'}/|\Delta|$ with that of Problem 10.7.

10.9. Chemiluminescence data are used to determine whether an exoergic potential surface is "attractive," "mixed," or "repulsive." Refer to the data of Table 6.3 and determine the following:

(a) Which reactions do not involve multiple collisions.

(b) Of these systems, which have surfaces that are (i) relatively more attractive, (ii) relatively more repulsive.

10.10. Consider the surface of Fig. 10.10 with an initial trajectory where $\theta = 0$. Modify the analysis used in Problem 10.2 and show that:

(i) all paths are reactive if $\pi/(2n + 1) \geq \beta \geq \pi/(2n + 1.5)$, $n \geq 1$;

(ii) no paths are reactive if $\beta = \pi/(2n)$, $n \geq 1$.

General References

P. J. Kuntz, in *Modern Theoretical Chemistry*, W. H. Miller, ed. (New York: Plenum Press, 1976), Vol. 2, Part B, Chapter 2.

R. D. Levine and R. B. Bernstein, *Molecular Reaction Dynamics* (Oxford: Clarendon Press, 1974), Chapter 6.

B. H. Mahan, *Acc. Chem. Res.* **8**, 55 (1975).

R. N. Porter, *Ann. Rev. Phys. Chem.* **25**, 317 (1974).

Physical Constants

Quantity	Symbol	Value
Avogadro's number	N_0	6.02252×10^{23} mol^{-1}
Boltzmann's constant	k	1.38054×10^{-23} $J\ K^{-1}$
Electron charge	e	1.60210×10^{-19} C
Electron mass	m_e	9.1091×10^{-31} kg
Faraday's number	F	9.64868×10^{4} $C\ mol^{-1}$
Gas constant	R	8.3143 $J\ K^{-1}\ mol^{-1}$
Planck's constant	h	6.6256×10^{-34} J sec
Velocity of light	c_0	2.997925×10^{8} $m\ sec^{-1}$

Index

Where new material or new concepts are introduced in problems, the page numbers are italic.